EXPLORATION AND VALORIZATION OF NATURAL RESOURCES FROM ARID ZONES

EXPLORATION AND VALORIZATION OF NATURAL RESOURCES FROM ARID ZONES

Edited by

Cristóbal Noé Aguilar, PhD

Norma Paola Meléndez-Rentería, PhD

Karen Nathiely Ramírez-Guzmán, PhD

AAP | APPLE ACADEMIC PRESS

First edition published 2025

Apple Academic Press Inc.
1265 Goldenrod Circle, NE,
Palm Bay, FL 32905 USA

760 Laurentian Drive, Unit 19,
Burlington, ON L7N 0A4, CANADA

CRC Press
2385 NW Executive Center Drive,
Suite 320, Boca Raton FL 33431

4 Park Square, Milton Park,
Abingdon, Oxon, OX14 4RN UK

Library and Archives Canada Cataloguing in Publication

Title: Exploration and valorization of natural resources from arid zones / edited by Cristobal Noé Aguilar, PhD, Norma Paola Meléndez-Rentería, PhD, Karen Nathiely Ramírez-Guzmán, PhD.
Names: Aguilar, Cristóbal Noé, editor. | Meléndez-Rentería, Norma Paola, editor. | Ramírez-Guzmán, Karen Nathiely, editor.
Description: First edition. | Includes bibliographical references and index.
Identifiers: Canadiana (print) 2024031297X | Canadiana (ebook) 20240312988 | ISBN 9781774915523 (hardcover) | ISBN 9781774915530 (softcover) | ISBN 9781032672236 (ebook)
Subjects: LCSH: Arid regions—Mexico, North. | LCSH: Natural resources—Mexico, North. | LCSH: Microorganisms—Mexico, North. | LCSH: Plants—Mexico, North. | LCSH: Animal products—Mexico, North
Classification: LCC GB618.16 .E97 2024 | DDC 333.73/609721—dc23

Library of Congress Cataloging-in-Publication Data

CIP data on file with US Library of Congress

ISBN: 978-1-77491-552-3 (hbk)
ISBN: 978-1-77491-553-0 (pbk)
ISBN: 978-1-03267-223-6 (ebk)

About the Editors

Cristóbal Noé Aguilar, PhD

Cristóbal Noé Aguilar, PhD, interconnects the development of bioprocesses (based on solid-state fermentation) with the investigation of issues within food science, combining subjects such as microorganism and chemistry with the generation of several bioproducts. His fermentation study integrates concerns from other disciplines, such as bioactive compounds and enzymes with product formation. Prof. Aguilar is the Director of Research at the Autonomous University of Coahuila (2018–2022), Mexico. He was Dean of the School of Chemistry (2014–2018). Prof. Aguilar is a level III member of the Mexican System of Researchers (SNI). He is a regular member of the Mexican Academy of Science (since 2014) and a member of the Central Council of IBA. He has received several prizes and awards which include the National Prize of Research 2010 of the Mexican Academy of Sciences, CCC Prize 2008 of the Mexican Society of Biotechnology and Bioengineering, National Prize AgroBio-2005, Mexican Prize in Food Science and Technology CONACYT-Coca Cola Mexico 2003, 2018 Outstanding Researcher Award from the International Association of Bioprocessing, and 2019 Coahuila-State Innovation, Science and Technology Award. Prof. Aguilar has led more than 45 research projects, including several international research projects (financed by ECOS-Nord, Alpha Network-EU, Marie Curie EU, AMEXID, and CONACYT). In 2022, Prof. Aguilar was listed among the top 10 scientists (Biology and Biochemistry) in Mexico (https://research.com/scientists-rankings/biology-and-biochemistry/mx) and, in 2021 he was listed in the World's top 2% by the SU Ranking (https://cutt.ly/oDZhnXB).

Norma Paola Meléndez-Rentería, PhD

Norma Paola Meléndez-Rentería, PhD, is a full-time Professor at the Biodiversity and Ecology of the Coahuila State Conservations and Research Center at the Autonomous University of Coahuila for the last two years. She is a graduate in chemistry, bacteriology, parasitology (2008) from the Autonomous University of Chihuahua; her MSc and PhD in Food Science and Technology (2013) are from the Autonomous University of Coahuila. She was working on Thermophilic enzyme production at Vigo University (2010), Vigo, Spain, and on the protective effect of oregano fermented wastes on *Lactobacillus* spp. at Foggia University, Foggia, Italy. She has worked with food safety/quality certifications, food industrialization, applied microbiology, and science communication. Dr. Melendez has published 13 papers in indexed journals, three book chapters, one patent, and has various contributions in scientific meetings. She is a member of the International Bioprocessing Association (IFIBiop). Finally, she has been an advisor for one PhD thesis, three MSc theses, and one BSc thesis. Her research areas are applied microbiology, food law and regulations, and science communication.

Karen Nathiely Ramírez-Guzmán, PhD

Karen Nathiely Ramírez-Guzmán, PhD, is a full-time Professor at the Center for Interdisciplinary Studies and Research of the Autonomous University of Coahuila for the last three years. She obtained her graduate degree in Pharmaceutical Chemistry (2014) from the Autonomous University of Coahuila, as well as her master's and doctorate degree in Food Science and Technology (2019) as well a doctorate in Biological Sciences with a concentration in Biotechnology from the Federal University of Pernambuco in Brazil (2019). She worked on topics related to the isolation of microorganisms, the production of metabolites of industrial interest, and biological control agents in France, Colombia, and Brazil. Prof. Ramírez is a level C member of the S.N.I. (Mexican Research System) since 2021. She has also published 11 articles in indexed journals, 13 book chapters, and has made various contributions to national and international scientific meetings, Finally, she has been an advisor to five doctoral students, eight master's students, and three bachelor's students. She teaches courses which include microbiology, food safety, biotechnology, and research methodology.

Contents

Contributors

Jaime Raúl Adame-Gallegos
Facultad de Ciencias Químicas, Universidad Autónoma de Chihuahua, Chihuahua, Mexico

Cristóbal Noé Aguilar
Bioprocesses and Bioproducts Research Group, Food Research Department, School of Chemistry, Universidad Autónoma de Coahuila, Saltillo, Mexico
Department of Research (DIA-UAdeC), School of Chemistry, University Autonomous of Coahuila, Saltillo, Coahuila, Mexico
Faculty of Chemistry, Autonomous University of Coahuila, Saltillo, Mexico

David Ramiro Aguillón-Gutiérrez
Laboratorio de Bioindicadores, Centro de Investigación y Jardín Etnobiológico, Universidad Autónoma de Coahuila, Viesca, Coahuila, Mexico

Nagamani Balagurusamy
Laboratory of Bioremediation, Faculty of Biological Sciences, Autonomous University of Coahuila, Torreón, Mexico

Cynthia Lizeth Barrera Martinez
Research Center for the Conservation of Biodiversity and Ecology of Coahuila (CICBEC), Autonomous University of Coahuila, Saltillo, Coahuila, Mexico

Arturo Bueno-Tokunaga
Centro de Investigación en Geociencias Aplicadas, Universidad Autónoma de Coahuila, Nueva Rosita, Coahuila, Mexico.

Luis Fernando Camacho-Guerra
Centro de Investigación y Jardín Etnobiológico del Semidesierto de Coahuila, Universidad Autónoma de Coahuila, Viesca, Coahuila, Mexico

Luis Fernando Camacho-Ortegón
Centro de Investigación en Geociencias Aplicadas, Universidad Autónoma de Coahuila, Nueva Rosita, Coahuila, Mexico

Karla Fabiola Chacón-Vargas
Facultad de Ciencias Químicas, Universidad Autónoma de Chihuahua, Chihuahua, Mexico

Dayane Correia Gomes
Centro de Investigações Biomédica, Universidad Veracruzana, Xalapa-Ver, Mexico

Miguel Ángel De León-Zapata
Research Center for the Conservation of Biodiversity and Ecology of Coahuila (CICBEC), Autonomous University of Coahuila, Saltillo, Coahuila, Mexico
CICBEC, Autonomous University of Coahuila, Cuatrocienegas, Mexico

Juan Josué Enciso-Cardenas
Centro de Investigación en Geociencias Aplicadas, Universidad Autónoma de Coahuila, Nueva Rosita, Coahuila, Mexico.

Mariana Delgado García
Tecnologico de Monterrey, Escuela de Ingenieria y Ciencias, Nuevo Mexico, Zapopan, Jalisco, Mexico

Dora Alicia García García
National Institute of Forestry, Agricultural and Livestock Research, Saltillo, Mexico

Fernando Hernández-Terán
Laboratory of Bioremediation, Faculty of Biological Sciences, Autonomous University of Coahuila, Torreón, Mexico

Javier Ulises Hernández-Beltrán
Laboratory of Bioremediation, Faculty of Biological Sciences, Autonomous University of Coahuila, Torreón, Mexico

Héctor Daniel Inurreta-Aguirre
National Institute of Forestry, Agricultural and Livestock Research, Saltillo, Mexico

David Francisco Lafuente-Rincon
Laboratory of Bioremediation, Faculty of Biological Sciences, Autonomous University of Coahuila, Torreón, Mexico

Elan Iñaky Laredo-Alcalá
CICBEC, Autonomous University of Coahuila, Cuatrocienegas, Mexico
Research Center for the Conservation of Biodiversity and Ecology of Coahuila (CICBEC), Autonomous University of Coahuila, Saltillo, Coahuila, Mexico

Eduardo Alberto Lara Reimers
Departamento de Forestal in Universidad Autonoma Agraria Antonio Narro

Liliana Londoño-Hernandez
Biotics Group, School of Basic Sciences, Technology and Engineering. Universidad Nacional Abierta y a Distancia – UNAD, Colombia

Diego Martinez-Carrillo
Centro de Investigación en Geociencias Aplicadas, Universidad Autónoma de Coahuila, Nueva Rosita, Coahuila, Mexico

Gloria A. Martínez Medina
Bioprocesses and Bioproducts Research Group, Food Research Department, School of Chemistry, Universidad Autónoma de Coahuila, Saltillo, Mexico.

Norma Paola Meléndez-Rentería
Research Center for the Conservation of Biodiversity and Ecology of Coahuila (CICBEC), Autonomous University of Coahuila, Saltillo, Coahuila, Mexico
CICBEC, Autonomous University of Coahuila, Cuatrocienegas, Mexico

J. Y. Méndez-Carmona
Bioprocesses and Bioproducts Research Group, Food Research Department, School of Chemistry, Universidad Autónoma de Coahuila, Saltillo, Mexico

Dayanira Morales-Corral
Facultad de Ciencias Químicas, Universidad Autónoma de Chihuahua, Chihuahua, Mexico

Anna María Polanía
MIBIA Group, Biology Department. Faculty of Natural and Exact Sciences. Universidad del Valle, Colombia

Guadalupe Virginia Nevárez-Moorillón
Facultad de Ciencias Químicas, Universidad Autónoma de Chihuahua, Chihuahua, Mexico
Faculty of Chemistry, Autonomous University of Chihuahua, Chihuahua, Mexico

Natalia Quintero-Pacheco
Biotics Group, School of Basic Sciences, Technology and Engineering, Universidad Nacional Abierta y a Distancia–UNAD, Colombia

Nathiely Ramírez-Guzmán
Center for Interdisciplinary Studies and Research, Universidad Autónoma de Coahuila, Saltillo, Mexico

Josefina Rodriguez
Center for Interdisciplinary Studies and Research, Universidad Autónoma de Coahuila, Saltillo, Mexico

Laihsa Rodríguez-González
Bioprocesses and Bioproducts Research Group, Food Research Department, School of Chemistry, Universidad Autónoma de Coahuila, Saltillo, Mexico

Raul Rodriguez-Herrera
Faculty of Chemistry, Autonomous University of Coahuila, Saltillo, Mexico

Salvador A. Saldaña-Mendoza
Bioprocesses and Bioproducts Research Group, Food Research Department, School of Chemistry, Universidad Autónoma de Coahuila, Saltillo, Mexico

Diana Uresti-Duran
National Institute of Forestry, Agricultural and Livestock Research, Saltillo, Mexico

Andrea Vásquez-García
Biotics Group, School of Basic Sciences, Technology and Engineering. Universidad Nacional Abierta y a Distancia – UNAD, Colombia

Rubi Viveros-Contreras
Centro de Investigações Biomédica, Universidad Veracruzana, Xalapa-Ver, Mexico

Abbreviations

ACC	1-aminocyclopropane-1-carboxylic acid
AHBA	3-amino-5-hydroxybenzoic acid
AHLs	acylated homoserine lactones
AI	aridity index
APC	allophycocyanin
BCFA	branched chain fatty acids
BS	biosurfactants
BSC	biological soil crusts
BWSE	black widow spider envenomation
cDPG	cyclic-2,3-bis(di)phosphoglycerate
CDS	crassicaule desert scrub
CH4	methane
CO_2	carbon dioxide
COD	chemical oxygen demand
CS	cysteine-stabilized
DGGGP	digeranylgeranylglycerophosphate
DGP	derivates as diglycerol phosphate
DHA	docosahexaenoic
ED	Entner–Duodoroff
EDTA	ethylenediaminetetraacetic acid
EM	Embden–Meyerhof
EO	essential oil
EPA	eicosapentaenoic
EPS	exopolysaccharides
ES	enzyme substrate
FDA	Food and Drug Administration
FOS	fructooligosaccharides
GA	gibberellic acid
GOS	galactooligosaccharides
GRAS	generally recognized as safe
HGT	horizontal transfer gene
HVA	humic vitamin agar
IAA	indole-3-acetic acid
ICK	inhibitor cystine knot

LC-PUFA	long-chain polyunsaturated fatty acids
LED	light-emitting diode
LOX	lipoxygenases
LP	lasso peptide
MAA	mycosporin-like amino acids
MAP	mean annual precipitation
MDS	microphyll desert scrub
MG	mannosylglycerate
MIC	minimum inhibitory concentration
MPE	measurement of the prebiotic effect
MRSA	multidrug-resistant bacteria
MUFA	monounsaturated fatty acids
N_2O	nitrous oxide
NDBPs	non-disulfide bridged peptides
NDGA	nordihydroguaiaretic acid
NRPSs	non-ribosomal peptide synthetases
PAD	phenolic acid decarboxylase
PAR	photosynthetic active radiation
PBP	phycobiliproteins
PBRs	photobioreactors
PC	phycocyanin
PDA	potato dextrose agar
PE	phycoerythrin
PEC	phycoerythrocyanin
PFD	prefoldin
PHAs	polyhydroxyalkanoates
PUFA	polyunsaturated fatty acids
QS	quorum sensing
RDS	rosette desert scrub
RiPP	ribosomally and post-translationally modified peptides
SAM	S-adenosyl-L-methionine
SCA	starch casein agar
SCFA	short-chain fatty acids
SEA	soil extract agar
SmF	submerged fermentation
SSF	solid-state fermentation
SYP	yeast peptone starch agar
TCA	tricarboxylic acid cycle
TPS	T6P synthase

UFA	unsaturated fatty acids
UNCCD	United Nations Committee on the Combat of Desertification
UNEP	United Nations Environment Program
YPD	yeast peptone dextrose agar

Preface

Mexico is a country with distinct landscapes and landforms. The north of the country features the arid zones as its principal landforms, boasting huge biodiversity that has been used by the population of the region to survive for centuries. This work presents the various technological, scientific, environmental, social, and weather challenges faced by these exotic landforms.

The book compiles works on the arid zones' bioprocessing uses of plants, animals, and microorganisms for added-value compounds production for the benefit of people. It also includes current information about taxonomy, botany, ecology, microbiology, functional foods, fermentation, innovative processes/bioprocesses for the utilization of by-products, nutraceuticals, extraction of natural products, nano- and microtechnology, alternative processes that reduce waste energy or water, among other topics.

All the topics discussed in this book make it an important reference for the sustainable use of arid zones' natural resources at all levels, with particular benefit for students, professors, investigators, and for professionals in the biotechnology industry. The editors, being members of the Autonomous University of Coahuila (Mexico) located in the Mexican arid zone, have made special efforts on the topics, sharing their extensive experience as teachers and investigators in the field, for the benefit of the academic as well as the general community.

The book explores important advances in the sustainable use of the arid zones' resources, focusing on the challenges that humanity faces in an era of a sedentary lifestyle, diseases associated with food consumption, stress, pandemic, social, economic, environmental, and cultural crisis consequences of drought and unpredictable weather changes.

The book directly addresses the objectives of Sustainable Development Article No. 13 of the 2030 Agenda for Sustainable Development to implement crucial measures to combat climate change and its effects. Likewise, the objective of Sustainable Development Article No. 15 aims to protect, restore, and promote the sustainable use of terrestrial ecosystems, sustainably manage forests, combat desertification, halt and reverse land degradation and halt biodiversity loss.

A detailed and up-to-date revision is a practical tool in teaching and research for arid zone's resource management. This book explains the potential and actual developments through current advances in the design and development of strategies to improve human life in the arid zones.

The book comprises 11 chapters which include relevant information about geographic localization, properties, and basic studies. Important topics covered in the book include economical resources of the arid zones, such as microorganisms, plants, lands, poisonous animals, microalgae, and more examples of biotechnological applications.

Due to its authentic and innovative content, this book will doubtlessly prove to be an important reference for food science and biotechnological and environmental research, for university students, and for professionals in the academic and research areas.

The book focuses to meet the demands and needs of people both in the field of science and technology, as well as for its practical applications for people, particularly in the arid zones.

We want to thank all the writers, who responded eagerly to our call by providing original and novel texts. The editors thank the writers for their time, effort and for communicating their experience for the benefit of the readers. In the end, we also recognize the value of the facilities granted by the institution for which we work, the Autonomous University of Coahuila.

Editors

Norma Paola Meléndez-Rentería, PhD
Karen Nathiely Ramírez-Guzmán, PhD
Cristóbal Noé Aguilar, PhD

CHAPTER 1

Arid Zone Description and Its Biodiversity

LAIHSA RODRÍGUEZ-GONZÁLEZ[1], GLORIA A. MARTÍNEZ MEDINA[1], J. Y. MÉNDEZ-CARMONA[1], SALVADOR A. SALDAÑA-MENDOZA[1], JOSEFINA RODRIGUEZ[2], CRISTÓBAL NOÉ AGUILAR[1], and NATHIELY RAMÍREZ-GUZMÁN[2]

[1]*Bioprocesses and Bioproducts Research Group, Food Research Department, School of Chemistry, Universidad Autónoma de Coahuila, Saltillo, Mexico*

[2]*Center for Interdisciplinary Studies and Research, Universidad Autónoma de Coahuila, Saltillo, Mexico*

ABSTRACT

Arid zones represent 47% of the entire land, which represents a high percentage. These are divided into: arid lands, hyper-arid, arid, semi-arid and dry subhumid regions, their main characteristic being that the average rainfall is less than the potential losses of humidity through evaporation, in addition to high climatic variety, these areas have a presence on the five continents, the main areas being Africa: North and Middle East, India, Eurasia, Australia northern region, North America, South America to mention a few.

Due to climate changes, arid areas have been affected by soil degradation and loss of diversity, which causes erosive processes due to low vegetation and the loss of native microorganisms in these soils; however, despite this, the Arid zones have an interesting flora, fauna and microbiota due to their ability to adapt and survive to hostile and adverse conditions, which is why they are capable of producing compounds, substances, and defense mechanisms that help them survive, which represents a niche, important for its interesting biological characteristics and its possible use in different areas; however these have not been widely explored, which represents an area of

Exploration and Valorization of Natural Resources from Arid Zones. Cristobal Noé Aguilar, Norma Paola Meléndez-Renteria, & Karen Nathiely Ramirez-Guzman (Eds.)

opportunity for research, to find processes that help achieve sustainable and sustainable use that does not affect or overexploit these areas or their natural resources, which is why it is important to see it from a multidisciplinary perspective taking into account all the variables involved.

1.1 INTRODUCTION

According to the National Commission of Arid Zones (CONAZA), the arid regions are distinguished by total annual precipitation values of 50–250 mm, resulting in a very sparse vegetation cover, while semiarid regions show a total precipitation of 250–500 mm/year and more abundant vegetation (steppe, savannah, and grassland). These arid lands are also characterized by persistent water scarcity, frequent droughts, high climate variability, and diverse forms of land degradation, among which are desertification and biodiversity loss (José, 2014). Arid terrestrial ecosystems play a very important role in global biophysical processes, as they reflect and absorb solar radiation and help to maintain the balance of atmospheric constituents (Pino-Vargas et al., 2013).

Arid zones have diverse characteristics, among which we find the precipitation characterized by short rains with high intensity, after a dry season, precipitation exerts a significant force on the soil, resulting in erosive processes, all these due to low vegetation and dry soil (Verbist et al., 2010). Whereas the soil in arid and semiarid areas is typically incipient or poorly developed (Marzuela, 2013), low in organic matter due to sparse vegetation cover and limited productivity, this results in microbial activity that is often limited by humidity; the periodic droughts characteristic of these areas result in a large amount of dead biomass of microorganisms that serve as reservoirs and thus make nutrients available for plants or other microorganisms (Celaya and Castellano, 2011). These areas occupy 41% of the planet's surface; according to the aridity index (AI), they are classified as follows: arid lands, hyperarid, arid, semiarid, and dry subhumid regions (Gaur and Squires, 2017).

In the different continents, arid climates dominate in five major zones, these five zones are North Africa and Eurasia; North America and South Africa; Australia; South America; and all zones have a central desert, partially surrounded by semiarid lands (UNESCO, 2015).

1.1.1 DISTRIBUTION

Arid ecosystems are spread between latitudes 50°N and 50°S and this divides them into cold and warm deserts and semideserts because they are found in latitudinally different situations (Archibold, 1995). Among the main hot deserts and semideserts, we find Africa: North and Middle East (Sahara, Libya, Arabia, Somalia), south and Madagascar (Namib, Karroo), and India (Thar). Australia: large sandy desert, Gibson, Simpson. North America: Mojave, Sonora, Chihuahua. South America: Sechura, Atacama, Guajira, Northwest Brazil, argentine mountain. For the case of cold deserts and semideserts, we find Occidental Asia: Dasht-j kavir, Karakum, Kizilikum. Oriental Asia: Takla Makan, Jungaria, Turpan, Beishan, Ala Shan, Ordos, Gobi. North America: Great Sand Desert, the Great Basin. South America: Patagonia (Ariza Alcaraz, 2012).

Nearly half of the Earth's land surface is made up of arid areas according to the World Atlas of Desertification (PNUMA, 1997), which is equivalent to some 6.45 billion hectares in all regions of the planet. One billion hectares are hyperarid (real deserts) like the Sahara Desert (Hernández, 2007). In contrast to the arid, semiarid, and dry subhumid regions, which occupy some 5.45 billion hectares and are the most populated by humans, the arid, semiarid, and dry subhumid regions are the most populated by humans, although in these regions, in particular, the soil is fragile, rare vegetation and climate, approximately 70% of the 5.2 trillion hectares of drylands used for agriculture around the world are degraded (Tapia, 2014).

1.1.2 ARID ZONES IN THE WORLD

The scarcity of water that characterizes arid zones makes them a vital part of the earth's human and physical environment. More than 6.1 billion hectares are arid and semiarid zones, corresponding to 40% of the planet's surface (Granados-Sanchez et al., 2013). These environments are found all over the world, among which we find the following significant zones:

1. In Mexico, arid and semiarid zones cover more than half of the territory and are located mainly in the northern and central regions of the country. They contain a varied vegetation and have an enormous wealth of endemic plants (Arámbula 2005), composed mainly of low-growing woody plants, as well as succulents and grasses that form thickets and pastures (Briones et al., 2018). In North America, one-third of its surface can be considered as arid or semiarid; Mexico is considered to have about 41% of arid and semiarid zones, of which about 18% of the national population lives in the country. The arid zone can be located in the lower part of the states of Sonora and Baja California, extending into the United States of America, and

is known as the Sonoran Desert. The other region is located in part of the states of San Luis Potosi, Zacatecas, Coahuila, Nuevo Leon, Tamaulipas, Durango, and Chihuahua, known as the Chihuahuan Desert (Francisco, 2012).

2. Africa's drylands are home to more than 268 million people comprising 43% of the continent's land area, excluding deserts (Mortimore, 2017), including the arid zone of Nigeria, which covers some 350,000 km^2, including the Sahel and Sudan semiarid regions (Fasola et al., 2004).
3. In North Africa, the margins of the Sahara provide examples of the transition to more humid climates with rainfall at contrasting seasons of the year. Desert and steppe, arid and semiarid, betray the combined effects of rainfall and evaporation, but there remains one additional basic distinction, namely, between the warm and temperate deserts with their associated steppe lands. Aridity in hot deserts depends not only on scarce and irregular rainfall but also on high temperatures and evaporation rates throughout the year. What distinguishes coastal deserts from inland deserts is the high relative humidity and fog produced by prevailing onshore winds (Walton, 2007).
4. The southern region of Peru, specifically, the Tacana region and northern Chile are located at the headwaters of the Atacama Desert, one of the driest places in the world, as mentioned above, there is little or no annual rainfall, all of this is reflected in the arid climates characteristic of Chile and this makes them very sensitive areas to climate change (Pino-Vargas et al., 2013).

1.1.3 CLIMATIC AND GEOGRAPHICAL CHARACTERISTICS OF ARID ZONES

Arid zones are geographical and ecological units where extreme drought and cover conditions prevail [Comisión Nacional Forestal (CONAFOR), 2022]. Arid zones are characterized by a higher rate of evaporation compared to the rate of precipitation. In this sense, it is common to associate an arid environment with drylands. Nevertheless, aridity is a climate phenomenon characterized mainly by a shortage of water. So, the arid zones also include cold climates since in these regions, the precipitation falls mainly in the form of snow (Joint Research Centre, 2019).

Arid zones are estimated to occupy 47.2% of the world's total land area according to the United Nations Convention to Combat Desertification (UNCCD) classification system.

The word "arid" derives from Latin *aridus* meaning "dry, arid, parched." The definition of an arid zone is based on several criteria, including the drainage pattern, erosion processes, vegetation types, and climate criteria based on plant growth (Gaur and Squires, 2017). The high variability of temperatures during the day and night, the lasting and frequent droughts, the irregular rains, and the accelerated wind speed are some of the climatic conditions that persist in these geographical regions.

There is an enumerable variety of concepts applied to identify "arid environment," which makes its definition a bit complicated. However, arid zones can be characterized climatically as areas where precipitation is less than the evapotranspiration of the permanent flora (Rodríguez-Núñez et al., 2020). (Gaur and Squires, 2017), expose the causality of an arid zone because of climatic, topographic, and oceanographic factors that prevent weather systems that support humidity from reaching a zone of the Earth's surface. Quality originates from general causes acting individually or in conjunction with other elements. Quantifying these climatic factors makes it possible to catalog arid zones. The AI represents the average water index available in the soil, that is, the ratio between mean average annual precipitation and mean annual evapotranspiration (Dave et al., 2019). Based on this index, the Food and Agriculture Organization of the United Nations classifies aridity. Also, (Rodríguez-Núñez et al., 2020) properly identify the geographic location of each type of arid zone:

1. Hyperarid zone (<0.03 AI): precipitation in those areas is quite unusual and irregular. Annually they receive less than 100 mm of rain, and sometimes the rains do not occur for several years. For this reason, they are largely unvegetated. Hyperarid areas include the Sahara Desert (North Africa), the Lut Salt Desert (Iran), the Gobi Desert (northern China and southern Mongolia), the Atacama Desert (South America), and the Antarctic. In total, the lands classified as hyperarid, also called deserts, occupy 6.6% of the Earth's surface.
2. Arid zone (0.03–0.2 AI): areas classified as arid normally receive less than 200 mm of winter rainfall or less than 400 mm of summer rainfall. Other references refer to an annual rainfall between 100 and 300 mm (Food and Agriculture Organization of the United Nations, 1993). They are also characterized by sparse vegetation that develops slowly, mainly small trees, shrubs, and herbaceous. Geographically these areas can be identified in Australia and Asia Minor. They are estimated to occupy 10.6% of the Earth's land surface.

3. Semiarid zone (0.2–0.5): In annual terms, rainfall ranges from 200 to 600 mm, including up to 800 mm. Nonetheless, rainfall in the semi-arid zones reports a greater variation according to the time of year. In winter, higher rainfall is reported, from 450 to 500 mm. Compared to reported rainfall during the summer season ranging from 200 to 250 mm. These areas also have the presence of dry gale-force winds and very high air temperatures, so high evaporation rates are maintained. The presence of vegetation is more evident compared to the other areas. Xerophytic vegetation prevails, and large trees. Geographically semiarid areas include regions in southern Spain, Chile (Coquimbo region), and Kazakhstan. These areas are estimated to extend across 15.2% of the Earth's surface.

Arid zones are now considered vulnerable ecosystems as they have been severely affected by the destabilizing effects resulting from anthropogenic climate change. The increase in the global temperature of 1–5°C estimated in the last decades brings with it more severe, intense, and lasting droughts to the arid zones, significantly affecting the probability of landscapes with a higher rate of degradation and desertification (Lewandrowski et al., 2021). In addition to projected increases in temperature and long droughts, arid ecosystems are extremely vulnerable to heatwaves and prone to rapid environmental and land degradation as a result of limited natural water resources (Spinoni et al., 2021).

In previous figures, it was estimated that about one billion people inhabit a large part of these arid and semiarid areas, a population classified among the most vulnerable on the planet (Malagnoux, 2007). With the current population increase, it is also estimated that the percentage of arid inhabited areas has increased significantly. Overpopulation poses an alarming risk to the preservation of arid ecosystems due to progressive natural degradation combined with overexploitation of available natural hybrid resources in these areas, deforestation, and degradation of flora due to inadequate human resource management (Lewandrowski et al., 2021).

Desert areas occupy about a third of the total land area. Nevertheless, one of the conditions of climate change currently affecting the planet is expected to be reflected in the gradual increase of drylands.

A comprehensive research study (Spinoni et al., 2021) analyzes the relationship between rapid population growth and possible gradual changes in drylands at specific levels of global warming resulting from land use. To estimate the possible changes in the land, the authors start from the analysis of a set of climate simulations of CORDEX that include three indicators: the AI established by the Food and Agriculture Organization of the United Nations-UN Environment Programme (FAO-UNEP), the Köppen–Geiger climatic classification, and the Holdridge life zones. The results are estimated at four projected global warming levels ranging from 1.5°C to 4°C. Concerning the above, about 1.8% of total global land will be converted into arid zones if global warming is limited to 2°C. On the contrary, approximately 4.6% of nonarid lands will become arid if global warming increases by 4°C. The affected areas show notable geographical differences, this drying is mainly strong in South America and Southern Europe.

The reality of climate change is imminent, but projected impacts on diverse ecosystems, including arid and semiarid lands, can happen alarmingly fast. Bringing about major changes in the natural resources available globally.

1.2 FLORA AND FAUNA OF ARID ZONES

The climatic, terrestrial, and environmental conditions characteristic of arid zones promote low vegetative growth. However, three main species of flora stand out in arid zones, which are discussed in the following subsections.

1.2.1 EPHEMERAL ANNUALS

They are characterized by growing in medium humidity conditions, so it is common for them to appear after the rains. They have a short life cycle governed during the short periods when humidity and temperature are favorable for their growth. During the dry stage of arid ecosystems, plants of this classification can survive for several years in the form of seeds. Therefore, it is considered that annual ephemeral plants do not present extraordinary adaptations to desert weather conditions. Usually, these plants have small sizes, their roots are shallow, and their physiological adaptation consists of their active growth.

Within this classification of ephemeral annuals, there is a subclassification of these species based on the seasonal conditions determined by temperature and humidity so that during the winter and spring months, winter annuals germinate and complete their life cycle. For their part, summer annuals complete their life cycle during the summer months and early autumn (Mulroy and Rundel, 2009; Sharma, 2021).

1.2.2 SUCCULENT PERENNIALS

Succulents have succulent tissue. This living tissue performs various functions in the plant, among others, is responsible for allowing the temporary storage of usable water, making the plant temporarily independent of the external water supply when soil conditions have deteriorated so that the root no longer absorbs the necessary water from the soil. In addition to this characteristic, succulents possess the quality of decreasing the transpiration rates of the plant (von Willert et al., 1990). The succulent organs are generally the stream or leaves. The water content of succulent organs can reach 90–95% storage. Many all-cell succulents are annual plants and use their leaves only as stores to extend the effective growing season (Griffiths and Males, 2017). Some plants of the family of succulents are cactus, spurge, milkweed, lily, and amaryllis.

1.2.3 NONSUCCULENT PERENNIALS

These types of plants occupy mostly arid lands. They are considered plants very resistant to the stress conditions of these areas, including grasses, woody herbs, shrubs, and some trees.

These plants have extensive roots that penetrate deep into the soil, establishing contact with subsoil water that allow them to withstand drought. During periods of water scarcity, they usually lose their leaves and begin to wither, and subsequently produce new leaves when the drought period ends.

One of the classifications of growth of nonsucculent perennials plants differs based on seasons: (1) Evergreen: those that are biologically active throughout the year, (2) Drought-deciduous: biologically dormant during the dry season, (3) Cold-deciduous: biologically dormant during the cold season.

The diversity of plant species that cover the landscapes of arid zones along the plant varies from region to region. The current reality of climate change that has been projected for decades and that affects any form of life as we know it, also triggers a clear series of changes in the plant species that are part of the life cycle including the preservation of the representative fauna of this type of land.

Arid zones are also home to a great diversity of fauna. Despite extreme conditions, the presence of animal life is imminent. From microorganisms, through underground life to mammals that have managed to adapt to the climatic conditions of arid lands.

In addition to the increase in temperature, climatic estimates for arid and semiarid areas project drier environmental conditions, including more frequent and extensive droughts. Such changes in habitats lead to direct changes in animals through increased environmental heat gain or increased water requirements to dissipate heat by evaporation (Fuller et al., 2014).

Exposure to thermal stress in combination with water scarcity in hot deserts triggers a series of pathophysiological processes that result in heat exhaustion, insolation, and kidney dysfunction. Exposure to extremely low temperatures in cold-arid deserts, or on winter nights in hot deserts, can lead to hypothermia. Therefore, it is likely that species adapted to the desert experienced a strong selection in metabolism and water retention (Rocha et al., 2021). These adaptations could harm species living in warmer environments. Adaptations to the thermoregulatory process have been examined mainly on species that inhabit warmer ecosystems belonging to the ectothermic group. Exothermic animals have an autonomous ability to cope better with environmental variations compared to endotherms. Most organisms are ectothermic (Kearney et al., 2009).

However, the effectiveness of thermoregulation is strongly linked to nature, in addition to the availability of flora. It has been shown that large mammals such as Oryx, Arabian sand gazelles, and Angora goats have had to modify and adapt their behaviors to reduce environmental thermal stress (Fuller et al., 2014). For its part, the large population of small desert mammals includes a diverse variety of animals, with rodents being most of these mammals. In addition to rodents, bats and birds could also be classified as arid ecosystem fauna despite the mobility of these animals (Walsberg, 2000). These small mammals also use several strategies that allow them to coexist in arid lands, such strategies include increasing water loss by evaporation to maintain a lower and stable corporal temperature, to carry out much of their daily activities at night (despite the increased risk of depopulation), to live in burrows and/or to move regularly to shady shelters under the vegetation where temperatures are lower, and adaptation to corporal temperatures above normal levels to conserve water (Alagaili et al., 2017).

Subterranean fauna also occupies a large part of the dryland animal ecosystem. This includes several vertebrate and invertebrate animals living in groundwater, as well as in unsaturated zones from depths of a meter or more below the ground surface to the water table. These habitats often serve as a refuge during periods of adverse conditions (Halse, 2018).

The adaptation of any form of life to ecosystems, a consequence of climate change, can be seen not only in large animals. The microbiological world has been present in climatologically extreme areas. Despite this, the specialized agencies are governed by a dry limit of life which, being exceeded by limited environmental conditions, is insufficient to maintain the metabolic functions of the organisms. In arid zones in addition to water limitations, microorganisms also need to withstand high solar fluxes, including lethal ultraviolet (UV) light, accelerated temperature variations, high rates of water evaporation, frequent and increasingly prolonged periods of droughts, oligotrophic conditions, and sometimes high levels of salinity (Wierzchos et al., 2012).

The environmental complexity that governs the arid zones allows a selective way of life concerning the flora and fauna of these ecosystems. Although both plants and animals have managed to adapt to recent changes in their habitats, the reality is that the excessive changes threaten the possible adaptations of species due to the speed at which these alterations are occurring. The risks projected on different ecosystems, including arid zones, are catastrophic for any form of life. Actions for the conservation of these extensive lands throughout the planet are imminent to ensure an adequate ecological balance.

1.3 MICROORGANISMS IN ARID ZONES

Arid zones represent a very particular ecosystem, which normally presents, an evaporation rate higher than precipitation. Arid zones are one of the bigger terrestrial ecosystems in the planet, with an extension of 5.36 million km^2, being zones with characteristics, as water scarcity, drought, constant climatic changes, high UV radiation, erratic and low rainfall, low atmospheric humidity, and high wind velocity; also, their soils possess special features as low water holding capacity, low organic content and low nitrogen and phosphorus content (Gaur and Squires, 2018). Despite the characteristics of arid zones, they possess a rich biodiversity, including animals, plants, and microbes (Osborne et al., 2020).

Arid zone microbiome, is also adapted to confront a set of abiotic parameters especially lack of water, changing temperatures, radiation, and salinity (Osborne et al., 2020). Normally arid zone microbiome could be animal- or plant- associated, but also funded in soils. Plants and animals generate mechanism that allows arid zones survival, and microbiome could help to generate diverse class of mechanism to support environmental conditions or could be key part to ameliorate these disadvantages. In dry lands, animals need to support lack of water and the high concentrated salts in plants that serve as food; Nouri et al. (2022) analyze the effect of water intake with different levels of salts in Mongolian gerbils, a small dessert mammal, and analyze their gut microbiome. The study demonstrates that elevated salts in animal body, generate a reduction in *Rikinella* and *Christensenella* but an increase in *Atopobium* genera, and at physiological level, the gerbils concentrate urine, which generate a negative energy balance and a systemic inflammation, also increase hypothalamic arginine vasopressin and intestinal aquaporin-2 to support water retention, all these symptoms was reversed via a fecal transplant from the control group (water without salts) which normalize the altered functions, this fecal sample reduce certain microorganisms in gerbils as *Parabacteroides distasonis* and *Prevolia copri* and increase *Lactobacillus reuteri*. Also, He et al. (2018) analyzed microbiome composition along gastrointestinal tract of Bactrian camel, where *Firmicutes* and *Bacteroidetes* are predominant in forestomach and *Firmicutes* and *Proteobacteria* are most abundant taxa in duodenum and jejunum, while *Firmicutes* and *Verrucomicrobia* were the most abundant in ileum and large intestine. The taxas found in forestomach belongs to *Clostridiales* and are unclassified, and also found in rumen and recognized as major polysaccharide degrading element in cattle; also unclassified *Ruminococcaceae* are extended funded in the ileum and large intestine in Bactriam camels, this could contribute to extent the food fermentation of usually poorly digestive forages. Also, the genera *Akkemansia* (from *Verrucomicrobia* phylum) is high extended in camel gastric tract and could be related to control the salt tolerance and glucose levels.

In the other hand, arid soils normally generate water and salinity stress for the plants, and salinity impacts in pH, nutrient availability, cation exchange and carbon input, shaping the microbial communities (Gamalero et al., 2020), plants could generate physiological adaptation to confront this group of stressful parameters, as vertical or steep layers, presence of wax or phenolic compounds, photosynthetic alterations, expression of heat shocking proteins, seed dormancy, longer roots, reduced gas levels, small leaves, and solutes accumulation (Alsharif et al., 2020). Plenty quantities of microorganism live in the rhizosphere of vegetation, that is the soil attached to the root of plants, where an important exchange between microbial–plant–soil components, exudates and metabolites occurs. They microbial organism could live in a free form and could since facilitate the nutrient acquisition (N, P, and other minerals) modulate hormone and, to suppress other microbial presence (Gamalero et al., 2020), also microbial presence has been detected in other parts of the plant as leaves or seeds. Mascot-Gómez et al., 2021 studied the microbiome present inside and outside the seed of different Cactus from Chihuahua's desert, and they observe that exist a group of dominant taxa, the final microbial community, is a combination between host and environmental selection. Among predominant taxa, *Stenotrophomonas* is selected in these cactus species group, due to the reason that these genera promote support against pathogens, phytohormones productions, and nitrogen fixation. *Bacilli* isolated from cactus species generate organic acids which solubilize rocks and generate phosphorus. Also, the study shows difference between internal and external microbes, specially by fungus, which could be related to the fact that fungi are more sensitive to environmental factors, also some genera as *Trichoderma* sp., *Penicillium* sp., and *Phoma* sp., was detected in low value and has been associated to the facilitation of the germination procedure specially into degrading testa.

In certain types of soils exits a very huge ecosystem composed by diverse microbial groups and called biocrusts. The biocrusts, are a complex biological community as bacteria, lichens, fungi, algae, cyanobacteria, and bryophytes that are normally founded in a top of centimeters of soil, they are present globally, nevertheless are predominant in arid and semiarid soils (Sosa-Quintero et al., 2022; Weber et al., 2022). One of main characteristics of biocrust are that the microbial components generate substances as polymeric substance or filamentous structures that aggregate surface soil particles, and this increases soil stability in the upper or down sides of soils, their microbial components also possess CO_2-fixation organism and are extremotolerant to UV-radiation, desiccation or salt concentrations (Weber et al., 2022); the biocrusts also could influence the hydrological distribution (Sosa-Quintero et al., 2022). Pombubpa et al. (2019) analyzes the composition of biocrusts along Mojave Desert in the USA they observe that one of most important features is de Geographic spot, followed by the soil depth, being strongly different the composition of biocrusts and subsurface communities; and finally the class of other microorganism where, cyanobacteria is the major center of biocrust networking, and the fungal taxa *Pleosporales* are the core for fungi–bacteria connectivity in biocrusts, also (Sosa-Quintero et al., 2022) mention that the composition of different biocrust from Mexican arid zones, includes 47 algae, 15 fungi, 61 cyanobacteria, 51 lichens, and 26 bryophytes, which could be linked to activities as to enhance nutrient dynamics in soils, germination promotion, soil stability maintenance and hydrology dynamics, whit also a very particular geographical – dependent composition.

Arid zones are the origin of multiple organisms with elevated importance since biotechnological point of view, with evolved features that allow their survival in a constraint environment, they usage and study could prepare to the humanity for future changes, especially those related to the adaptation and mitigation in the constant environmental issues.

1.4 ARID ZONE AND CLIMATIC CHANGE

Climate change is considered as global alteration in climate arrangement caused by greenhouse gases. Greenhouse gases are considered carbon dioxide (CO_2), methane (CH_4), Nitrous oxide (N_2O), and fluorinated gases, like hydrofluorocarbons, perfluorocarbons, and sulfur hexafluoride; all of them produced by two main sources: natural systems and human activities, the first includes fires, earthquakes, oceans, permafrost phenomena, wetlands, and volcanoes, while the seconds are related almost to industrial activities, energy production, land use, and land use change. These gases generate change in certain environmental factors as, temperature, precipitation, sea-level rise, and extreme weather conditions (Fawzy et al., 2020).

Dry lands have been also affected for climate change, and actually, this class of ecosystem could be very sensitive to climate changes (Huang et al., 2005). As multiple researches and analyzes highlights, and as previous mention, climate change support diverse climatological adverse effects including global warming, whit this, an increase in moisture content in atmosphere, resulting in the predicted behavior of dry gets drier, wet gets wetter (Yu et al., 2019); nevertheless, this trend could be inversed and being and some arid zone became wetter (Huang et al., 2016; Yu et al., 2019). Some authors predict an expansion of global territory of Arid zones, between 11 and 26% at the end of the century, being global arid land covering to 50 at 56% planet surface, principally in East Siberia, Northeast China, West Asia, Central Africa, and Canada, the behavior of becoming wetter is predicted to appears in the west half of continents, being a reduced area, compared with the desertification process (Huang et al., 2005).

The biodiversity in dry lands could be affected by climate change, mammals for example, in a dryer arid zone could confront several problems, besides they possess their natural adaptations. Climate change could trigger higher temperatures in arid zones, and as consequence reduce the water a food availability; when the heat is higher and without the enough intake of water is difficult to maintain the correct temperature, generating elevated temperatures in their bodies during days, while in the night, became harder to maintain the temperature needed and demand more energy, this and the low quantity of food make harder their survival, their change in habits from day to night or the migration to new zones, could make it suitable to new predators or increase the competition in certain areas or diminishing the patron of the dispersion of seeds, changing dynamics of entire environment (Fuller et al., 2021).

Some authors also report the possible effects in arid zone biocrusts, specifically in biophytes, during normally situations with moderate rate water absorption, they could modulate the water absorption by the other participants in the biocrust, but in the contrary situation the lack of water could promote an intense competition between all biocrust members, also, being the major Carbon fixators in biocrust mosses with extended period of dissection could be affected morphologically and due to other important characteristics as water condensation and distribution, (Ladrón de Guevara and Maestre, 2022), while the other microbial components in soil also could be affected by climate change making their adaptation and lag phases longer, also experts suggests that during a warm environment first the microbials in soil promote a soil carbon rapid utilization, and followed by a degradation of more recalcitrant carbon, while at long term, the organic soil carbon could be depleted where sustainability in soil being compromised; on the contrary, droughts could cause a lower carbon utilization, and change completely the equilibrium in soil microbiome (Jansson and Hofmockel, 2020).

Also, vegetation in arid zone could be affected by climate change. The climate shift could bring several or extended droughts, and this could alter the plant composition in diverse ecosystems, and could be specie- and recover functionality facilities-dependent, they main mechanisms that allows vegetation being drought-resistant, they are the hydraulic composition and gas exchange (Creek et al., 2018). Salinity is other factor that could be affected in soil and as consequence in plant growth, climate change could make higher the evaporation rate in soil, generating more salt-concentrated soil, or as well, they could promote a sea water intrusion by high sea levels, other key comportment is the changes in soil microbiomes that could break the synergy between microbes and plant; this expected behaviors, affect the class of plant, the availability, and also the quality of plant-based food for some arid zones animals making a noncontrolled cycle and creating a unbalanced ecosystem (Applequist et al., 2020; Corwin, 2021).

Authors as (Zhu et al., 2019) generate a deep analysis in how climate change could affect vegetation dynamics, focused in the multiple subhumid, semiarid, arid, and hyperarid of North China, during the period of 1982–2015. The response of vegetation during this period differs across the different desert sites, and the group formed by trees as *Pinus sylveestris* var. *mogolica* and *Ulmus* spp. and tall shrubs *Caragana* spp. Was most sensitive to climatic variation than the drought-tolerant shrubs *Artemisia deserorum* and *Artemisia ordosica.*

Climate change could promote deeper change in a sensitive ecosystem as arid zones, perturbing since their microbiome to the fauna and flora of the place, if well they are adapted to constant changes a constraint variation could even generate a big stress in their bodies or environments.

1.5 ARID ZONES OPPORTUNITIES

Arid zones are environments where the entry of water by precipitation is lower than its loss by evaporation, high temperatures reside even reaching extreme points and the scarcity of nutrients in the soil is present. However, despite the bleak appearance of these landscapes, they are actually ecosystems with high biological diversity (Briones et al., 2018). These areas have multiple forest resources that can be exploited, generating economic and social benefits. The productive potential of timber and nontimber resources of desert plants such as lechuguilla, maguey, gobernadora, mesquite, and oregano, exercised through their transformation into high-value products with a wide range of applications, is a profitable business opportunity (Comisión Nacional Forestal, 2018).

Among the most notable desert products, we can find candelilla wax. This is a natural wax produced by *Euphorbia antisyphilitica Zucc*, which grows in northern Mexico's arid and semiarid zones. This natural product is recognized by the United States Food and Drug Administration (FDA) as a generally recognized as safe (GRAS) substance and has physicochemical properties that make it a raw material of interest for use as an additive in the preparation of formulations in the food industry (Aranda-Ledesma et al., 2022; De Leãn-Zapata and Aguilar, 2018).

To date, research has been carried out on the application of candelilla wax nanocoatings in fruits such as apples in combination with phytomolecules, achieving an increase in shelf life, maintaining firmness, and reducing moisture loss (De León-Zapata et al., 2018). Its application has also been reported in the development of formulations based on candelilla wax enriched with protein and plant extracts of *Flourensia cernua* to prolong the shelf life of tomatoes, contributing to an increase in protection against phytopathogenic fungi such *as Colletotrichum gloeosporioides*, *Botrytis cinerea*, and *Fusarium oxysporum* (Ruiz-Martínez et al., 2020). In addition to its food application, candelilla wax is used in other areas, such as the development of hydrophobic coatings for textiles; since being a coating of natural origin, its application is an ecological alternative with a safe and skin-friendly application (Sachan and Purwar, 2021). In the same way, it is used in the production of cosmetic products such as lipstick (Mawazi et al., 2022). as well as a stabilizer in formulations of topical application products (Botega et al., 2021).

Oregano (*Lippia graveolens*) is a nontimber forest species that grows in the arid and semiarid zones of Mexico. This plant stands out for its use as a strong flavoring in traditional Mexican cuisine, as well as for its antimicrobial, anti-inflammatory, and antioxidant properties associated with the polyphenolic content of its essential oils (Bautista-Hernández et al., 2021). The applications of the essential oil of oregano are diverse. Its use as an additive in emulsions has been reported by implementing technologies such as micro and nanoencapsulation, as a green solution to the use of synthetic additives for food preservation (Plati and Paraskevopoulou, 2022). It has also been used as a feed additive in the production of Nile tilapia (*Oreochromis niloticus*) to combat the adverse effects of intensive agriculture, achieving an increase in fish growth, an improvement in its hematobiochemical indices and immune condition (Shourbela et al., 2021). Thanks to its antimicrobial properties, oregano essential oil has also been used in the design of pH indicator food packaging produced with cellulose nanofibers and purple sweet potato anthocyanins, achieving up to 99.99% inhibition against foodborne pathogens (*Escherichia coli* and *Listeria monocytogenes*) (Chen et al., 2020). Similarly, its use has been reported in the production of sweet potato starch-based packaging in combination with octenylsuccinate, improving its hydrophobic characteristics and achieving an inhibitory effect against *Staphylococcus aureus* and *E. coli* (Li et al., 2018). Likewise, the potential application of monoterpenoids obtained from oregano leaves, such as carvacrol and thymol, for the prevention of oral diseases has been reported (Vu and Bose, 2020).

Another interesting plant with a wide distribution in the arid zones of Mexico and the USA is the gobernadora (*Larrea tridentata*). This bush has multiple pharmacological properties and is one of the best-known medicinal plants in the region for its effectiveness in relieving problems of arthritis, rheumatism, diabetes, inflammation, and infertility, among others (Morales-Ubaldo et al., 2021). Some of the most important biological activities that have been attributed to *L. tridentata* are antioxidants (Skouta et al., 2018), antibacterial (Núñez-Mojica et al., 2021), antifungal (Aguirre-Joya et al., 2018), and antiparasitic (Bashyal et al., 2017), which have allowed the development of research on the application of gobernadora extracts for the biological control of phytopathogens of agricultural importance such as *Fusarium*, *Botrytis*, *Colletotrichum,* and *Alternaria* (Asael et al., 2018). It has also been reported that the methanolic extracts of gobernadora constitute a renewable resource that acts as an alternative to the use of synthetic antioxidants used in the production of biodiesel, contributing to an improvement in its stability to oxidative stress and prolonging its shelf life (Sagaste et al., 2019). Likewise, research has been carried out on the production of cellulose hydrogels enriched with *L. tridentata* with biomedical applications for tissue regeneration (Tovar-Carrillo et al., 2020). The composition of the gobernadora is rich in resins and organic acids to which antiviral and antimicrobial properties are attributed that allow its use in the development of interesting biotechnological products. Such is the case of the gobernadora-based surface sanitizer developed by the Department of Food Research of the Faculty of Chemical Sciences in collaboration with the Research Center for the Care of Biodiversity in Cuatro Ciénegas and the Research Center and Ethnobiological Garden in Viesca, institutions belonging to the Autonomous University of Coahuila in Mexico, which was born from the search for economical and environmentally friendly alternatives to meet the strong demand for sanitizers generated during the international contingency produced by COVID-19 (Universidad Autónoma de Coahuila Coordinación de Comunicación Institucional Oficina Virtual de Prensa, 2020).

Another plant of great importance is the lechuguilla (*Agave lechuguilla*), one of the most collected species from the arid and semiarid zones of the Mexican highlands. Its harvest represents one of the fundamental elements of the economy of numerous peasant families for at least a third of the year, since multiple national and foreign companies use its fiber (ixtle) to elaborate diverse products. There are vestiges of its use since pre-Hispanic times, since sandals made with ixtle, approximately 8000 years old, have been found in archaeological zones in northern Mexico (Reyes-Agüero et al., 2017). It is currently known that lechuguilla pulp contains bioactive compounds of interest, such as saponins, which have pharmacological application properties, and other compounds that serve as precursors in the production of cortisone, estrogen, and progesterone (Hernández et al., 2005). The main residue from lechuguilla processing is known as guishe and represents about 85% of the crop. Usually, this residue is discarded or incinerated. However recent research has discovered its potential application as a food additive, finding that the incorporation of crude extracts of gishe in the diet of juvenile shrimp (*Litopenaeus vannamei*) improves their health and nutritional status (Peña-Rodríguez et al., 2020).

Among the timber resources that we can find in semiarid areas, the mesquite (*Prosopis juliflora*) is present, which has been used for generations as a source to produce firewood, food, and charcoal (Instituto Nacional de Ecologia y Cambio Climático, 2007). Nowadays, mesquite has been incorporated into the development of research in the production of renewable energies through the use of mesquite seed oil for the production of biodiesel (Venkatesan et al., 2020). In addition to that, recent studies confirm the potential of mesquite flour as a relevant source of polyphenolic compounds and protein for food production (Cavalcante et al., 2022).

These are just some of the examples of how the resources provided by arid zones can be used through biotechnological processes to generate products with a wide variety of applications. However, to carry out its exploitation, it is vitally important to highlight that it is necessary to consider respect for the biological cycles that preserve and restore plant populations, thus ensuring the subsistence of this type of activity as well as of the communities that depend on it economically from its use.

1.6 CONCLUSION

Now that we have seen that the arid and semiarid zones globally occupy a large part of the earth making these ecosystems an important part of biodiversity conservation processes and the maintenance of present ecological processes, but they are also a vital part of the earth's human and physical environments, as many arid areas are densely populated and one of the main sources of food crops. These ecosystems are fragile in the face of climate change, but even so, little use is made of the flora and fauna characteristic of these ecosystems, due to the lack of water and frequent droughts, high climate variability and various forms of land degradation. All these characteristics make the arid zones interesting ecosystems for study, and little exploited in all senses.

KEYWORDS

- **arid zones**
- **biodiversity**
- **erosive processes**
- **semi-deserts**
- **desertification**
- **arid zones opportunities**

REFERENCES

Aguirre-Joya, J. A.; Pastrana-Castro, L.; Nieto-Oropeza, D.; Ventura-Sobrevilla, J.; Rojas-Molina, R.; Aguilar, C. N. The Physicochemical, Antifungal and Antioxidant Properties of a Mixed Polyphenol Based Bioactive Film. *Heliyon* **2018,** *4* (12), e00942. https://doi.org/10.1016/j.heliyon.2018.e00942

Alagaili, A. N.; Bennett, N. C.; Mohammed, O. B.; Zalmout, I. S.; Boyles, J. G. Body temperature Patterns of a Small Endotherm in an Extreme Desert Environment. *J. Arid Environ.* **2017,** *137*, 16–20. https://doi.org/10.1016/j.jaridenv.2016.10.010

Alsharif, W.; Saad, M. M.; Hirt, H. Desert Microbes for Boosting Sustainable Agriculture in Extreme Environments. *Front. Microbiol.* **2020,** *11*, 1666. https://doi.org/10.3389/fmicb.2020.01666

Applequist, W. L.; Brinckmann, J. A.; Cunningham, A. B.; Hart, R. E.; Heinrich, M.; Katerere, D. R.; Van Andel, T. Scientists Warning on Climate Change and Medicinal Plants. *Planta Medica* **2020,** *86* (1), 10–18. https://doi.org/10.1055/a-1041-3406

Aranda-Ledesma, N. E.; Bautista-Hernández, I.; Rojas, R.; Aguilar-Zárate, P.; Medina-Herrera, N. del P.; Castro-López, C.; Guadalupe Martínez-Ávila, G. C. Candelilla Wax: Prospective Suitable Applications Within the Food Field. *Lwt* **2022,** *159*, 113170. https://doi.org/10.1016/j.lwt.2022.113170

Archibold, O. W. Las Regiones Áridas. In *Ecología de La Vegetación Mundial*; Springer: Dordrecht, 1995.

Ariza Alcaraz, J. F. Desiertos y Semidesiertos. *Geobotánica* **2012,** *24*, 1–17. https://www.um.es/docencia/geobotanica/ficheros/tema24.pdf.

Asael, R. G. H.; Guevara-Gonzalez, R. G.; De Jesus, R. G. S.; Angelica, F. P. A. In *Antifungal Activity of Mexican Endemic Plants on Agricultural Phytopathogens: A Review*, 14th International Engineering Congress, CONIIN 2018, 2018. https://doi.org/10.1109/CONIIN.2018.8489793

Bashyal, B.; Li, L.; Bains, T.; Debnath, A.; LaBarbera, D. V. *Larrea tridentata*: A Novel Source for Anti-Parasitic Agents Active Against Entamoeba histolytica, Giardia Lamblia and Naegleria Fowleri. *PLoS Negl. Trop. Dis.* **2017,** *11* (8), 1–19. https://doi.org/10.1371/journal.pntd.0005832

Bautista-Hernández, I.; Aguilar, C. N.; Martínez-ávila, G. C. G.; Torres-León, C.; Ilina, A.; Flores-Gallegos, A. C.; Kumar Verma, D.; Chávez-González, M. L. Mexican Oregano (*Lippia graveolens* kunth) as Source of Bioactive Compounds: A Review. *Molecules* **2021,** *26* (17), 1–19. https://doi.org/10.3390/molecules26175156

Botega, D. C. Z.; Nogueira, C.; de Moura, N. M.; Martinez, R. M.; Rodrigues, C.; Barrera-Arellano, D. Influence of Aqueous Matrices into Candelilla Wax Organogels Emulsions for Topical Applications. *JAOCS, J. Am. Oil Chem. Soc.* 2021, *98* (3), 317–328. https://doi.org/10.1002/aocs.12438

Briones, O.; Búrquez, A.; Martínez-Yrízar, A.; Pavón, N.; Perroni, Y. Biomass and Productivity in Mexican Arid Lands. *Madera Bosques* **2018,** *24* (Special Issue). https://doi.org/10.21829/myb.2018.2401898

Cavalcante, A. M. de M.; de Melo, A. M.; da Silva, A. V. F.; Neto, G. J. da S.; Barbi, R. C. T.; Ikeda, M.; Silva, G. B.; Steel, C. J.; Silva, O. S. da. Mesquite (*Prosopis juliflora*) Grain Flour: New Ingredient with Bioactive, Nutritional and Physical-Chemical Properties for Food Applications. *Future Foods* **2022,** *5*. https://doi.org/10.1016/j.fufo.2022.100114

Celaya-Michel, H.; Castellano, A. Nitrogen Mineralization on Arid and Semi-Arid Land Soil. *Terra Latinoamericana* **2011,** *29*, 343–56. http://www.scielo.org.mx/pdf/tl/v29n3/2395-8030-tl-29-03-00343.pdf%0Ahttp://www.cabdirect.org/abstracts/20123049100.html.

Chen, S.; Wu, M.; Lu, P.; Gao, L.; Yan, S.; Wang, S. Development of pH Indicator and Antimicrobial Cellulose Nanofibre Packaging Film Based on Purple Sweet Potato Anthocyanin and Oregano Essential Oil. *Int. J. Biol. Macromol.* **2020,** *149*, 271–280. https://doi.org/10.1016/j.ijbiomac.2020.01.231

Comisión Nacional Forestal. *Las zonas áridas son más que desierto |Comisión Nacional Forestal | Gobierno | gob.mx*. 2018.

Comisión Nacional Forestal (CONAFOR). *Zonas Aridas* 2022.

Corwin, D. L. Climate Change Impacts on Soil Salinity in Agricultural Areas. *Eur. J. Soil Sci.* 2021, *72* (2), 842–862. https://doi.org/10.1111/ejss.13010

Creek, D.; Blackman, C. J.; Brodribb, T. J.; Choat, B.; Tissue, D. T. Coordination Between Leaf, Stem, and Root Hydraulics and Gas Exchange in Three Arid-Zone Angiosperms During Severe Drought and Recovery. *Plant Cell Environ.* **2018,** *41* (12), 2869–2881. https://doi.org/10.1111/pce.13418

Dave, V.; Pandya, M.; Ghosh, R. Identification of Desertification Hot Spot Using Aridity Index. *Ann. Arid Zone* **2019,** *58*, 1–2.

De Leãn-Zapata, M. A; Aguilar, C. N. Extraction Methods and Common Uses of Candelilla Wax. In *Food Process Engineering and Quality Assurance*; Mohan, C. O., Carvajal-Milan, E., Ravishankar, C. N., Haghi, A. K., Eds.; Apple Academic Press, **2018;** pp 505–524. https://doi.org/10.1201/9781315232966-18

De León-Zapata, M. A.; Ventura-Sobrevilla, J. M.; Salinas-Jasso, T. A.; Flores-Gallegos, A. C.; Rodríguez-Herrera, R.; Pastrana-Castro, L.; Rua-Rodríguez, M. L.; Aguilar, C. N. Changes of the Shelf Life of Candelilla Wax/Tarbush Bioactive Based-Nanocoated Apples at Industrial Level Conditions. *Sci. Hortic.* **2018,** *231*, 43–48. https://doi.org/10.1016/j.scienta.2017.12.005

Fasola, T. R.; Ogunshe, A. A. O.; Onyeachuchim, H. D. Ethnobotanical Importance of Endangered Species in the Arid Zones of Nigeria. *Zonas Áridas* **2004,** *8* (1013-445X), 57–61.

Fawzy, S.; Osman, A. I.; Doran, J.; Rooney, D. W. Strategies for Mitigation of Climate Change: A Review. *Environ. Chem. Lett.* **2020,** *18* (6), 2069–2094. https://doi.org/10.1007/s10311-020-01059-w

Food and Agriculture Organization of the United Nations. In *Key Aspects of Strategies for the Sustainable Development of Drylands*, 1993.

Francisco, G. M. In *Las Zonas Áridas y Semiáridas de Mexico y Su Vegetación*, 2012.

Fuller, A.; Hetem, R. S.; Maloney, S. K.; Mitchell, D. Adaptation to Heat and Water Shortage in Large, Arid-Zone Mammals. *Physiology* **2014,** *29* (3), 159–167. https://doi.org/10.1152/physiol.00049.2013

Fuller, A.; Mitchell, D.; Maloney, S. K.; Hetem, R. S.; Fons ca, V. F. C.; Meyer, L. C. R.; Ven, T. M. F. N. V.; Snelling, E. P. How Dryland Mammals will Respond to Climate Change: The Effects of Body Size, Heat Load and a Lack of Food and Water. *J. Exp. Biol.* **2021,** *224*, 1–11. https://doi.org/10.1242/jeb.8113

Gamalero, E.; Bona, E.; Todeschini, V.; Lingua, G. Saline and Arid Soils: Impact on Bacteria, Plants, and Their Interaction. *Biology* **2020,** *9* (6), 1–27. https://doi.org/10.3390/biology9060116

Gaur, M. K.; Squires, V. R. In *Climate Variability Impacts on Land Use and Livelihoods in Drylands*, 2017; pp 1–348. https://doi.org/10.1007/978-3-319-56681-8

Gaur, M. K.; Squires, V. R. Geographic Extent and Characteristics of the World's Arid Zones and Their Peoples. In *Climate Variability Impacts on Land Use and Livelihoods in Drylands*; Gaur, M. K., Squires, V. R., Eds.; 2018; pp 3–20. https://doi.org/10.1007/978-3-319-56681-8

Granados-Sanchez, D.; Hernández-García, M. A.; Vázquez-Alarcón, A.; Ruiz-Puga, P. The Processes of Desertification and Arid Regions. *Revista Chapingo serie ciencias forestales y del ambiente* **2013,** *19* (1), 45–66. https://doi.org/10.5154/r.rchscfa.2011.10.077.

Griffiths, H.; Males, J. Succulent plants. *Curr. Biol.* **2017,** *27* (17), R890–R896. https://doi.org/10.1016/j.cub.2017.03.021

Halse, S. A. Subterranean Fauna of the Arid Zone. In *On the Ecology of Australia's Arid Zone*, 2018; pp 215–241. https://doi.org/10.1007/978-3-319-93943-8_9

He, J.; Yi, L.; Hai, L.; Ming, L.; Gao, W.; Ji, R. Characterizing the Bacterial Microbiota in Different Gastrointestinal Tract Segments of the Bactrian Camel. *Sci. Pharm.* **2018,** *8* (654), 1–9. https://doi.org/10.1038/s41598-017-18298-7

Hernández, H. M. La Vida En Los Desiertos Mexicanos. In *Acta Bot. Mex.* 2007; vol *80*, p 188.

Hernández, R.; Lugo, E. C.; Díaz, L.; Villanueva, S. Extraction and Indirect Quantification of Saponins from the *Agave lechuguilla* Torrey. *E- Gnosis Revista Digital Cientifica y Tecnologica* **2005,** *3*, 3–12.

Huang, J.; Ji, M.; Xi, Y.; Wang, S.; He, Y.; Ran, J. Global Semi-Arid Climate Change Over Last 60 Years. *Clim. Dyn.* **2016,** *46*, 1131–1150. https://doi.org/10.1007/s00382-015-2636-8

Huang, D.; Ou, B.; Prior, R. L. The Chemistry behind Antioxidant Capacity Assays. *J. Agric. Food Chem.* **2005,** *53*, 1841–1856.

Instituto Nacional de Ecologia y Cambio Climático. *Usos y propiedades del mezquite*, 2007.

Jansson, J. K.; Hofmockel, K. S. Soil Microbiomes and Climate Change. *Nat. Rev. Microbiol.* **2020,** *18* (1), 35–46. https://doi.org/10.1038/s41579-019-0265-7

Joint Research Centre. In *World Atlas of Desertification*; Patterns of Aridity, 2019.

Kearney, M.; Shine, R.; Porter, W. P. The Potential for Behavioral Thermoregulation to Buffer "Cold-Blooded" Animals Against Climate Warming. *Proc. Natl. Acad. Sci. U. S. A.* **2009,** *106* (10), 3835–3840. https://doi.org/10.1073/pnas.0808913106

Ladrón de Guevara, M.; Maestre, F. T. Ecology and Responses to Climate Change of Biocrust-Forming Mosses in Drylands. *J. Exp. Bot.* **2022,** *1*, 1–16.

Lewandrowski, W.; Stevens, J. C.; Webber, B. L.; Dalziell, E. L.; Trudgen, M. S.; Bateman, A. M.; Erickson, T. E. Global Change Impacts on Arid Zone Ecosystems: Seedling Establishment Processes are Threatened by Temperature and Water Stress. *Ecol. Evol.* **2021,** *11* (12), 8071–8084. https://doi.org/10.1002/ece3.7638

Li, J.; Ye, F.; Lei, L.; Zhao, G. Combined Effects of Octenylsuccination and Oregano Essential Oil on Sweet Potato Starch Films With an Emphasis on Water Resistance. *Int. J. Biol. Macromol.* 2018, *115*, 547–553. https://doi.org/10.1016/j.ijbiomac.2018.04.093

Malagnoux, M. Afforestation and Sustainable Forests as a Means to Combat Desertification Arid Land Forests of the World Global Environmental Perspectives. *Forestry* **2007,** 1–14.

Marzuela, P. Agricultura En Zonas Áridas y Semiáridas. *Idesia* 2013, *31* (2), 3–4.

Mascot-Gómez, E.; Flores, J.; López-Lozano, N. E. The Seed-Associated Microbiome of Four Cactus Species from Southern Chihuahuan Desert. *J. Arid Environ.* **2021,** *190* (104531). https://doi.org/10.1016/j.jaridenv.2021.104531

Mawazi, S. M.; Redzal, N. A. B. A.; Othman, N.; Alolayan, S. O. Lipsticks History, Formulations, and Production: A Narrative Review. *Cosmetics* **2022,** *9* (1), 1–19. https://doi.org/10.3390/cosmetics9010025

Morales-Ubaldo, A. L.; Rivero-Perez, N.; Avila-Ramos, F.; Aquino-Torres, E.; Prieto-Méndez, J.; Hetta, H. F.; Batiha, G. E. S.; Zaragoza-Bastida, A. Bactericidal Activity of *Larrea tridentata* Hydroalcoholic Extract Against Phytopathogenic Bacteria. *Agronomy* **2021,** *11* (5), 1–11. https://doi.org/10.3390/agronomy11050957

Mortimore, M. La Transición Agrícola En Las Zonas Áridas Africanas. *LEISA - Revista de agroecología* **2017,** *22* (2).

Mulroy, T. W.; Rundel, P. W. *Annual Plants: Desert Environments Adaptations to*. **2009,** *27* (2), 109–114.

Nouri, Z.; Zhang, X.; Khakisahneh, S.; Degen, A. A.; Wang, D. H. The Microbiota-Gut-Kidney Axis Mediates Host Osmoregulation in a Small Desert Mammal. *Biofilm Microbiomas* **2022,** *8* (16), 1–10. https://doi.org/10.1038/s41522-022-00280-5

Núñez-Mojica, G.; Vázquez-Ramírez, A. L.; García, A.; Rivas-Galindo, V. M.; Garza-González, E.; Cuevas González-Bravo, G. E.; Toscano, R. A.; Moo-Puc, R. E.; Villanueva-Toledo, J. R.; Marchand, P.; Camacho-Corona, M. del R. New Cyclolignans of *Larrea tridentata* and their Antibacterial and Cytotoxic Activities. *Phytochem. Lett.* **2021,** *43*, 212–218. https://doi.org/10.1016/j.phytol.2021.04.013

Osborne, P.; Hall, L. J.; Kronfeld-schor, N.; Thybert, D.; Haerty, W. A Rather Dry Subject ; Investigating the Study of Arid-Associated Microbial Communities. *Environ. Microbioma* **2020,** *15* (20), 1–14.

Peña-Rodríguez, A.; Pelletier-Morreeuw, Z.; García-Luján, J.; Rodríguez-Jaramillo, M. del C.; Guzmán-Villanueva, L.; Escobedo-Fregoso, C.; Tovar-Ramírez, D.; Reyes, A. G. Evaluation of *Agave lechuguilla* by-Product Crude Extract as a Feed Additive for Juvenile Shrimp *Litopenaeus vannamei*. *Aquacult. Res.* **2020,** *51* (4), 1336–1345. https://doi.org/ 10.1111/are.14497

Pino-Vargas, E.; Montalvan-Díaz, I.; Avendaño-Jihuallanga, C. "La Disponibilidad Hídrica Futura En Los Ecosistemas de Zonas Áridas En El Sur de Perú y Norte de Chile. *Agroind. Sci.* **2013,** *2* (2), 147–54.

Plati, F.; Paraskevopoulou, A. Micro- and Nano-encapsulation as Tools for Essential Oils Advantages' Exploitation in Food Applications: The Case of Oregano Essential Oil. *Food Bioproc. Technol.* **2022,** *2021*, 1–29. https://doi.org/10.1007/S11947-021-02746-4

Pombubpa, N.; Pietrasiak, N.; De Ley, P.; Stajich, J.; Hawkes, C. V.; States, J. S.; Christensen, M. Insights Into the Desert Living Skin Microbiome: Geography, Soil Depth, and Crust Type Affect Biocrust Microbial Communities and Networks in Mojave Desert, USA. *BioRxiv* **2019,** 810002. https://doi.org/10.21973/N3S942

Reyes-Agüero, J. A.; Aguirre-Rivera, J. R.; Peña-Valdivia, C. B. Biología y aprovechamiento de *Agave lechuguilla* Torrey. *Bot. Sci.* **2017,** *88* (67), 75. https://doi.org/10.17129/botsci.1626

Rocha, J. L.; Godinho, R.; Brito, J. C.; Nielsen, R. Life in Deserts: The Genetic Basis of Mammalian Desert Adaptation. *Trends Ecol. Evolut.* **2021,** *36* (7), 637–650. https://doi.org/10.1016/j.tree.2021.03.007

Rodríguez-Núñez, K.; Rodríguez-Ramos, F.; Leiva-Portilla, D.; Ibáñez, C. Brown Biotechnology: A Powerful Toolbox for Resolving Current and Future Challenges in the Development of Arid Lands. *SN Appl. Sci.* **2020,** *2* (7). https://doi.org/10.1007/s42452-020-2980-0

Ruiz-Martínez, J.; Aguirre-Joya, J. A.; Rojas, R.; Vicente, A.; Aguilar-González, M. A.; Rodríguez-Herrera, R.; Alvarez-Perez, O. B.; Torres-León, C.; Aguilar, C. N. Candelilla Wax Edible Coating With *Flourensia cernua* Bioactives to Prolong the Quality of Tomato Fruits. *Foods* **2020,** *9* (9), 1303. https://doi.org/10.3390/foods9091303

Sachan, R.; Purwar, R. Silica Modified Candelilla Wax/Thermoplastic Polyurethane Blend Coatings for Hydrophobic Textiles. *J. Text. Inst.* **2021,** *0* (0), 1–7. https://doi.org/10.1080/00405000.2021.1926128

Sagaste, C. A.; Montero, G.; Coronado, M. A.; Ayala, J. R.; León, J.; García, C.; Rojano, B. A.; Rosales, S.; Montes, D. G. Creosote Bush (*Larrea tridentata*) Extract Assessment as a Green Antioxidant for Biodiesel. *Molecules* **2019,** *24* (9), 1–15. https://doi.org/10.3390/molecules24091786

Sharma, P. K. Ecology of Arid Zone Plants with Special Reference to North Eastern Region of Ecology of Arid Zone Plants with Special Reference to North Eastern Region of Rajasthan and Methods to Conserve Them. *Int. J. Mod. Trends Sci. Technol.* **2021,** *7 (11)*, 162–171. https://doi.org/10.46501/IJMTST0711027

Shourbela, R. M.; El-Hawarry, W. N.; Elfadadny, M. R.; Dawood, M. A. O. Oregano Essential Oil Enhanced the Growth Performance, Immunity, and Antioxidative Status of Nile Tilapia (*Oreochromis niloticus*) Reared Under Intensive Systems. *Aquaculture* **2021,** *542*, 736868. https://doi.org/10.1016/j.aquaculture.2021.736868

Skouta, R.; Morán-Santibañez, K.; Valenzuela, C. A.; Vasquez, A. H.; Fenelon, K. Assessing the Antioxidant Properties of *Larrea tridentata* Extract as a Potential Molecular Therapy Against Oxidative Stress. *Molecules* **2018,** *23* (7), 1–17. https://doi.org/10.3390/molecules23071826

Sosa-Quintero, J.; God´nez-Alvarez, H.; Camargo-Ricalde, S. L.; Gutiérrez-Guitiérrez, M.; Huber-Sannwald, E.; Jiménez-Aguilar, A.; Maya-Delgado, Y.; Dinorah, M. A.; Montaño, N. M.; Pando-Moreno, M.; Rivera-Aguilar, V. Biocrust in Mexican Deserta and Semideserts: A Review of Their Species Composition, Ecology and Ecosystem Function. *J. Arid Environ.*. **2022,** *199*, 104712.

Spinoni, J.; Barbosa, P.; Cherlet, M.; Forzieri, G.; McCormick, N.; Naumann, G.; Vogt, J. V.; Dosio, A. How Will the Progressive Global Increase of Arid Areas Affect Population and Land-Use in the 21st Century? *Glob. Planet. Change* **2021,** *205*, 103597. https://doi.org/10.1016/j.gloplacha.2021.103597

Tapia, A. Geografía de Zonas Áridas y de Montañas 2014; pp 1–138. http://sb.uta.cl/libros/DossierGeogafíaZonasÁridasyMontañas-1.pdf.

Tovar-Carrillo, K. L.; Saucedo-Acuña, R. A.; Ríos-Arana, J.; Tamayo, G.; Guzmán-Gastellum, D. A.; Díaz-Torres, B. A.; Nava-Martínez, S. D.; Espinosa-Cristóbal, L. F.; Cuevas-González, J. C. Synthesis, Characterization, and in Vitro and in Vivo Evaluations of Cellulose Hydrogels Enriched with *Larrea tridentata* for Regenerative Applications. *BioMed Res. Int.* **2020,** *2020*. https://doi.org/10.1155/2020/1425402

UNESCO. La Ciencia Hace Retroceder al Desierto. *Gastronomía ecuatoriana y turismo local.* **2015,** *1* (69), 5–24.

Universidad Autónoma de Coahuila Coordinacion de Comunicación Institucional Oficina Virtual de Prensa. *Crea UAdeC Solución Sanitizante Natural a Base de la Planta "Gobernadora."* 2020.

Venkatesan, S. P.; Ganesan, S.; Prabhahar, J. J.; Kaveti, V. R.; Anoop, A.; Andrew, A. Performance and Emission Test on Diesel Engine Using *Prosopis juliflora* Seed Oil. *Int. J. Ambient Energy* 2020, *43* (1), 1701–1706. https://doi.org/10.1080/01430750.2020.1719887

Verbist, K.; Santibañez, F.; Gabriels, D.; Soto, G. *DocumentoTécnico No 25. Atlas de Zonas Áridas de América Latina y El Caribe*, 2010.

von Willert, D. J.; Eller, B. M.; Werger, M. J. A.; Brinckmann, E. Desert Succulents and Their Life Strategies. *Vegetatio* **1990,** *90* (2), 133–143. https://doi.org/10.1007/BF00033023

Vu, A. A.; Bose, S. Natural Antibiotic Oregano in Hydroxyapatite-Coated Titanium Reduces Osteoclastic Bone Resorption for Orthopedic and Dental Applications. *ACS Appl. Mater. Interf.* **2020,** *12* (47), 52383–52392. https://doi.org/10.1021/acsami.0c14993

Walsberg, G. E. Small Mammals in Hot Deserts: Some Generalizations Revisited. *BioScience* **2000,** *50* (2), 109–120. https://doi.org/10.1641/0006-3568(2000)050[0109:SMIHDS]2.3.CO;2

Walton, K. *Las Zonas Áridas*, 1ra ed.; 2007.

Weber, B.; Belnap, J.; Büdel, B.; Antoninka, A. J.; Barger, N. N.; Chaudhary, V. B.; Darrouzet-nardi, A.; Eldridge, D. J.; Faist, A. M.; Ferrenberg, S.; Havrilla, C. A.; Huber-sannwald, E.; Issa, O. M.; Maestre, F. T. What is a Biocrust? A Refined, Contemporary Definition for a Broadening Research Community. *Biol. Rev.* 2022, 000–0000. https://doi.org/10.1111/brv.12862

Wierzchos, J.; de los Ríos, A.; Ascaso, C. Microorganisms in Desert Rocks: The Edge of Life on Earth. *Int. Microbiol.* **2012,** *15* (4), 173–183. https://doi.org/10.2436/20.1501.01.170

Yu, Y.; Pi, Y.; Yu, X.; Ta, Z.; Sun, L.; Disse, M.; Zeng, F.; Li, Y.; Chen, X.; Yu, R. Climate Change, Water Resources and Sustainable Development in the Arid and Semi-Arid Lands of Central Asia in the Past 30 Years. *J. Arid Land* **2019,** *11* (1), 1–14. https://doi.org/10.1007/s40333-018-0073-3

Zhu, Y.; Zhang, J.; Zhang, Y.; Qin, S.; Shao, Y.; Gao, Y. Responses of Vegetation to Climatic Variations in the Desert Region of Northern China. *Catena* **2019,** *175*, 27–36.

CHAPTER 2

Geographical Description of the Arid Zone in the Chihuahuan Desert

LUIS FERNANDO CAMACHO-ORTEGÓN[1],
LUIS FERNANDO CAMACHO-GUERRA[2], ARTURO BUENO-TOKUNAGA[1],
JUAN JOSUÉ ENCISO-CARDENAS[1], and DIEGO MARTINEZ-CARRILLO[1]

[1]*Centro de Investigación en Geociencias Aplicadas, Universidad Autónoma de Coahuila, Nueva Rosita, Coahuila, Mexico*

[2]*Centro de Investigación y Jardín Etnobiológico del Semidesierto de Coahuila, Universidad Autónoma de Coahuila, Viesca, Coahuila, Mexico*

ABSTRACT

As a result of the constant discussion about the importance of the conservation of our natural resources and to update the knowledge about our ecoregion, we synthesize in a geographic description updated aspects of the Chihuahuan Desert, which is located between two crucial elongated orographic walls parallel to the coasts, the Sierra Madre Occidental to the west and the Sierra Madre Oriental to the east. This arid ecoregion is farther away from moisture sources, which has generated the vegetation and climate gradients typical of this large ecogeographic unit. The climate and topography of the Chihuahuan Desert are the main factors that create these patterns by limiting plant productivity, density, and species composition. The Chihuahuan Desert harbors an essential reserve of biological diversity. It possesses an unparalleled and rich native diversity, being the only ecoregion classified for its terrestrial and aquatic importance since its lakes, springs, rivers, and streams harbor a great variety of freshwater species. Specifically, in Coahuila, Mexico, we locate the Cuatrociénegas region, where a group of pools converges, fed by abundant subway water sources, where freshwater

Exploration and Valorization of Natural Resources from Arid Zones. Cristobal Noé Aguilar, Norma Paola Meléndez-Renteria, & Karen Nathiely Ramirez-Guzman (Eds.)

fish and invertebrates are extinct in the rest of the planet are still alive. In contrast, in the Viesca region, today, considered a moorland with a minimum humidity of 11% and an average of 32%, we can find biodiversity of endemic cacti that amaze for their excellent resistance to survival in a highly arid region of the Chihuahuan Desert.

2.1 INTRODUCTION

Arid zones are characterized mainly by their climate and a deficit of humidity in the environment, with highly irregular rainfall cycles, and by endemic herbaceous vegetation, rarely wooded, very irregularly distributed, and excessively sandy soils. These soils are frequently saline (Aubert, 1962). Under such severe conditions, its grounds show various characteristics, such as low abundant and superficial organic matter, high porosity, stable clay colloids, and saline nodules or calcareous crusts.

Aridity is complex and challenging for many life forms (Thomas, 2011), as moisture is a fundamental requirement for many of them, so the conservation of physical landscapes in arid areas against wind erosion and soil degradation from overgrazing becomes more challenging to restore, because of the extreme climatic conditions, as well as the relationships between people and their environment.

Globally, arid and semi-arid ecosystems occupy at least one-third of the Earth's surface (Pointing and Belnap, 2012) and are likely to expand due to human-induced desertification (Asner et al., 2004; Reynolds et al., 2005). In Mexico, these ecosystems are widely recognized for their plant diversity of more than 6000 endemic species, as well as for the variety of functional processes that distinguish them and for occupying nearly 60% of the national territory (Challenger, 1998; Balbontin et al., 2009; Montaño et al., 2016).

Besides being scarce in arid ecosystems, water input through precipitation is highly variable in time and space (Noy-Meir, 1973; Schwinning et al., 2004). These inputs determine the species composition of plant communities, the rate and seasonality of vegetation growth, the size and distribution of animal populations, nutrient cycles, and the cultural habits, survival strategies, and quality of life of the region's inhabitants (Noy-Meir, 1973). In this context, plants' loss of water utilization capacity becomes a critical process in desertification processes (Villagra et al., 2013). As a consequence, one of the most general ideas in arid zone ecology is that rainwater is mainly used by vegetation preventing deep drainage, runoff, and local recharge of subsurface aquifers (Noy-Meir, 1973; Jobbágy et al., 2011).

This work aims to geographically describe the arid zone of northeastern Mexico, particularly the arid-semiarid ecosystems within the Chihuahuan Desert (Figure 2.1).

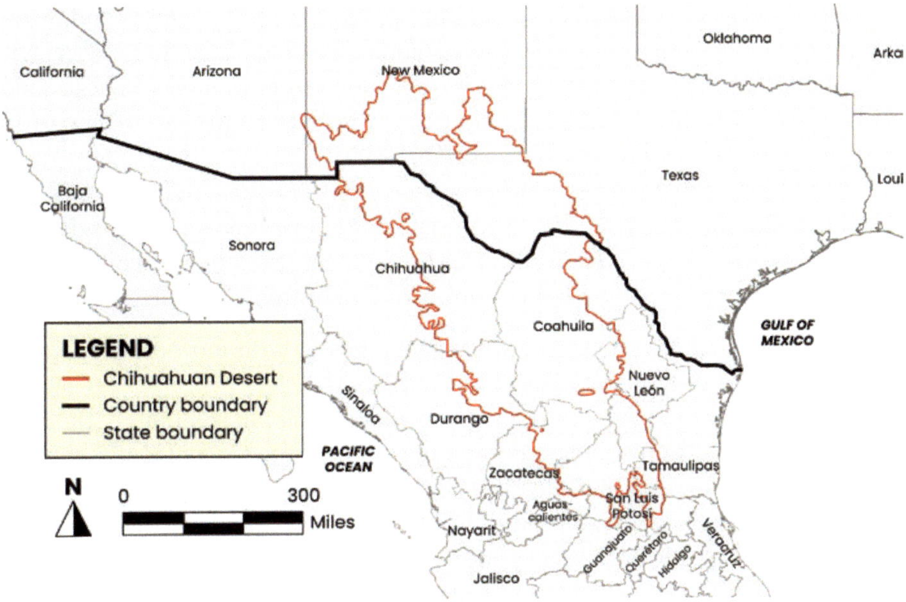

FIGURE 2.1 Delineation of Chihuahuan Desert ecoregion (red line) within Mexico and USA as defined by Schmidt (1979) and recently based on the wide distribution of endemic cactus species (see Henrickson and Straw, 1976; Hernández et al., 2008; Granados-Sánchez, et al., 2011; Adapted from Unnasch et al., 2017). The black line is the Country boundary, and the grey line is the state boundary.

Source: Adapted.

2.2 GEOGRAPHY OF ARID AND SEMI-ARID ZONES

According to Udvardy (1975), arid and semi-arid lands include forested and non-forested grasslands and savannas, warm deserts and semi-deserts, temperate zone grasslands, tundra communities, and cold desert biomes. These lands cover about one-third of the Earth's surface and are home to more than 900 million people.

Arid and semi-arid lands are notable for their genetic diversity within species rather than for variation between species or species richness. However, they contain a remarkable endowment of plant and animal species, including microorganisms. Dryland species have a very restricted geographic distribution (endemism) and a wide range of morphological, physical, and chemical adaptation to their harsh environment. These ecosystems provide critical habitats for wildlife and diverse ecosystems, including wetlands for migratory species.

Grove (1977) determined that, during the last twenty thousand years, semi-arid lands were sometimes wetter and sometimes drier than today, with significant consequences for today's soil and water resources. These lands were the scene of the beginnings of pastoralism, cereal cultivation, and urban life between 10,000 and 5000 years ago.

Considering that there are differences in the physical environment of the Chihuahuan Desert, Schmidt (1983) defines that, in general, it is relatively homogeneous, with varied differences at the local level, thus specifying four main characteristics of the general uniformity for this arid zone:

1. The mountain massifs surround the desert with uniform heights and similar orographic effects.
2. Most of the Chihuahuan Desert is located at approximately the same distance from sources of moisture, such as the Gulf of California, the Pacific Ocean, and the Gulf of Mexico.
3. The desert's position on the continent, with its relatively low latitudinal location and surrounding mountain masses, means that little moisture is derived from frontal activity; as a result, this arid zone is characterized by a very high percentage of its precipitation falling as rain during the summer months.
4. Very few flowing rivers exist, and few eroded lowland areas exist. Consequently, most of the surface area of this arid zone is characterized by basin and mountain range topography.

2.2.1 CHIHUAHUAN DESERT ECOREGION

The Chihuahuan Desert ecoregion covers an average of 647,500 square kilometers. It is the largest in the Americas and extends for 1500 km from the southeastern tip of Arizona, New Mexico, and the transpacific region of Texas in the United States of America, to portions of the states of Chihuahua, Coahuila, Durango, Nuevo León, San Luis Potosí, Tamaulipas, and Zacatecas, in Mexico (Figure 2.1).

Two nations share this ecoregion of the Chihuahuan Desert. It is flanked by a mountain range that blocks the passage of humid winds, made up of the Rocky Mountains (in the United States), the Sierra Madre Oriental, the

Trans-Mexican Volcanic Belt, and the Sierra Madre Occidental (in Mexico). About 9000 years ago, this area was wetter, and forests dominated the mountain slopes. As the area became drier, species experienced isolation, differentiation, and for some species, extinction, resulting in today's unique Chihuahuan flora and fauna (Granados-Sánchez et al., 2011; NPS, 2020).

The Chihuahuan Desert is considered the most diverse desert in the western hemisphere and one of the most varied arid regions in the world. Unfortunately, the Chihuahuan Desert ecoregion is also one of the most threatened regions in the world. Overgrazing, water depletion, detour, changes in fire regime, urbanization, increased agricultural and resource extraction activities, invasive exotic species, and overharvesting of native plants and animals are some of the greatest threats to the ecoregion's biodiversity (NPS, 2020).

In a recent delineation, Schmidt (1979) established a definitive boundary for the "true" Chihuahuan Desert, a frontier for the lesser-known desert in North America to arrive at something more definitive than the previous vague interpretations of the Chihuahuan Desert. The methodology to identify the Chihuahuan Desert's regional limits was mapped to mean annual temperature and precipitation values according to De Martonne (1926) aridity index (I_a). The indices were derived in eq 2.1.

$$Ia = \frac{P}{T°C + 10} \tag{2.1}$$

Where P is the annual amount of precipitation (in millimeters), and $T°C$ is the mean annual air temperature (in degrees Celsius) (see De Martonne, 1926; Schmidt, 1979; Croitoru et al., 2013).

2.3 VEGETATION AND SOILS

Following Rzedowski (1965, 1978), three basic vegetation types are recognized in the Chihuahuan Desert: microphyll desert scrub (MDS) on alluvial soils, rosette desert scrub (RDS) on limestone rocks and well-drained soils, and crassicaule desert scrub (CDS) linked to rocks and soils of an igneous nature. The MDS as alluvial desert scrub. For example, Shreve (1942) defines it as a microphyllous desert, and Leopold (1950) is a creosote bush. Gentry (1957) classified it as a Chihuahuan Desert scrub (see Granados-Sanchez et al., 2011).

Rzedowski (1965) defined a few terms to refer to the vegetation in the Chihuahuan desert. The MDS can be recognized by the predominance of small-leaved shrubby elements and is typical of flat terrain and the lower parts of the mountains. The RDS is distinguished by the predominance of shrub or subshrub species with elongated and narrow leaves, which take on the appearance of a rosette and correspond to what Rzedowski (1965, 1978) called calcareous desert scrub. Miranda and Hernández (1963) named yucca (izotal), maguey and Lechuguilla areas. Leopold (1950) defined the Nopalera areas of central Mexico as mesquite–grassland vegetation. Rzedowski (1978) had initially named it cactus–mesquite scrub, and Miranda and Hernández (1963) designated it as Nopalera (see Granados-Sanchez et al., 2011; Figure 2.2).

FIGURE 2.2 Photographic exhibition of the vegetation of the Chihuahuan Desert. Images of ten species of shrubs were captured, located mainly from the Centro de Investigación y Jardín Etnobiológico del Semidesierto de Coahuila (CIJE-UAdeC), in Viesca, Coahuila, Mexico. The images show in bold letters their common name, and their scientific name is slim letters. For more information see Rzedowski, 1978; Dinerstein, 2000; Cotera et al., 2004; Granados-Sánchez et al., 2011.

Source: Original.

Eight primary subdivisions of Chihuahuan Desert scrub include classification by Henrickson and Johnston (1986): (1) Chihuahuan Desert Scrub; (2) Lechuguilla Scrub; (3) Yucca Woodland; (4) Prosopis-Atriplex Scrub; (5) Alkali Scrub; (6) Gypsophilous Scrub; (7) Arorescent Cactus Scrub; and (8) Riparian Woodland.

Each classification is a floristic unit of the Chihuahuan Desert. At the regional scale, Granados-Sánchez et al. (2011) subdivide the Chihuahuan

Desert scrub into five types: (1) Larrea Tridentata Scrub; (2) Mixed Desert Scrub; (3) Sand Scrub; (4) Canyon Scrub; and (5) Sanddune Scrub. Henrickson and Johnston (1986) recognized that some of these scrub types are part of the thermal azonal vegetation of the riparian forest community and the dune phase of the Chihuahuan Desert scrub.

According to Reynolds (1986) and McAuliffe (1994), in the extension of the Chihuahuan Desert, three main geomorphological processes dominate the evolution of the landscapes: (1) The mechanical weathering caused by the growth of salt crystals (Figure 2.3), (2) The fracturing of the rocks by thermal expansion and contraction during the daily cycle of heating and cooling (Figure 2.4), and (3) The chemical weathering of the silicates exposed to the environment (Figure 2.5). We observe in the desert a unique effect of the exposed rock surfaces; these are darkened by the thin films of clay, iron oxides, and manganese that cover them (desert varnish).

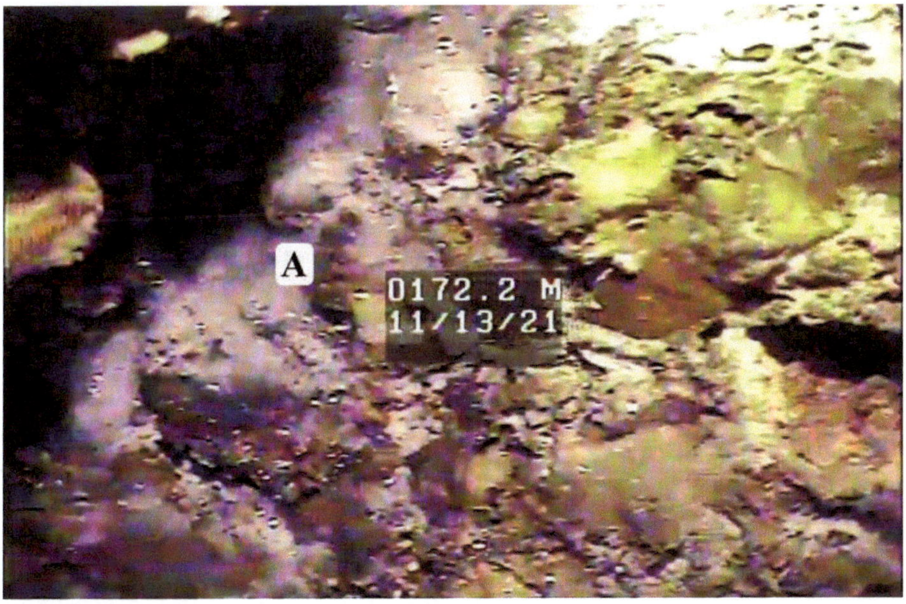

FIGURE 2.3 This photography is an example of the first geomorphological processes dominating the evolution of the landscapes; here, one observes the mechanical weathering caused by the growth of salt crystals (A) in this Chihuahuan Desert Mountain. The picture was captured on November 13, 2021, at a depth of 172 m, inside a natural cavern located by accident in the exploratory drilling of the Manuelita 1 well in Ejido Tomás Garrido, Viesca, Coahuila.

Source: Original.

FIGURE 2.4 The fracture of rocks by thermal expansion and contraction during the daily heating and cooling cycle is a natural phenomenon in the Chihuahuan desert area. This photo shows rock fragments affected by this phenomenon, creating a favorable environment for the endemic flora that grows around it. The photography was captured in the year 2022 in Ejido Tomás Garrido, Viesca, Coahuila. At the center of the image mysteriously coexists a Bishop's cap cactus (*Astrophytum myriostigma* subsp. *Coahuilense)* among hundreds of small rocks. This image demonstrates how desert endemism is in harmony.

Source: Original.

Information in published literature (e.g., Godínez, 1998) estimates that the evaporation of water from the soil and the transport of groundwater to the surface by capillarity produce a great variety of crusts on the rocks. According to Sánchez and Granados (2003), the surface layer of the sandstone layer is formed by a hard crust caused by the deposition of carbonates. According to Godínez (1998) and Sánchez and Granados (2003), in desert soils (aridisols), the petrocalcic horizon, so-called "caliche" when this eroded, gives rise to reliefs similar to cliffs, plateaus, and hills (see Granados-Sánchez et al., 2011).

As Schlesinger et al. (1996) and Ugalde et al. (2008) state, the soil, and vegetation are closely integrated; they affect the nutrient cycle and the fitness of organisms. The poor soils of the higher elevations are exposed due to the scarce vegetation cover weathering and the transport of materials to the lower elevations, generating vegetation chains along this gradient.

FIGURE 2.5 The arched folds (anticlines) are formed when movements in the crust compress rocks. These tectonic movements are perfectly observed in the folded sedimentary rock strata in the Tomás Garrido Canyon in Viesca, Coahuila. For example, this image shows the weathering of the arched strata exposed rock surfaces, obscured by chemical weathering of the silicates exposed to the environment (desert varnish).

Source: Original.

The distribution of vegetation is usually widely spaced except in wet sites, such as streams. Shrub plants generate microhabitats (oasis) for other plants through leaf fall and root input (Valiente-Banuet and Ezcurra, 1991; Miranda et al., 2004; see Granados-Sánchez et al., 2011).

2.3.1 CHIHUAHUAN DESERT GRASSLANDS

Natural grasslands or zacatales currently cover less than 15% of the Chihuahuan Desert, often in the form of mosaics of grasses and shrubs, distributed mainly towards the west and north of this ecoregion. In this ecoregion, there are several types of grassland such as (1) medium open grassland with dominant species such as *Bouteloua curtipendula*, *Bouteloua eriopoda*, and *Bouteloua gracilis*; (2) toboso grasslands with dominant species such as *Pleuraphis mutica*; (3) alkali grassland with dominant species such as *Sporobolus wrightii* "Windbreaker" and *Sporobolus airoid*es; and (4) open grassland with agave, sotol, yucca, and nopal (Dinerstein et al., 2000; Cotera et al., 2004; Guzman-Aranda et al., 2012).

2.4 HYDROLOGY

The availability of surface water, groundwater, and air humidity is the main factor limiting the natural distribution of trees in arid zones, climate (precipitation, temperature, and wind), and soil quality (Malagnoux et al., 2007).

Considering that rainfall is the indicator that defines land aridity, Noy-Meir (1973) determined three main attributes for arid ecosystems, one almost by definition, two others by correlation with the first: (1) precipitation is so low that water is the dominant controlling factor for biological processes; (2) precipitation is highly variable throughout the year and occurs in infrequent and discrete events; (3) precipitation variation has a significant (unpredictable) component, defining desert ecosystems as "water-controlled ecosystems with infrequent, discrete, and largely unpredictable water inputs."

Brisson and Reynolds (1994) infer that the topography of arid zones is modeled mainly by moving water, although there are places where rainfall occurs once every few years; when it does, watercourses play an essential role as agents of erosion, transport, and sedimentation. In a dry climate, rivers flow along significant accumulations of permeable alluvium, and stream water infiltrates the bed and contributes to groundwater collection. Without ample vegetation cover to protect the terrain and slow the descent of water downstream, large quantities of eroded material are transported to the valleys, forming river fans.

Alluvial fans are low cones of sand and gravel whose profile resembles that of an open Japanese fan. The fan's apex, or center point, is located at the mouth of a canyon or gully and opens onto the adjacent plain. Alluvial fans are highly variable in size; some can reach several kilometers. The rivers that originate these fans carry large loads of material from the higher elevations (e.g., Figure 2.6).

FIGURE 2.6 Map of the study area showing the subaerial Alluvial fan plain with vegetation zones in Viesca lagoon. Google Earth captured this image and clearly shows the fan's apex, or center point, located at the mouth of a canyon or gully and opens onto the adjacent plain. This alluvial fan is located exactly in the Tomás Garrido Mountains, Ejido Tomás Garrido, Viesca Coahuila. Reprinted with permission from Google Earth Pro, Apache License, Version 2.0, January 2004, https://www.apache.org/licenses/. (See https://earth.google.com/web/@25.25255783,-102.70396079,1146.49770763a,12697.93199029d,35y,0h,0t,0r)

Source: Reprinted.

2.5 CLIMATE IN ARID AND SEMI-ARID ZONES

Precipitation and temperature are the main factors that determine the structure and functioning of an ecosystem. Other factors include long-term weather patterns, such as the timing, duration, and intensity of extreme weather conditions (such as frosts and violent downpours), and seasonal variations between years.

The important differences in aspect and topography of the Chihuahuan Desert Ecoregion make the climate spatially variable. For example, Schmidt

(1979), according to De Martonne's (1926) model to delimit the Chihuahuan Desert ecoregion, considers that an aridity index value (I*a*; see eq 2.1) of less than five generally characterizes true deserts, around ten corresponds to dry steppes, values of about twenty to grasslands, and above thirty dominates the forest. These factors greatly influence the presence of plant species, the suitability of the habitat for wildlife, surface and subsurface hydrology, nutrient cycling, and the relationships between soil, plant, and water availability. They can also affect the sensitivity or resilience of an ecosystem to disturbance.

The Earth's variable climate is a function of the uneven distribution of solar radiation. At the same time, the moisture content in the atmosphere is a function of air temperature and proximity to water sources. It has been shown that warm tropical air rises, expands, cools adiabatically, condenses, forms clouds, and produces precipitation over much of the equatorial zone. This, related to the Earth's rotation, causes air masses to migrate towards the poles, where the air descends, compresses, warms adiabatically, increases its tendency to retain moisture, and eventually results in warm and dry air at the surface.

This heating of the middle troposphere inhibits the deep convection necessary for forming heavy rainfall. The arid climatic conditions mean that many of the world's deserts are located in these high-pressure belts surrounding the planet.

2.5.1 CLIMATE OF THE CHIHUAHUAN DESERT

The Chihuahuan Desert is a large and relatively isolated arid zone in the central subtropical latitudes of North America, ranging in a relatively high-elevation basin and range region (ranging approximately 400–3100 m above sea level) generally south of the Colorado Plateau and between the Sierra Madre Oriental and the Sierra Madre Occidental ranges and is bisected by the USA–Mexico border (e.g., Baddock et al., 2011). As a result, it tends to have a slightly milder climate during the summer compared to the Sonoran Desert to the west. However, temperatures usually range between 35°C and 40°C during the day.

Winter weather varies from moderately mild to very cold, depending on altitude. Precipitation in the form of snow is scarce except in regions located at higher elevations (Godínez, 1998; Huerta-Martínez and García-Moya, 2004). Precipitation is more abundant compared to the Chihuahuan Desert with most of the Great Basin Desert and the Sonoran and Mojave deserts;

however, it is less than 250 mm per year, with most of the rain falling during the monsoon season in late summer (see Granados-Sanchez et al., 2011).

2.6 CONCLUSIONS

The Chihuahuan Desert lies between two primary elongated orographic walls parallel to the coasts. These mountains are the Sierra Madre Occidental to the west and the Sierra Madre Oriental to the east. The high intermountain plateau that occupies the area between the sierras is a source of high heat that triggers instability and lifts moist air on the slopes of the sierras to the east and west of the Chihuahuan Desert. Thus, the air currents release most moisture by forcing elevation toward the sea on the mountain slopes. To the north, between the vast plains of the Chihuahuan Desert, there are isolated mountains, where the sierras become more fragmented. Still, the desert is farther away from the sources of moisture, which has generated the vegetation and climate gradients typical of this large ecogeographic unit.

The climate and topography of the Chihuahuan Desert are the main factors that create these patterns by limiting plant productivity, density, and species composition, which indirectly influence the soil formation processes of the ecoregion.

By their very name, deserts are rarely considered essential reserves of biological diversity, but some deserts are extraordinarily rich in species, plants, fauna, and endemic natural communities. For this reason, the Chihuahuan Desert, home to 350 of the 1500 species of cacti known worldwide, possesses an unequaled and rich native diversity, including 333 species of birds, 23 species of fish, and 76 species of reptiles and amphibians. It is the only ecoregion classified for its terrestrial and aquatic importance, as its lakes, springs, rivers, and streams are home to many freshwater species. In addition, in the Viesca ecoregion, there is an endemic abundant biological diversity that is being recognized and is little studied, which will allow us to discover more information about such biodiversity in such an extremely arid environment in the coming decades.

The climatic conditions and low rainfall in the Chihuahuan Desert ecoregion make water a crucial resource. The Río Bravo, fed by the Pecos and Conchos rivers, among other smaller rivers, is considered the heart of the only central river system in the ecoregion. In Coahuila, we find in the region of Cuatrociénegas a group of pools fed by abundant subway water sources, where fish and freshwater invertebrates that are not found anywhere else on Earth are alive.

KEYWORDS

- **ecoregion**
- **natural resources**
- **geographic description**
- **hydrology**
- **Chihuahuan Desert**

REFERENCES

Asner, G. P.; Elmore, A. J.; Olander, L. P.; Martin, R. E.; Harris, A. T. Grazing Systems, Ecosystem Responses, and Global Change. *Ann. Rev. Environ. Resour.* **2004,** *29*, 261–299. https://doi.org/10.1146/annurev.energy.29.062403.102142

Aubert, G. Les Sols de la Zone Aride Etude de leur formation, de leurs caractères, de leur utilisation et de leur conservation. UNESCO, Colloque général sur les problèmes de la zone aride : Paris 1960. *Actes du colloque* **1962**, 127–150.

Baddock, M. C.; Gill, T. E.; Bullard, J. E.; Dominguez-Acosta, M.; Rivera-Rivera, N. I. Geomorphology of the Chihuahuan Desert Based on Potential Dust Emissions. *J. Maps* **2011,** *7* (1), 249–259. https://doi.org/10.4113/jom.2011.1178

Bailey, H. P. Semi-Arid Climates: Their Definition and Distribution. In *Agriculture in Semi-Arid Environments*; Hall, A. E., Cannell, G. H., Lawton, H. W., Eds.; Ecological Studies, Springer, 1979; vol *34*, pp 73–97. https://doi.org/10.1007/978-3-642-67328-3_3

Balbontin, C.; Cruz, C. O.; Paz, F.; Etchevers, J. D. Soil Carbon Sequestration in Different Ecoregions of Mexico. In *Soil Carbon Sequestration and the Greenhouse Effect*; Rattan L.; Ronald F. F., Eds.; SSSA Special Publication: Madison, WI, USA, 2009; vol *57*, pp 71–96. https://doi.org/10.2136/sssaspecpub57.2ed.c5

Brisson, J.; Reynolds, J. F. The Effect of Neighbors on Root Distribution in a Creosotebush (Larrea Tridentata) Population. *Ecology* **1994,** *75*, 1693–1702. https://doi.org/10.2307/1939629

Challenger, A. Utilización, y conservación de los ecosistemas terrestres de Mexico: pasado, presente y futuro. CONABIO: Mexico, DF, 1998; p 848.

Cotera, M.; Guadarrama, E.; Brenner, J.; Arango, A. M.; et al. Ecoregional Conservation Assessment of the Chihuahuan Desert; *Pronatura Noreste, The Nature Conservancy, WWF*, 2004, p 706. www.worldwildlife.org/wildplaces/cd/science.cfm

Croitoru, A. E.; Piticar, A.; Imbroane, A. M.; Burada, D. C. Spatiotemporal Distribution of Aridity Indices Based on Temperature and Precipitation in the Extra-Carpathian Regions of Romania. *Theor. Appl. Climatol.* **2013,** *112*, 597–607. https://doi.org/10.1007/s00704-012-0755-2

Cruz-Aragón, A.; Cruz-Medina, J.; Melgarejo, E. D.; Martinez de la Vega, G.; Flores-Rivas. J. D.; et al. La biodiversidad en San Luis Potosí – Estudio de Estado. *CONABIO, Secretaria de Ecología y Gestión Ambiental*, 2019; p 295.

De Martonne, E. Areisme et Indice d'aridite, *Comptes Rendus de L'Academie des Sciences de Paris*. **1926,** *182*, 1395–1398.; (in English) Regions of Interior Basin Drainage. *Geogr. Rev.* **1927,** *17*, 397–414. https://doi.org/10.2307/208323

Dinerstein, E.; Olson, D.; Atchley, J.; Loucks, C.; Contreras-Balderas, S.; Abell, R.; Iñigo, E.; Enkerlin, E.; Williams, C.; Castilleja, G. Ecoregion-Based Conservation in the Chihuahuan Desert; *WWF/TNC/CONABIO/PRONATURA/ITESM*, Mexico, **2000**; p 376. http://awsassets.panda.org/downloads/wwfbinaryitem2757.pdf

Gentry, H. S. In *Los pastizales de Durango*, Instituto Mexicano de Recursos Naturales Renovables: Mexico, D.F., 1957; p 361.

Godínez, H. Los desiertos mexicanos, sus características e importancia. *Ciencia y Desarrollo*. **1998,** *143*, 17–22.

Granados-Sánchez, D.; Sánchez-González, A.; Granados-Victorino, R. L.; Borja de la Rosa, A. Ecología de la vegetación del Desierto Chihuahuense. *Rev. Chapingo Ser. Cienc. For. Ambient.* **2011,** *17* (spe), 111–130. https://doi.org/10.5154/r.rchscfa.2010.10.102

Grove, A. T. The Geography of Semi-Arid Lands. *Phil. Trans. R. Soc. Lond.* **1977,** *278*, 457–475. https://doi.org/10.1098/rstb.1977.0055

Guzmán-Aranda, J. C.; Hoth, J.; Berlanga, H. Plan Maestro de la Alianza Regional para la Conservación de los Pastizales del Desierto Chihuahuense; *Comisión para la Cooperación Ambiental*; Montreal, 2012, p 64. http://rmbo.org/v3/Portals/0/Documents/International/Master_Plan_Chihuahua_jan2012.pdf

Henrickson, J. Y.; Johnston, M. C. In *Vegetation and Community Types of the Chihuahuan Desert*, Second Symposium on Resources of the Chihuahuan Desert Region; Barlow, J. C., Powell, A. M., Timmermann, B. N., Eds.; The Chihuahuan Desert Research Institute: Chihuahuan Desert, U.S. and Mexico, 1986; vol II, pp 20–39.

Huerta-Martínez, F. M.; García-Moya, E. Diversidad de especies perennes y su relación con el ambiente en un área semiárida del centro de Mexico. *Interciencia* **2004,** *29* (8), 435–441.

Jobbágy, E. G.; Nosetto, M. D.; Villagra, P. E.; Jackson, R. B. Water Subsidies from Mountains to Deserts: Their Role Sustaining Groundwater-Fed Oases in a Sandy Landscape. *Ecol. Appl.* **2011,** *21* (3), 678–694.

Leopold, A. S. Vegetation zones of Mexico. *Ecology* **1950,** *31*, 507–518.

Malagnoux, M.; Sène, E. H.; Atzmon. N. Les forêts, les arbres et l'eau dans les terres arides : un équilibre précaire. *Unasylva* **2007,** *229* (58), 24–29.

McAuliffe, J. R. Landscape Evolution, Soil Formation, and Ecological Patterns and Processes in Sonoran Desert Bajadas. *Ecol. Monogr.* **1994,** *64* (2), 11–48.

Miranda, F.; Hernández, X. E. Fisiografía y vegetación. In: Las zonas áridas del centro y noreste de Mexico, Ed.; *IMRNR*: Mexico, DF, 1963; pp 1–27.

Miranda, J. D.; Padilla, F. M.; Pugnaire, F. I. Sucesión y restauración en ambientes semiáridos. *Ecosistemas* **2004,** Año XIII, N°1, Enero–Abril, 1–5. http://iis1.eeza.csic.es/eeza/documentos/Miranda%20et%20al_Ecosistemas%202004_1.pdf

Montaño, N. M.; Ayala, F.; Bullock, S. H.; Briones, O.; García-Oliva. F.; et al. Almacenes y flujos de carbono en ecosistemas áridos y semiáridos de Mexico: Síntesis y perspectivas. *Terra Latinoamericana*. **2016,** *34* (1), 39–59.

Noy-Meir, I. Desert Ecosystems: Environment and Producers. *Ann. Rev. Ecol. Syst.* **1973,** *4* (1), 25–51. https://doi.org/10.1146/annurev.es.04.110173.000325

NPS. Chihuahuan Desert Inventory & Monitoring Network, National Park Service, U.S. Department of the Interior, **2020**. https://www.nps.gov/im/chdn/index.htm

Pointing, S.; Belnap, J. Microbial Colonization and Controls in Dryland Systems. *Nat. Rev. Microbiol.* **2012,** *10*, 551–562. https://doi.org/10.1038/nrmicro2831

Reynolds, J. Adaptative Strategies of Desert Shrubs with Special Reference to the Creosotebush. In *Pattern and process in desert ecosystems*; University of New Mexico Press, 1986; p 139.

Reynolds, J. F.; Maestre, F, T.; Huber-Sannwald, E.; Herrick, J.; Kemp, P. R. Aspectos socioeconómicos y biofísicos de la desertificación. Ecosistemas. *Revista científica y técnica de ecología y medioambiente* **2005,** *14* (3), 3–21.

Rzedowski, J. Vegetación del Estado de San Luis Potosí. *Acta Científica Potosina* **1965,** *1* (2), 5–291.

Rzedowski, J. Vegetación de Mexico, Limusa: Mexico, D. F.; 1978, p 432.

Sánchez, G. A.; Granados, S. D. Ordenación de la Vegetación de la Sierra Catorce, San Luis Potosí, a lo largo de gradientes ambientales. *Rev. Terra.* **2003,** *21* (3), 311–319.

Schlesinger, W. H.; Raikes, J. A.; Hartley A. E.; Cross, A. F. On the Spatial Pattern of Soil Nutrients in Desert Ecosystems. *Ecology* **1996,** *77*, 364–374.

Schmidt, R. H. A Climatic Delineation of the "Real" Chihuahuan Desert. *J. Arid Environ.* **1979,** *2* (3), 243–250. https://doi.org/10.1016/S0140-1963(18)31774-9

Schmidt, R. H. Climate and the Chihuahuan Desert, Chapter 4. In *Natural Resources and Development in Arid,* Regions (1st ed.); Campos-López, E., Anderson, R. J., Eds.; Routledge, 1983; pp 35–52. https://doi.org/10.4324/9780429049378

Schwinning, S.; Sala, O. E.; Loik, M. E.; Ehleringer, J. R. Thresholds, Memory, and Seasonality: Understanding Pulse Dynamics in Arid/Semi-Arid Ecosystems. *Oecologia* **2004,** *141*, 191–193. https://doi.org/10.1007/s00442-004-1683-3

Shreve, F. The Desert Vegetation of North America. *Bot. Rev.* **1942,** *8* (4), 195–246. https://www.jstor.org/stable/4353268

Thomas, D. S. G. In *Arid Zone Geomorphology: Process, Form and Change in Drylands*; Third Ed.; John Wiley & Sons, 2011; p 656. https://doi.org/10.1002/9780470710777

Udvardy, M. D. F. A Classification of Biogeographical Provinces of the World. *UNESCO, IUCN* Occasional Paper no. 18, Morges, Switzerland, 1945, p 49.

Ugalde, A. J.; Granados-Sánchez D.; Sánchez-González, A. Sucesión en el matorral desértico de Larrea tridentata (DC.) Cov. en la Sierra de Catorce, San Luis Potosí, Mexico. *Terra Latinoamericana* **2008,** *26*, 153–160.

Unnasch, R.; Braun, D.; Young, K. *Chihuahuan Desert Rapid Ecoregional Assessment Pre-Assessment Report*; Batcher, M., Fogarty, F., Marty, J., Salo, C., Seamster, V., Welch, N., Whittier, T., Sound Science Technical Report to the U.S. Department of the Interior Bureau of Land Management, Rapid Ecoregional Assessment Program, 2017; p 485.

Valiente-Banuet, A.; Ezcurra, E. Shade as a Cause of the Association Between the Cactus Neobuxxaunia Tetetzo and the Nurse Plant Mimosa Luisana. *J. Ecol.* **1991,** *79*, 961–971. https://doi.org/10.2307/2261091

Vaughn, D. M. Arid Climates. In *Encyclopedia of World Climatology*; Oliver, J. E., Ed.; Encyclopedia of Earth Sciences Series; Springer: Dordrecht, 2005; pp 85–89. https://doi.org/10.1007/1-4020-3266-8_16

Villagra, P. E.; Pablo. A.; Meglioli, F. I.; Pugnaire, B. V.; et al. La regulación de la partición del agua en zonas áridas y sus consecuencias en la productividad del ecosistema y disponibilidad de agua para los habitantes, In *Servicios ecosistémicos hídricos: estudios de caso en América Latina y el Caribe*; Lara, A., Laterra, P., Manson, R., Barrantes. G., Eds.; Red ProAgua CYTED, Imprenta América: Valdivia, Chile, 2013; pp 111–125.

CHAPTER 3

Physiological Requirements of Microorganisms for Growth in Extreme Temperatures

MARIANA DELGADO GARCÍA

Tecnologico de Monterrey, Escuela de Ingenieria y Ciencias, Nuevo Mexico, Zapopan, Jalisco, Mexico

ABSTRACT

Currently, the study of extremophile microorganisms had increased and some of them such as Pyrococcus sp., Thermothoga sp. Thermococcus sp. and others have been identified and studied, being an important subject to understand the different strategies of life on Earth. However, the microbial life in the extreme conditions is complex, interesting, and little understood. The first step to understand and take advantage of these microorganisms, is to change the perception and know the difference compared to others. Particularly, in extreme temperatures, the microorganisms are able to survive, have special physiology and metabolic machinery, and are different from such psychrophiles as thermophiles. To adapt, the cell has a robust composition such as the lipid cell wall; particularly, UFAs, PUFAs, BFCAs, and SFACs are the most predominant ones but it also depends of the temperature to which the cell is exposed, giving it greater fluidity and integrity in front of that environmental changes. Their great capacity to contrast the oxidative stress, the heat shock proteins, and the synthesis of thermoprotectants make them a model for the design and construction of improved microorganisms. Their extraordinary molecular capacity to express a protein with different chemical and structural properties makes them an object to study for the generation of new and robust biofactories. Not only their cell properties are

Exploration and Valorization of Natural Resources from Arid Zones. Cristobal Noé Aguilar, Norma Paola Meléndez-Renteria, & Karen Nathiely Ramirez-Guzman (Eds.)

very interesting in these microorganisms, their behavior and cooperation with other microorganisms, which has allowed us to understand that the most important communications are Autoinducer-1 (AI-1 system), Peptide based system, Autoinducer-2 (AI-2), and AHL signaling between them for their survival.

3.1 INTRODUCTION

Until past few decades, it was known that in the Earth there were hostile places to live. However, with several studies, this is erroneous and now known that limits of life in the Earth are not well defined. Due to this, the nature has shown us that is very wise and is an instructor of survival for humanity for the distinct and impressive life styles of the biologic systems where some of them have endured through millions of years. One of them is the life in extreme conditions, pressure, drought, salinity, pH, and temperature where much of the biologic systems are sensible to drastic conditions, but there are biologic systems particularly microorganisms that have shown tolerance not only toward these conditions, but they are so adapted, that is a necessity to live: they are called extremophiles (Rampelotto, 2010).

Described above, there are great varieties of extreme conditions present in different zones or ecosystems around the Earth. One of them is the arid zones that occupy as much as 33% of the global terrestrial surface. In this area, there are some abiotic characteristics like strong winds that cause erosion and dunes formation (in some places for the evotranspiration due to the high temperatures) and are characterized by 100–300 mm of annual rainfall (Powell et al., 2013). In these places, there is presence of organisms that are able to grow at extreme temperatures, thermophiles that grow between 45 and 80°C. However, in the case of arid zones, not only are high temperatures, but the fluctuation of the temperatures during the day makes possible cold temperatures (0–37°C in hot deserts), so it can present the psycrophiles that grow below 0–20°C. To understand the presence of extremophiles, particularly in arid zones, it is important to mention that one extreme microorganism can live and adapt to one more extreme condition, called poliextremophiles (prokaryotic microorganisms are most frequently present) (Rampelotto, 2010; Wang et al., 2013). In this section, some characteristics of their wide and versatile metabolic diversity coupled with the extraordinary physiological capacities to survive at extreme temperatures are described, showing the unlimited amount of habitats on Earth that are suitable for microbial life leading to the search of life inside and outside of it.

This leads to understanding a little, one of the theories of the life origin and the unique ancestral survival strategies.

3.2 TYPICAL CELL CHARACTERISTICS FOR EXTREME TEMPERATURES SURVIVAL

3.2.1 LIPID MEMBRANE COMPOSITION

All cells have different components that constitute it, with different functions and chemical composition. The cellular membrane, a component that has been known for his role as a barrier for extracellular environment, also is usually permeable for several ions and solutes to maintain a cellular homeostasis. Although the cellular membrane is very complex, an important characteristic is the lipid composition and its interactions that allow a matrix with a "fluidity and permeability" that it is possible that it changes chemically due to the external environment like extreme temperatures (Konings et al., 2002). Commonly, the lipid composition of the cellular membrane is lipids of head polar groups oriented into a water phase, while having a hydrocarbon chain composed of the hydrophobic section positioned in the interior part of the membrane, generally in crystalline phase. However, one of the differences between chemical composition of the lipid membrane of the three life domains is associated with Darwanian evolution, LUCA, and cenancestor as some hypothesis mentioned, although this distinction is not so sharp (Fitch and Upper, 1987; Konings et al., 2002; Lombard et al., 2012). It is known that since several millions of years during the origin of life, the presence of hydrothermal systems was noted, which is largely transient in nature and crucial to understand the survival processes of microorganisms in extreme environments, especially for thermophiles and hyperthermophiles (Rampelotto, 2010; Lombard et al., 2012).

Another important factor of the difference between lipid composition of the three life domains is the physochemical parameters, like temperature, pH, hydrostatic pressure influencing fluidity and permeability of membrane. Normally, while Bacteria and Eukarya have two fatty acyl group chains linked by ester bond with glycerol (sn-glycerol-3-phosphate) and this is linked to phospho- or glycol- headgroups, Achaea has two phytanyl chains linked by ether bonds with glycerol or other alcohols (sn-glycerol-1-phosphate) and the acyl groups are fully saturated isoprenoids (Konings et al., 2002; Koga, 2012). For this, the core lipids are different between three life domains named "lipid divided," phosphatidic acid, and archeatidic acid, more evidenced and studied between Bacteria and Archaea, playing an important role in membrane functionality and they are who suffer major structural and composition changes (Koga 2011, 2012).

3.2.2 PARTICULAR CHEMICAL COMPOSITION OF LIPID MEMBRANE UNDER EXTREME TEMPERATURES

The conditions to survive to temperature changes or extreme temperatures that microorganisms can be exposed and maintain the membrane integrity and fluidity, had been documented and investigated a particular chemical-structural lipid membrane modification describe below (Table 3.1), not only at high temperature, also for cold temperatures. Some peculiarities are that archaeal membrane is more robust than bacterial membrane, bacteria dominate low temperatures and Archaea high temperatures, the core lipids arrangement of bacteria allows a better permeability and in archaeal core lipids confers more rigidity (Koga, 2012).

In cold temperatures the fatty acids are low disorder and the density is reduced, but allow a transition of liquid to gel phase to maintain the fluidity given principally by SCFAs, or/and the ratio of BFCA's anteiso (C-15 around 50%) with methyl groups in the case of Bacteria. The polyunsaturated fatty acids (LC-PUFAs) are the principal to cold adaptation, principally eicosapentanoic, docosahexanoic, and arachidonic that are in high content. These (LC-PUFAs) (LC-PUFAs) have an antioxidant activity and are able to form hydrophobic interfaces preventing ROS inside of cell (Feng et al., 2014; Yoshida et al., 2016; Collins and Margesin, 2019). Also are genes for fatty acid denaturation (desaturases), production of branched chain fatty acids (KAS-II and KAS III) cis-isomerisation and genes for membrane rigidity (De Maayer et al., 2014; Gordial et al., 2016; He et al., 2015; Koga, 2012; Collins and Margesin, 2019).

The more the temperature decreases, more levels of SCFAs, UCFAs, carotenoids, and glycolipids, this showed in (T_{opt}: 4°C), *Colwellia psychrerythraea* (T_{opt}: 8.5°C) as examples (Koga, 2012; Guan et al., 2013; Wan et al., 2016). In addition, polar carotenoids create a "balance" of the fluidizing effect and allow functionality of the transport of oxygen and ion inside the cell. Also, these participate in the modulation of the membrane fluidity in the homeosviscous adaptation creating a balance between the fatty acids stabilizing it; the pigmentation is very common in these microorganisms for the photoprotection, antioxidant activity (protection against to ROS), and cryoprotectant in front to freeze cycles (Pandey et al., 2018; Shen et al., 2018; Collins and Margesin, 2019). On the other hand, in the case of Archea, there are more unsaturations given by a high number of diethers, more fluidity is the archeal lipidic membrane, shown in *Thermococcus kodakarensis* (Matsuno et al, 2009). The presence of squalene or lycopene as principal poly-isoprenoids acts as packers and organizer of polar lipids; squalene can promote this rigidity and necessary permeability to maintain the cell integrity shown in *Halobacterium salinarum*. As a cold-shock response, Archaea has decrement in tetraether and pentacycle number and can maintain the liquid crystalline phase in temperatures below 0°C (Koga, 2012).

TABLE 3.1 Chemical Composition of Microbial Lipid Membrane Under Extreme Temperatures (Matsuno et al., 2009; Koga, 2012, Gilmore et al., 2013; Guan et al., 2013; Wan et al., 2016).

Environment	Microorganisms lipid membrane
Cold Temperatures	**Bacteria**
	1. Degree saturation acyl chain: Unsaturated fatty acids (UFAs), short chain fatty acids (SCFAs), monounsaturated fatty acids (MUFAs), and long chain polyunsaturated fatty acids (LC-PUFAs), these last as the most modification incorporated
	2. Branched chain fatty acids (BCFAs) (ratio iso/anteiso)
	3. Carotenoids
	4. Glycolipids and uncommon polar lipids
	Archaea
	1. Unsaturated diethers
	2. Isoprenoid hydroxylation
	3. Diether/tetraether ratio
	4. Pentacicly presence
High Temperatures	**Bacteria**
	1. Branched chain iso-fatty acids
	2. Saturated fatty acids
	3. Long-chain fatty acids
	4. Polar carotenoid content
	Archaea
	1. Tetraether amount
	2. Pentacicly presence
	3. Core lipids modifications

Contrarily, when microorganisms of Bacteria are in high temperatures, the chemical structural modification (Table 3.1) allows a hexagonal and cubic liquid crystalline phase, generating a more rigidity and ordering of hydrocarbon chains. There are more amount of BCFAs, particularly, *iso* and the modification in fatty acid chain composition allow the correct permeability in the cell (Konings et al., 2002; Koga, 2012). In Archaea, the presence of tetraether in thermophiles is important, confers high stability at monolayers and impact on the motility of hydrocarbon chains. In addition, the pentacycli particularly of C40 isoprenoids stabilize the membrane and modified archaeols (macrocyclics) that are cross-linked and can be packed more tightly generating a retiringly motion. The glycerol by ether bonds confers a thermal stability (Konings et al., 2002; Koga, 2011, 2012) These structural changes can show a major presence of water molecules trapped in the membrane layer, allowing a major proton permeability (Konings et al., 2002). Actually, there are a typical named "thermophilic lipids" for their prevalence in lipid membrane in the species of Bacteria an Archaea which are: archeol, cyclic archeol, digeranylgeranylglycerophosphate (DGGGP), caldarcheol, cyclopentane-containing caldarcheol, 15,16-dimethyl-30-glyceryloxytriacontanoic acid, 1,2-di-hydroxynonadecane, palmitic acid, *cis*-vaccenic acid (MUFA), iso-C17 fatty acid, anteiso-C17 fatty acid, 15,16-dimethyltriacontandioic acid (diabolic acid), 11-cyclohexylundecanoic acid (Koga, 2011).

3.3 METABOLISM ASSOCIATED TO SURVIVE IN EXTREME TEMPERATURES

Change of environment conditions can produce metabolic disorders as reduced growth, extended lag phase, metabolic cessation, or death. However, when the microorganisms are exposed gradually to extreme conditions, their adaptation generally to more than one condition is possible (Guan et al., 2017). To understand the survival metabolic strategies of microorganism under extreme temperatures, it is not only important to know their special characteristics, but their biotechnological potential of special metabolites, systems biology, synthetic biology or metabolic engineering applied like a new way to solve the humanity and environmental actual problems.

3.3.1 OXIDATIVE STRESS

Molecular oxygen is an important element for different metabolic processes specifically for aerobic microorganisms during respiration chain and energy generation, is a small and nonpolar molecule able to diffuse through the membranes. During different cell reactions, the generation of ROS (reactive oxygen species), superoxide (O_2^-), hydrogen peroxide (H_2O_2), or hydroxyl group (OH^-) is possible which react with other O_2^- molecules or nitrogen that makes very toxic for lipid membrane. H_2O_2 can be a precursor of free radicals when the UV radiation is present, while OH^- can react with metals like Fe^{2+}, Fe^{3+}, or Cu^+ and the reaction into the cells is also possible (Bokare and Choi, 2014).

The results of these different reactions are the lipid peroxidation reactions where causing and accumulation decreasing the permeability, fluidity, damage DNA and disruption of carbohydrates of nitrogen bases generating cell death (Cabiscol et al., 2010; Uchida, 2015; Gambino and Cappitelli, 2016; Guan et al., 2017). Specifically, the microorganisms able to grow in extreme temperature have some strategies to counter the negative effect due to this environmental condition causing a high oxidative stress. During these processes, there are three important strategies: lipid membrane enzymes, catalytic enzymes, and the nonenzymatic antioxidants production for ROS detoxification and the production of biofilms and exopolysaccharides. First, the lipid membrane composition or structural conformation changes becoming a gross membrane, then the unsaturated fatty acid increasingly reduces the fluidity, and the fatty acid desaturase participates in this process (Guerzoni et al., 2001; Guan et al., 2017).

Second, catalytic enzymes like superoxide reductase, superoxide dismutase, catalase, and hydrogen peroxide are produced by thermophiles in defense of oxidative stress. The peroxiredoxin in most systems is used for ROS scavenge, reported in *Hydrogenobacter thermophilus*. This enzyme has a domain swapping which is highly conserved in thermophiles having a NADPH oxidoreductase (FNR)-dependent peroxidase activity and hydrogen and peroxidase reductase activities (Sato et al., 2012; Ranawat and Rawat, 2017). Other enzymes, like superoxide dismutase, are used to avoid the oxidative stress forming hydrogen peroxide and molecular oxygen, where the form reported in some thermophiles is Fe as atom in the active site of SOD (García et al., 2015; Imlay et al., 2008). For DNA damage caused by ROS, thermophiles have repair mechanism using DNA glycosylase gene encoding for an enzyme for guanine to thymine transversion. This enzyme had been found in *T. thermophilus* and is stable at 85°C and in low temperatures (Ranawat and Rawat, 2017). Also, the transient control of DNA topology is given by a DNA topoisomerase or gyrase B as a regulation of stress response. During the ROS balance into the cell, the production of non-antioxidants like ascorbate, flavonoids, carotenoids, thioredoxin, and glutathione is possible.

Third, one of the most important mechanisms to avoid the negative effect of ROS is the production of biofilms and exopolysaccharides. These compounds are large, complex, highly diverse structurally, but they are composed principally of carbohydrates, but it can have proteins, lower concentration of nucleic acids, lipids, phenols, and humic substances (Gambino and Cappitelli, 2016; Collins and Margesin, 2019). Some conditions to activate this mechanism is obviously, the extreme temperature, but the possibility of others conditions present such as elevated pH, high salt, UV light, etc. is also known. These metabolites are in response to continuous increase of oxygen in the environment regulated by different metabolic pathways, which are sensitive to concentration, quality and durability of ROS (Ewert and Deming, 2013; Gambino and Cappitelli, 2016; Deming and Young, 2017). The oxidative stress and biofilm formation have the same regulators; OxyR protein senses H_2O_2 active genes related to peroxide scavengers, thiol redox and involved in the biofilm in minimal medium, RpoS activate under limited condition of nutrient, as some common regulators. The other regulator of biofilm production is the quorum sensing in interspecies interaction or coculture, where thermophilic Bacteria and Archaea are able to produce biofilms using ABC-dependent pathway (Gambino and Cappitelli, 2016; Kambourova et al., 2016). Specifically, the exopolysaccharides (EPS) are composed of different monosaccharides linked by 1,4-β-, 1,3-β- or 2,6-β or more flexible 1-2-α- or 1,6-α- bonds with repeated units of glucose, galactose, uronic acid, and manose, this last is found in *Thermococcus litoralis* and *Geobacillus thermotoleran;* other sugars reported are glucose, fructose, galactose, arabinose, these are found in other *Geobacillus* sp., *Sulfolobus* sp., *Thermotoga maritima* (Nicolaus et al., 2003; Johnson et al, 2005; Koerdt et al., 2010). In addition, the EPS have helical aggregates by N-acetyl-glucosamine conferred rigidity, this composition is found in *Thermus aquaticus*. A special characteristic of biofilms produced by thermophiles is their high molecular weight around 300–1000 kDa (Gambino and Cappitelli, 2016). In psycrophiles, EPS gel matrix generates a protective shell acting as a diffusion barrier to solutes and physical barrier to ice formation due the role in the reduction of free energy for ice nucleation impeding the ice crystal growth and increasing the liquid space (Casillo et al., 2017; Collins and Margesin, 2019).

3.3.2 COLD STRESS AND HEAT SHOCK PROTEINS

Temperatures above 20°C can cause a negative effect in the microorganism. However, some are able to grow at low temperatures. TtecspC gene had been found in some microorganisms like *Thermoanaerobacter tengcongensis*. MB4 is associated in DNA replication, recombination and repair, transcriptional and translational, cell membrane, cell fluidity, carbohydrate metabolism strategies for cold shock response and acts as chaperones (Lui et al., 2014). Also, the energy production during cold shock is low and genes *CinA* and *RecA* decreased impacting DNA transport (Joe Shaw et al., 2010). Cold shock proteins families present in psycrophiles are: CspA, CspB, CspC, CspD, CspE and CspD in addition to fatty acid desaturase, ice nucleation protein and trehalose synthase (Koo et al., 2017). Other strategies of psycrophiles are the antifreeze proteins and noncolligative biological molecules with the capacity to bind to ice and inhibit ice growth recrystallization, blocking secondary nucleation events and shaping unique ice morphology. It had been associated with stabilization of cell membranes, integrity protectant and less frequent adhesion domains and ice-binding function allowing a specific position of the cell in the ice to enhance the oxygen and nutrient access (Bar Dolev et al, 2016; Lorv et al, 2014; Voets et al., 2017; Collins and Margesin, 2019).

One of the interesting and more studied strategies that thermophiles have is the production of heat shock proteins for their survival to high temperatures and growth. The principal collaborations are receptor regulation, cytoskeleton stabilization and protein translocation, folding, and repair (Gao et al., 2016). The principal heat shock proteins are Hsp14 (ClpB), Hsp90 (HtpG), Hsp70 (DnaK), Hsp60 (chaperonin GroEL), Hsp40 (DnaJ), Hsp27 (sHsp), and Hsp10 (GroES). In the case of thermophilic archaea, the chaperonins are six: chaperonin group II, prefoldin (PFD), small heat-shock, peptidyl-prolyl-cis-trans isomerases, AAA proteins, and NAC (Sahlan and Yohda, 2013). The function is like proteases that protect the cell, in the regulation of translation, maintain the integrity of proteins, participate in the degradation of abnormal proteins, becoming HSP70s the most important, so, when HSP70s decreased, the thermal effect was high on the cell. Other HSP104s act as ATPase that enhance the thermotolerance for the proteins aggregation solubilization (Okuda et al., 2015; Guan et al., 2017). These proteins are known as chaperones and found in *P. furiosus*, *Sulfolobus acidocaldarius*, *Thermococcus kodakarensis*; in this last, the presence of HSP60 (CpkA and CpKB) is highly expressed during logarithmic and stationary phase growth (Fujiwara et al., 2008; Ranawat and Rawat, 2017). Specifically, the chaperonins have the function of folding of new synthesized and stress-denatured proteins and grouped I (GroEL) and II groups present in bacteria and archaea cytosol and eukaryotic organelles (mitochondria) (Sahlan and Yohda, 2013). There exists a chaperonins III located in the operon DnaK together with cochaperonins (DnaJ and GrpE) with activity of refold of denatured proteins in a GroES-independent manner (Techtmann and Robb, 2010).

The cell induction of the chaperonins is performed by global factor ϵ, sigma factor, and distinct genes to express these heat shock proteins. There are constitutive genes of Hsp60 expressing an optimum temperature of thermophiles associated with specific regulatory mechanism. The major expression of these chaperonins is involved with different carbon metabolism pathways activated by high temperatures (Li and Xu, 2017).

3.3.3 FERMENTATION PATHWAYS

In the case of psycrophiles, metabolism like glycolysis, the pentose phosphate pathway, tricarboxylic acid cycle (TCA), and electron transport chain are depressed. In substitution, are present the glyoxylate, methylglyoxal or 2-methylcitrate cycles, the ethanol oxidation, acetate metabolism or propionyl-CoA catabolism (Ayala-del Río et al, 2010; Piette et al. 2012; Tribelli et al., 2015; Collins and Margesin, 2019). Particularly, glyoxylate enzymes had been recognized in *Nesterenkonia* sp. AN1 at 5°C, in *Exiguobacterium sibiricum* the presence of methylglyoxal synthase. The degradation of glyoxal and methylglyoxal allows the formation of important metabolic products and the removal of electrophilic species through the recycling of reactive carbonyls to lactate, used for cellular metabolism. Also, the metabolism of fatty acids influences the energy generation, due to their role in the propionate pathway, where propionyl-CoA is the central metabolic intermediate obtained to different catabolism of some amino acids and can be carboxylated to S-methylmalonyl-CoA and then isomerize to succinyl-CoA and possibly participate in the TCA pathway; this is observed in *Sphingopyxis alaskaensis* in growth at 4–10°C (Ting et al., 2010; Tribelli and Lopez, 2018).

Contrarily, thermophiles are able to use different carbohydrates, especially from complex polysaccharides, mono- and disaccharides. There are some thermophiles able to grow in different carbohydrates with some peculiarities, e.g. *Thermotogales* is able to grow under ≥80°C and metabolize glucans, hemicellulose, pentoses, hexoses, disaccharides are chemoheterotrophs and have ABC-transporters. In addition, *Thermotoga* species are able to use Embden-Meyehorf (EM) and Entner-Duodoroff (ED) as principal pathways, and produce large complexes for carbohydrate degradation (cellulosomes). The importance of *Thermotoga* sp. is the production of H_2 using enzymes phosphate acetyl-transferase and acetate kinase, and other unique enzymes are able to optimize the use of reducing power and energy generated from biomass utilization as carbon source (Conners et al., 2006; Blumer-Schuette et al., 2008; Lin and Xu, 2013). Another case, *Caldicellulosiruptor* sp., is a thermophile very versatile because it has a multi-modular carbohydrate degrading enzymes like glycoside hydrolases and transporters with the capacity to degrade various polysaccharides like cellulose and xylan simultaneously (Dam et al., 2011). In the same way, *Caldicellulosiruptor* species can form cellulosomes in response to different growth conditions. This genus is considered "biohydrogen factory" (Counts et al., 2017). *Pyrococcus furiosus*, a hyperthermophile, is able to grow until 100°C can use disaccharides and glucans, buy not pentoses because they are susceptible to Maillard reactions to high temperatures. Also, use Embden-Meyehorf (EM) and Entner-Duodoroff (ED) as principal pathways, but in the case of EM glucokinase and phosphofructokinase use ADP as phosphoryl group donor, generating AMP. The conversion of glyceraldehyde-3-phosphate to 3-phosphoglycerate resulting in a reduced ferredoxin. *P. furiosus* can use elemental sulfur to produce hydrogen sulfide; a proton gradient is formed by membrane-bound oxidoreductase reducing sulfur to use NADPH (Counts et al., 2017).

3.3.4 ACCUMULATION OF THERMOPROTECTANS

Although it is highly reported that some microorganisms are able to produce small-molecular weight compounds to avoid an osmotic stress, there are some compounds with additionally acting to avoid heat shock, scavenging free radicals (possible exclusion from proteins surface and/or water entrapment), compact the proteins unfolded, prevention of protein aggregation (Collins and Margesin, 2019). Trehalose is a disaccharide widespread in the microorganisms. The presence of this is given by cellular accumulation, enhancing the thermotolerance due to the capacity to stabilize the proteins, the structure of membranes gives cell protection and increases the strength of hydrogen-bonding interactions (Reina-Bueno et al., 2012; Gibney et al., 2015; Glatz et al., 2016; Guan et al., 2017). Four pathways can carry the synthesis of the trehalose; however, the most common is the TSP/TPP pathway, where one glucose is transferred from NDP-glucose to glucose-6-phosphate to form trehalose-6-phosphate (T6P) by T6P synthase (TPS). Then this intermediate is dephosphorylated to form trehalose. Other pathways are TreS (participation of trehalose synthase), TreYZ (involves maltooligotrehalose synthase and maltooligosyltrehalose tetrahydrolase), and others where there is a conversion of ADP-glucose and glucose instead of glucose-6-phosphate, all these pathways had been found in *T. thermophilus, Sulfolobus acidocalcarius*, and *Pyrococcus horikoshii* (Empadinhas and Costa, 2006; Iordachescu and Imai, 2011). During the heat shock, the cell starts to have a response to survival and the trehalose mobilized rapidly but can be affected by Hsp40 (Guan et al., 2017). When trehalose binds to Mn and phosphate is able to exhibit antioxidant properties for avoiding the radiation damage and oxidation stress (Webb and Diruggiero, 2012). In the case of psycrophiles, it can in addition produce trehalose, clycine, glycerol, sucrose, sarcosine, mannitol and sorbitol, participating obviously in the restoring osmotic balance but also in counteracts water loss and cell shrinkage during freezing (Fonseca et al., 2016).

Phosphate derivates are other osmolytes accumulated in response to supraoptimal temperatures, principally DIP [Di-myo-inositol-1,1-′(3,3′)-phosphate] and derivates as diglycerol phosphate (DGP) reported in *Pyrococcus* sp. and *Thermococcus* sp. DIP is present in microorganisms that are able to grow at 80°C and is the osmolyte predominant. Cyclic-2,3-bis(di) phosphoglycerate (cDPG), other derivate, had been detected in methanogens where it acts as an intermediate in gluconeogenic pathway and as a phosphate reserve. Other osmolyte, mannosylglycerate (MG), also is common in bacteria, marine thermophilic bacterium, *Pyrococcus* sp. and *Thermococcus* sp acting as osmoprotectant, but it is important to mention that these osmolytes can have protection for osmotic stress and radiation resistance, freezing, desiccation (Empadinhas and Costa, 2006; Ranawat and Rawat, 2017).

3.3.5 POLYHYDROXYALKANOATES (PHAS) PRODUCTION

This metabolism and interesting products are produced and degraded by psycrophiles. These are polymers accumulated under unbalanced growth conditions like dynamic reserves of carbon, nitrogen, or phosphorus, reducing equivalents and energy in the cell, associated with functions such as cryoprotectant, avoiding oxidative stress, redox balance, and cell motility. The production is associated with multiple gene duplications of acyl-CoA dehydrogenase and enoyl-CoA dehydrogenase for PHA synthesis in *C. psychrerytraea*, while in *S. alaskensis* it had been detected PhaP phasin as biosynthesis protein and secreted by 3-hydroxybutyrate dehydrogenase (Ting et al., 2010; Nunn et al. 2015; Collins and Margesin, 2019). There are a great diversity of psycrophiles to produce PHAs in a range of 5–20°C, and in the case of Antarctic bacteria Alpha, Beta and Gammaproteobacteria are able to produce them (Ciesielski et al., 2014).

3.4 MOLECULAR RESPONSE TO ADAPTATION

One of the central parts of the adaptation of microorganisms to extreme temperatures is the molecular system that leads to having extraordinary elements to survive. The "adaptation information" is stored in genes and realized in proteins regulated by a RNA. It is easy to mention, inside of these three important steps, occurring several changes. It is important consider microorganisms as system where there are interaction between several parts, which aren't yet fully studied. Below, some particular elements and molecular phenomes occurred in extreme microorganisms will be mentioned (thermophiles and psycrophiles).

3.4.1 DNA

Microbial adaptation to extreme temperatures is largely due to the DNA transfer ways carrying out the evolution. There are three common mechanisms to transfer DNA in bacteria and archaea (natural transformation, conjugation, and horizontal transfer gene), but it has been reported others such as gene transfer agents, membrane vesicles, and nanotubes (van Wolferen et al., 2013). The more important aspect of the common mechanism to transfer DNA is described in Table 3.2.

TABLE 3.2 Elements to Transfer DNA for Microorganisms Able to Grow in High Temperatures (van Wolferen et al., 2013).

Mechanism to transfer DNA	Aspects to microbial thermostability
Horizontal transfer gene	• Use of reverse gyrase to increase the melting temperature of double-stranded DNA • Integrases transposon and insertion of sequences • Homologous recombination • Low spontaneous but specifically DNA mutations
Conjugation	Plasmids, megaplasmids, and integrating conjugative elements (pTT27, pNOB8, pING1)
Natural transformation	Formation of Pili o pseudopili (conserved genes: PilA1-4, PilD, F,C,Q in *Thermus* sp. and *Sulfulobus* sp. reported) and participation of DNA translocator proteins (conserved genes ComEA, ComEC, DprA in *Thermus* sp. and *Sulfulobus* sp. reported).

Three important mechanisms are already described above to transfer DNA. First, horizontal transfer gene (HGT) from cell mom to daughters, where bacteria and archaea can obtain DNA from the environment and incorporate; this aspect leads to genomic evolution. Above 20% and 40% of the genome in bacteria and archaea respectively is horizontally transferred contributing to evolution. The use of integrases, transposons or insertion of the sequences facilitates the HGT; this had been found in *P. furiosus* (Diruggiero et al., 2000). Reverse gyrase is the other strategy to give thermostability to DNA due to the capacity to introduce positive supercoiling demonstrated in *Thermococcus kodakarensis* (Vetone et al., 2014). This gyrase had been found in a megaplasmid in *T. thermophilus* that contains several genes to thermostability and DNA repair system associated principally as protector. Homologous recombination is other mechanism of transfer to repair the double-stranded DNA breaks caused by UV-light e.g. and is consider as SOS response is several bacteria an archaea. It is reported the presence of spontaneous mutations in thermophilic bacteria and archaea as a mechanism to repair the DNA and thermal adaptability; however, this mutation level is low compared with non-thermophile microorganisms (van Wolferen et al., 2013; Wang et al., 2015).

Second, conjugation is the other mechanism to transfer unidirectional of DNA by cell-cell contact, plasmid, and integrating conjugative elements. From this way can be mentioned some plasmids reported in some thermophilic bacteria and archaea such as pTT27 found in *Thermus* species associated to have an encoded region of nitrate reductase for anaerobic growth and hygromycin resistance (Álvarez et al, 2011). In some Solfolobus strains, pNOB8 is well characterized plasmid that contained gene encoding virB4/traU ATPase, TCP4, a relaxase, and MPF-specific proteins. Third, natural transformation for competence between species where pili and pseudopili participate in DNA uptake. In *Thermus* genus this mechanism is one of the most efficient due to the DNA uptake has a speed of around 40kb/s per cell, and has conserved pili genes (type IV pili-related proteins related with pilin subunits:PilA1, PilA2, PilA3 and PilA4, PilD, F,C,Q). The genes ComEA and ComEC also are conserved and have participation in the DNA binding to membrane and transport to cytosol. Together they allow the DNA transfer PilA1-4 associated to pili built. PilC acts as anchor and PilD facilities the entry of DNA. Finally, DNA is bound to comEA and comEC transport the DNA to cytoplasm (van Wolferen et al., 2013).

The genomes sizes of microorganisms adapted to extreme temperature are small, >60°C the size is smaller than 4 Mb, while 6 Mb in lower 45°C. This shows a lower complexity in the genome, in some cases there is no presence of TCA-cycle related genes (e.g. *Thermoanaerobacter tengcongensis*). The other aspect is the high content of guanine and cytosine with the values of 69%, 52%, and 68% in some thermophiles; also, the pattern of synonymous codon usage to protein codification is unique in thermophiles, e.g. for isoleucine and arginine the codons are AGG, ATA, and AGA frequently. These are some evidence of the differences of microorganisms able to grow in high temperatures and mesophilic (Wang et al., 2015).

3.4.2 RNA AND PROTEINS

In microorganisms able to grow in high temperatures, the presence of high content of GC in tRNAs and mRNAs is common, and mRNAs are enriched with purines and clusters of these are high in ratio of purines/pyrimides, especially in genes encoding elements of transcriptional and translational machinery (Wang et al., 2015). mRNA can be edited by ADAR (Adenosine Deaminase Acting on RNA) specifically using a deamination reaction to adenosine to inosine and, this is important for some aspect to cold adaptation and acclimation. In the activation of this system the temperature is the major factor, and to activate, ADAR can recognize complex tertiary structures in mRNA to edit it, regulated by alternative splicing cellular compartmentalization and posttranslational modifications. This edition allows a structural rearrangement in mRNA, e.g. in some bacteria heat shock protein mRNAs have specialized hairpins in 5′unstraslated regions that in low temperatures obstruct Shine-Dalgarno sequence blocking the translation but in high temperatures the structure destabilizes and the translation occurs (Garret and Rosenthal, 2012). Operons is another characteristic to adaptation in extreme temperatures, in thermophiles the structure is more robust and in high presence. In *Thermotoga maritima*, ~15 operons are expressed under high temperatures (hrcA-grpE-dnaJ (TM0851-TM0850-TM0849), groES-groEL (TM0505-TM0506), dnaK-sHSP (TM0373-TM0374)) allowing a gene regulation "economical" (Wang et al., 2015).

The microbial proteins exposed to extreme temperatures, also have an interesting peculiarity. Outside the optimal range of temperature, the proteins suffer unfolded states, but, there are some strategies to protein stabilization carried out by amino acids interactions. Per example, the cold enzymes are disordered and flexible, reduce the activation energy and activation enthalpy accompanied by more negative activation entropy, suggesting a reorder during the activation, their Km values are lower (more disorder enzyme substrate (ES)), but the flexibility is more in active region (Collins and Margesin, 2019). It is important to mention that particular amino acids participate strongly in thermal adaptation, generally the thermostability of the proteins is given by a major content of lysine, isoleucine, and glutamate and low of glycine, this last, in thermophiles arginine and valine, in addition lysine has a large side basic chain, provides a great stabilization. Glycine is in low content because it has highly dihedral angles around N-C increasing the degree of freedom of an unfolded polypeptide backbone. Also there is a decrease of thermos-labile amino acids (histidine, glutamine, and threonine). Glutamate, a large acidic chain, is present in salt bridges formation, but if it is in key positions it can generate a protein destabilization, but in cold adapted enzymes it provides a great flexibility due to its participation in packed hydrophobic cores and surface salt (Garret and Rosenthal, 2012). The amino acids are key to thermal adaptation of the proteins due to their interactions and dynamism to stabilize and avoid losing their activity, these are described below.

Protein interactions (Pucci and Roman, 2017):

i. Hydrophobicity: this is a mechanism to give force of protein folding, in high temperatures it is possible a great conformational fluctuation while in cold temperature there is weakening hydrophobicity, observed in aromatic and hydrophobic residues in high *T* above 25° and below 25°C, respectively. The polar amino acids are important to maintain a stabilization and in high temperatures it is formed a hydrophobic core, while in cold enzyme nonpolar amino acids residues on the surface (vanDijk et al., 2015; Sharma et al., 2019).
ii. Salt bridges: major presence in thermophiles, principally on the surface mediated by arginines (or proline and lysine) that also can form H-bonds and are associated with increase of hydrophobicity of charges residues. Contrary in cold proteins, avoid it to enhance the flexibility for a major water penetration (Parvizpour et al., 2016).
iii. Other interactions: surface charges participate in stabilization of proteins but it depends on substitution and environment to create these other strategies. First, π-π interaction is given for aromatic residues, geometry and number of these, growth temperature, increasing at high temperatures and are less in cold temperatures (Lanzarotti et al., 2011). Second, cation-π related to arginine given in a delocalized charge on guanidinium group allows a stabilization of the protein close to melting temperature, is also low in cold proteins (Tang et al., 2014). Cold enzymes have few substitutions in nonconserved charged residues generating an electrostatic potential difference (Sharma et al., 2019).

iv. Disulfide bridges: favor flexible regions to a thermostability and in cold proteins promote a special conformation in active regions and basic amino acids participate to form it (Ogawa et al., 2007; Sharma et al., 2019).

Structural dynamism (Pucci and Roman, 2017):

i. Loops: They are related to flexibility or rigidity, more short favor the thermostability, proline insertion give more rigidity, non-local effect in adjacent region; the loops also promotes flexibility. In cold enzymes, a substitution of proline by small amino acids like alanine and an increase of glycine, enhance the flexibility of connecting region of adjacent secondary structure. Proline can be in surface loops introducing kink/turn in the main chain H-bind increasing the rigidity in thermophile proteins (Sharma, 2019).
ii. Surface hydration: the generation of large and stronger shell given by hydrogen bonds is important for thermostability but, in cold proteins, are more exposed non-polar residues and the water around of the surface is de-established (Paredes et al., 2011).
iii. Core packing: in thermostable proteins are less cavities in core region that favor the optimization of the backbone and side chains, while cold proteins are larger cavities that accommodate more to a better water position (Paredes et al., 2011; Sammond et al., 2016). Alanine has a participation in the thermophile and glycine in cold enzymes to provide a reducing packing density at core and in the case of cold enzymes more flexible (present a reduction of strength entropic and enthalpic contribution) (Sharma et al., 2019).
iv. Internal wetting: in thermophiles, contribute to major stabilization of the cavities for the special hydration and internal solvation favors the flexibility in cold proteins (Paredes et al., 2011; Chakraborty et al., 2015).

3.5 QUORUM SENSING: A CONTRIBUTION TO SURVIVAL

During microbial growth in extreme environments, the objective of the cell is survived and niche colonization. In the ecosystems like arid zones, are distinct type of microorganisms so the competence, natural selection, and evolution processes are imminent. During this survival, one of the most trends of study is quorum sensing. Quorum sensing (QS) is a typical microbial communication that regulates and expresses genes according to density population and hence physiology, this is detected by the production of different signal molecules called autoinducers (Kaur et al., 2019). QS is direct to study the regulation bioluminescence, cell competency and horizontal gene transfer, virulence, motility, the formation of biofilms, and the production of antibiotics, growth inhibition, and other secondary metabolites (Montgomery et al., 2013). Actually, the QS in extreme environments is still unclear and a little studied detail, however with strong bioinformatics studies had contributed the QS relation with the survival strategies of the microorganisms in extreme environments and especially to produce "untypical molecules" impacting in their biotechnology importance (Abbamondi et al., 2019).

To explain the QS in extremophiles, there are different systems produced and reported by different microorganisms explained below.

i. Autoinducer-1 (AI-1 system): uses acylated homoserine lactones (AHLs) as signals (N-acylated homoserine-lactone ring consisting of 4-18 carbon acyl chain). Produced by LuxI (AI-1 synthase) using a receptor LuxR (AI-1 receptor), the lactone is obtained from S-adenosinexylmethionine and acyl chain by fatty acid biosynthesis. Associated with virulence, swimming motility and biofilm formation.
ii. Peptide-based system: peptides of 5-34 amino acids long synthesized as inactive pro-peptides and exported by ABC transporters. These are impermeable to membranes using as transporters. Peptide signals are activated by cleavage using membrane-bound histidine kinase receptor, cognate cytoplasmatic response regulator, or peptide transporter complexes. Associated with virulence.
iii. Autoinducer-2 (AI-2): a furanosyl borate diester, produced by LuxS associated in activated methyl cycle for the generation of S-adenosyl-L-methionine (SAM) related with major methyl group donor in cell. Also is present between interspecies communication and biofilm formation (Rasamiravaka et al., 2012; Kim et al., 2017; Kaur et al., 2019).

In thermophiles and psicrophyles, different QS systems had been explored in some microorganisms. *T. maritima* coculture with *M.. jannaschi* produces EPS during mid phase and it generates cellular aggregates and promotes the H_2 production at interspecies level and the production of a signal peptide gene (TM0504). A1-2 system is reported in *T. maritima* with *P. furiosus* by series of biotic and abiotic steps and also was detected SAH hydrolase associated with methyl cycle which catalyzes the cleavage of SAH to adenosine, then converted to A1-2. It had been proposed that A1-2 in *T. maritima* play a role in the heat shock response (Nichols et al., 2009; Kaur et al., 2019). In *Thermus* sp. GH5 had been detected in the overexpression of AHL signaling, particularly were produced acetyl-transferase and S-adenosylmethioninsynthetase as a response of cold shock to produce biofilm (Yousefi-Nejad, et al., 2011). *Geobacillus stereathermophilus* T-6 produce a xynA (extracellular xylanase) and it has been suggested their regulation in a QS system, due to the overexpression during exponential growth, but when the culture was diluted, the expression also decreased. It has been associated with a xylanase density factor, a small peptide that acts a signal to QS (Shulami et al., 2014).

On the other hand, in the case of psycrophiles, some QS systems reported DiKetoPiperazines activers of LuxR and are produced de novo or from catabolic products of constituents of nutrient rich medium. Also, it had been detected cyclic di-guanylate to regulate genes for motility, biofilm, and adhesion factors, indicating that QS is very important to survival at cold temperatures, and it had been exploited to produce metabolites, like alginate from *Pseudomonas mandelii* by the genes expression of *algU* and *mucA* (Mitova et al., 2005; Vasquez-Ponce et al., 2017; Kaur et al., 2019). The reports about QS in psycrophiles are associated with biofilm formation as resilience mechanism to sea-ice, suggesting that biofilm production allows the availability of water to grow. Actually, there are few reported about QS in extremophiles, and is unclear the presence of this mechanism in them, especially in Archaea there is not clear evidence of these processes (Riley et al., 2008; Montgomery et al., 2013). Although QS information in extremophiles is limited, the use of bioinformatics can be of help to understand QS genes/proteins implicated in the survival in extreme conditions and use to develop future industrial applications or improve those already studied.

3.6 CONCLUSION

The efforts about the study of the origin life have led to exploring hostile place, finding that arid zones have a great microbial diversity with especial and extraordinary physiological, metabolic, and molecular machinery. Through the development of different tools, it is possible to understand the cell adaptation of microorganisms in front of extreme temperatures, one of the conditions present in arid zones extended around the world, allowing a conservator them and showing the great potential that has. The intense search of this knowledge allows the use of these as an astrobiology contribution, as instructors that how life in the Earth is present nowadays or how can survive to change of the environment that actually we suffer, but specially to see them as cell factories or part of them to biotechnological purposes and mimic strategies to resolve humanity problems.

KEYWORDS

- **thermophiles**
- **psycrophiles**
- **adaptation**
- **metabolism**
- **proteins**
- **quorum sensing**

REFERENCES

Abbamondi, G. R.; Kambourova, M.; Poli, A.; Finore, I.; Nicolaus, B. Quorum Sensing in Extremophiles. In *Quorum Sensing: Molecular Mechanism and Biotechnological Application*; Academic Press, 2019.

Ayala-Del-Río, H. L.; Chain, P. S.; Grzymski, J. J.; Ponder, M. A.; Ivanova, N.; Bergholz, P. W.; Tiedje, J. M. The Genome Sequence of Psychrobacter Arcticus 273-4, A Psychroactive Siberian Permafrost Bacterium, Reveals Mechanisms for Adaptation to Low-Temperature Growth. *Appl. Environ. Microbiol.* **2010,** *76* (7), 2304–2312.

Alvarez, L.; Bricio, C.; Gómez, M. J.; Berenguer, J. Lateral Transfer of the Denitrification Pathway Genes Among Thermus Thermophilus Strains. *Appl. Environ. Microbiol.* **2011,** *77* (4), 1352–1358.

Bar Dolev, M.; Braslavsky, I.; Davies, P. L. Ice-Binding Proteins and Their Function. *Annurev. Biochem.* **2016,** *85,* 515–542.

Blumer-Schuette, S. E.; Kataeva, I.; Westpheling, J.; Adams, M. W.; Kelly, R. M. Extremely Thermophilic Microorganisms for Biomass Conversion: Status and Prospects. *Curr. Opin. Biotecnhol.* **2008,** *19* (3), 210–217.

Dam, P.; Kataeva, I.; Yang, S. J.; Zhou, F.; Yin, Y.; Chou, W.; Adams, M. W. W. Insights into Plant Biomass Conversion from the Genome of the Anaerobic Thermophilic Bacterium Caldicellulosiruptor Bescii DSM 6725. *Nucleic Acids Res.* **2011,** *39* (8), 3240.

Cabiscol Català, E.; Tamarit Sumalla, J.; Ros Salvador, J. Oxidative Stress in Bacteria and Protein Damage by Reactive Oxygen Species. *Int. Microbiol.* **2000,** *3* (1), 3–8.

Casillo, A.; Parrilli, E.; Sannino, F.; Mitchell, D. E.; Gibson, M. I.; Marino, G.; Corsaro, M. M. Structure-Activity Relationship of the Exopolysaccharide from a Psychrophilic Bacterium: A Strategy for Cryoprotection. *Carbohyd. Polym.* **2017,** *156,* 364–371.

Chakraborty, D.; Taly, A.; Sterpone, F. Stay Wet, Stay Stable? How Internal Water Helps the Stability of Thermophilic Proteins. *J. Phys. Chem.* **2015,** *119* (40), 12760–12770.

Ciesielski, S.; Górniak, D.; Możejko, J.; Świątecki, A.; Grzesiak, J.; Zdanowski, M. The Diversity of Bacteria Isolated from Antarctic Freshwater Reservoirs Possessing the Ability to Produce Polyhydroxyalkanoates. *Curr. Microbiol.* **2014,** *69* (5), 594–603.

Collins, T.; Margesin, R. Psychrophilic Lifestyles: Mechanisms of Adaptation and Biotechnological Tools. *Appl. Microbiol. Biotechnol.* **2019,** *103* (7).

Conners, S. B.; Mongodin, E. F.; Johnson, M. R.; Montero, C. I.; Nelson, K. E.; Kelly, R. M. Microbial Biochemistry, Physiology, and Biotechnology of Hyperthermophilic Thermotoga Species. *FEMS Microbiol. Rev.* **2006,** *30* (6), 872–905.

Counts, J. A.; Zeldes, B. M.; Lee, L. L.; Straub, C. T.; Adams, M. W. W.; Kelly, R. M. Physiological, Metabolic and Biotechnological Features of Extremely Thermophilic Microorganisms. *Wiley Interdiscip. Rev. Syst. Biol. Med.* **2017,** *9* (3).

De Maayer, P.; Anderson, D.; Cary, C.; Cowan, D. A. Some Like it Cold: Understanding the Survival Strategies of Psychrophiles. *EMBO Rep.* **2014,** *15* (5), 508–517.

Deming, J. W.; Young, J. N. The Role of Exopolysaccharides in Microbial Adaptation to Cold Habitats. In *Psychrophiles: From Biodiversity to Biotechnology*; Springer Cham, 2017; pp 259–284.

Diruggiero, J.; Dunn, D.; Maeder, D. L.; Holley-Shanks, R.; Chatard, J.; Horlacher, R.; Weiss, R. B. Evidence of Recent Lateral Gene Transfer Among Hyperthermophilic Archaea. *Mol. Microbiol.* **2000,** *38* (4), 684–693.

Empadinhas, N.; da Costa, M. S. To be or Not to be a Compatible Solute: Bioversatility of Mannosylglycerate and Glucosylglycerate. *Syst. Appl. Microbiol.* **2008,** *31* (3), 159–168.

Ewert, M.; Deming, J. W. Sea Ice Microorganisms: Environmental Constraints and Extracellular Responses. *Biology* **2013,** *2* (2), 603.

Fitch, W. M.; Upper, K. The Phylogeny of tRNA Sequences Provides Evidence for Ambiguity Reduction in the Origin of the Genetic Code. *Cold Spring Harb. Symp. Quant. Biol.* **1987,** *52*, 759–767.

Fonseca, F.; Meneghel, J.; Cenard, S.; Passot, S.; Morris, G. J. Determination of Intracellular Vitrification Temperatures for Unicellular Micro Organisms under Conditions Relevant for Cryopreservation. *PLoS One* **2016,** *11* (4), e0152939.

Fujiwara, S.; Aki, R.; Yoshida, M.; Higashibata, H.; Imanaka, T.; Fukuda, W. Expression Profiles and Physiological Roles of Two Types of Molecular Chaperonins from the Hyperthermophilic Archaeon Thermococcus Kodakarensis. *Appl. Environ. Microbiol.* **2008,** *74*(23), 7306–7312.

Gambino, M.; Cappitelli, F. Mini-Review: Biofilm Responses to Oxidative Stress. *Biofouling* **2016,** *32* (2), 167–178.

Gao, Y.; Dai, J.; Peng, H.; Liu, Y.; applied, Xu, T. Isolation and Characterization of a Novel Organic Solvent-Tolerant Anoxybacillus sp. PGDY12, a Thermophilic Gram-Positive Bacterium. *J. Appl. Microbiol.* **2011,** *110* (2), 472–478.

García, C. A.; Alcaraz, E. S.; Franco, M. A.,De Rossi, B. N. P. Iron is a Signal for Stenotrophomonas Maltophilia Biofilm Formation, Oxidative Stress Response, OMPs Expression, and Virulence. *Front. Microbiol.* **2015,** *6* (SEP).

Garrett, S. C.; Rosenthal, J. J. C. A Role for A-to-I RNA Editing in Temperature Adaptation. *Physiology (Bethesda)* **2012,** *27* (6), 362–369.

Gilmore, S. F.; Yao, A. I.; Tietel, Z.; Kind, T.; Facciotti, M. T.; Parikh, A. N. Role of Squalene in the Organization of Monolayers Derived from Lipid Extracts of Halobacterium Salinarum. *Langmuir* **2013,** *29* (25), 7922–7930.

Guan, Z.; Tian, B.; Perfumo, A.; Goldfine, H. The Polar Lipids of Clostridium Psychrophilum, an Anaerobic Psychrophile. *Biochim. Biophys. Acta* **2013,** *1831* (6), 1108–1112.

Guan, N.; Li, J.; Shin, H. dong, Du, G.; Chen, J.; Liu, L. Microbial Response to Environmental Stresses: From Fundamental Mechanisms to Practical Applications. *Appl. Microbiol. Biotechnol.* **2017,** *101* (10), 3991–4008.

Guerzoni, M. E.; Lanciotti, R.; Cocconcelli, P. S. Alteration in Cellular Fatty Acid Composition as a Response to Salt, Acid, Oxidative and Thermal Stresses in Lactobacillus Helveticus. *Microbiology* **2001,** *147* (8), 2255–2264.

He, J.; Yang, Z.; Hu, B.; Ji, X.; Wei, Y.; Lin, L.; Zhang, Q. Correlation of Polyunsaturated Fatty Acids with the Cold Adaptation of Rhodotorula Glutinis. *Yeast* **2015,** *32* (11), 683–690.

Iordachescu, M.; Imai, R. Trehalose and Abiotic Stress in Biological Systems. In *Abiotic Stress in Plants—Mechanisms and Adaptations*; IntechOpen, 2011.

Joe Shaw, A.; Hogsett, D. A.; Lynd, L. R. Natural Competence in Thermoanaerobacter and Thermoanaerobacterium Species. *Appl. Environ. Microbiol.* **2010,** *76* (14), 4713–4719.

Johnson, M. R.; Montera, C. I.; Conners, S. B.; Shockley, K. R.; Bridger, S. L.; Kelly, R. M. Population Density-Dependent Regulation of Exopolysaccharide Formation in the Hyperthermophilic Bacterium Thermotoga Maritima. *Mol. Microbiol.* **2005,** *55* (3), 664–674.

Kambourova, M.; Radchenkova, N.; Tomova, I.; Bojadjieva, I. Thermophiles as a Promising Source of Exopolysaccharides with Interesting Properties. In *Biotechnology of Extremophiles; Grand Challenges in Biology and Biotechnology*; Springer Cham, **2016.**

Kaur, A.; Capalash, N.; Sharma, P. Communication Mechanisms in Extremophiles: Exploring their Existence and Industrial Applications. *Microbiol. Res.* **2019,** *221,* 15–27.

Koerdt, A.; Gödeke, J.; Berger, J.; Thormann, K. M.; Albers, S. V. Crenarchaeal Biofilm Formation under Extreme Conditions. *PLoS One* **2010,** *5* (11), e14104.

Koga, Y. Early Evolution of Membrane Lipids: How did the Lipid Divide Occur? *J. Mol. Evol.* **2011,** *72* (3), 274–282.

Koga, Y. Thermal Adaptation of the Archaeal and Bacterial Lipid Membranes. *Archaea* **2012,** *2012,* 789652.

Konings, W.; Albers, S.; Koning, S.; Driessen, A. J. M. The Cell Membrane Plays a Crucial Role in Survival of Bacteria and Archaea in Extreme Environments. *Antonie Van Leeuwenhoek* **2002,** *81* (1–4), 61–72.

Imlay, J. A. Diagnosing Oxidative Stress in Bacteria: Not as Easy as You Might Think. *Curr. Opin.* **2015,** *24,* 124–131.

Lanzarotti, E.; Biekofsky, R. R.; Estrin, D. A.; Marti, M. A.; Turjanski, A. G. Aromatic-Aromatic Interactions in Proteins: Beyond the Dimer. *J. Chem. Inf. Model.* **2011,** *51* (7), 1623–1633.

Lin, L.; Xu, J. Dissecting and Engineering Metabolic and Regulatory Networks of Thermophilic Bacteria for Biofuel Production. *Biotechnol. Adv.* **2013,** *31* (6), 827–837.

Liu, B.; Zhang, Y.; Zhang, W. RNA-Seq-Based Analysis of Cold Shock Response in Thermoanaerobacter Tengcongensis, a Bacterium Harboring a Single Cold Shock Protein Encoding Gene. *PLoS One* **2014,** *9* (3), e93289.

Lombard, J.; López-García, P.; Moreira, D. The Early Evolution of Lipid Membranes and the Three Domains of Life. *Nat. Rev. Microbiol* **2012,** *10* (7), 507–515.

Lorv, J. S. H.; Rose, D. R.; Glick, B. R. Bacterial Ice Crystal Controlling Proteins. *Scientifica* **2014,** 1–20.

Matsuno, Y.; Sugai, A.; Higashibata, H.; Fukuda, W.; Ueda, K.; Uda, I.,Fujiwara, S. Effect of Growth Temperature and Growth Phase on the Lipid Composition of the Archaeal Membrane from Thermococcus kodakaraensis. *Biosci. Biotechnol.* **2009,** *73* (1), 104–108.

Mitova, M.; Tutino, M. L.; Infusini, G.; Marino, G.; De Rosa, S. Exocellular Peptides from Antarctic Psychrophile Pseudoalteromonas Haloplanktis. *Mar. Biotechnol.* **2005,** *7* (5), 523–531.

Nichols, J. D.; Johnson, M. R.; Chou, C. J.; Kelly, R. M. Temperature, Not LuxS, Mediates AI-2 Formation in Hydrothermal Habitats. *FEMS Microbiol. Ecol.* **2009,** *68* (2), 173–181.

Nunn, B. L.; Slattery, K. V.; Cameron, K. A.; Timmins-Schiffman, E.; Junge, K. Proteomics of Colwellia Psychrerythraea at Subzero Temperatures—A Life with Limited Movement, Flexible Membranes and Vital DNA Repair. *Environ. Microbiol.* **2015,** *17* (7), 2319–2335.

Ogawa, K.; Sonoyama, T.; Takeda, T.; Ichiki, S. I.; Nakamura, S.; Kobayashi, Y.; ... Sambongi, Y. Roles of a Short Connecting Disulfide Bond in the Stability and Function of Psychrophilic Shewanella Violacea Cytochrome c (5). *Extremophiles* **2007,** *11* (6), 797–807.

Okuda, M.; Niwa, T.; Taguchi, H. Single-Molecule Analyses of the Dynamics of Heat Shock Protein 104 (Hsp104) and Protein Aggregates. *J. Biol. Chem.* **2015,** *290* (12), 7833–7840.

Pandey, N.; Jain, R.; Pandey, A.; Tamta, S. Optimisation and Characterisation of the Orange Pigment Produced by a Cold Adapted Strain of Penicillium sp. (GBPI_P155) Isolated from Mountain Ecosystem. *Mycology* **2018,** *9* (2), 81.

Paredes, D. I.; Watters, K.; Pitman, D. J.; Bystroff, C.; Dordick, J. S. Comparative Void-Volume Analysis of Psychrophilic and Mesophilic Enzymes: Structural Bioinformatics of Psychrophilic Enzymes Reveals Sources of Core Flexibility. *BMC Struct. Biol.* **2011,** *11* (1), 1–9.

Parvizpour, S.; Razmara, J.; Shamsir, M. S.; Illias, R. M.; Abdul Murad, A. M.; The Role of Alternative Salt Bridges in Cold Adaptation of a Novel Psychrophilic Laminarinase. *J. Biomol. Struct. Dyn.* **2017,** *35* (8), 1685–1692.

Piette, F.; D'Amico, S.; Mazzucchelli, G.; Danchin, A.; Leprince, P.; Feller, G. Life in the Cold: A Proteomic Study of Cold-Repressed Proteins in the Antarctic Bacterium Pseudoalteromonas haloplanktis TAC125. *Appl. Environ. Microbiol.* **2011,** *77* (11), 3881.

Powell, J. T.; Chatziefthimiou, A. D.; Banack, S. A.; Cox, P. A.; Metcalf, J. S.; Desert Crust Microorganisms, their Environment, and Human Health. *J. Arid Environ.* **2015,** *112* (PB), 127–133.

Pucci, F.; Rooman, M.; Physical and Molecular bases of Protein Thermal Stability and Cold Adaptation. *Curr. Opin. Struct. Biol.* **2017,** *42,* 117–128.

Rampelotto, P. H. Resistance of Microorganisms to Extreme Environmental Conditions and Its Contribution to Astrobiology. *Sustainability* **2010,** *2* (6), 1602–1623.

Ranawat, P.; Rawat, S. Stress Response Physiology of Thermophiles. *Arch. Microbiol.* **2017,** *199,* 391–414.

Rasamiravaka, T.; Vandeputte, O. M.; Pottier, L.; Huet, J.; Rabemanantsoa, C.; Kiendrebeogo, M.; El Jaziri, M. Pseudomonas aeruginosa Biofilm Formation and Persistence, along with the Production of Quorum Sensing-Dependent Virulence Factors, Are Disrupted by a Triterpenoid Coumarate Ester Isolated from Dalbergia trichocarpa, a Tropical Legume. *PLoS One* **2015,** *10* (7).

Reina-Bueno, M.; Argandoña, M.; Salvador, M.; Rodríguez-Moya, J.; Iglesias-Guerra, F.; Csonka, L. N.; Vargas, C. Role of Trehalose in Salinity and Temperature Tolerance in the Model Halophilic Bacterium Chromohalobacter salexigens. *PLoS One* **2012,** *7* (3), e33587.

Riley, M.; Staley, J. T.; Danchin, A.; Zhang, T. Z.; Brettin, T. S.; Hauser, L. J.; Thompson, L. S. Genomics of an Extreme Psychrophile, Psychromonas Ingrahamii. *BMC Genom.* **2008,** *9* (1), 1–19.

Sharma, M. Thermophiles vs. Psychrophiles: Cues from Microbes for Sustainable Industries. *Environ. Biotech.* **2019,** 323–340.

Sahlan, M.; Yohda, M. Molecular Chaperones in Thermophilic Eubacteria and Archaea: Thermophilic Microbes in Environmental and Industrial Biotechnology. In *Biotechnology of Thermophiles*; Springer Netherlands, 2013.

Sato, Y.; Kameya, M.; Fushinobu, S.; Wakagi, T.; Arai, H.; Ishii, M.; Igarashi, Y. A Novel Enzymatic System against Oxidative Stress in the Thermophilic Hydrogen-Oxidizing Bacterium Hydrogenobacter thermophilus. *PLoS One* **2012,** *7* (4), e34825.

Shen, L.; Liu, Y.; Wang, N.; Jiao, N.; Xu, B.; Liu, X. Variation with Depth of the Abundance, Diversity and Pigmentation of Culturable Bacteria in a Deep Ice Core from the Yuzhufeng Glacier, Tibetan Plateau. *Extremophiles* **2018,** *22* (1), 29–38.

Shulami, S.; Shenker, O.; Langut, Y.; Lavid, N.; Gat, O.; Zaide, G.; Shoham, Y. Multiple Regulatory Mechanisms Control the Expression of the Geobacillus Stearothermophilus Gene for Extracellular Xylanase. *J. Biol. Chem.* **2014,** *289* (37), 25957.

Siliakus, M. F.; van der Oost, J.; Kengen, S. W. M. Adaptations of Archaeal and Bacterial Membranes to Variations in Temperature, pH and Pressure. *Extremophiles* **2017,** *21* (4), 651–670.

Tang, M. A. K.; Motoshima, H.; Watanabe, K. Cold Adaptation: Structural and Functional Characterizations of Psychrophilic and Mesophilic Acetate Kinase. *Protein J.* **2014,** *33* (4), 313–322.

Techtmann, S. M.; Robb, F. T. Archaeal-Like Chaperonins in Bacteria. *Proc Natl Acad Sci U S A* **2010,** *107* (47), 20269–20274.

Tribelli, P. M.; Venero, E. C. S.; Ricardi, M. M.; Gómez-Lozano, M.; Iustman, L. J. R.; Molin, S.; López, N. I. Novel Essential Role of Ethanol Oxidation Genes at Low Temperature Revealed by Transcriptome Analysis in the Antarctic Bacterium Pseudomonas Extremaustralis. *PLoS One* **2015,** *10* (12), e0145353.

Uchida, K. Aldehyde Adducts Generated During Lipid Peroxidation Modification of Proteins. *Free Radic. Res.* **2015,** *49* (7), 896–904.

van Dijk, E.; Hoogeveen, A.; Abeln, S. The Hydrophobic Temperature Dependence of Amino Acids Directly Calculated from Protein Structures. *PLoS Comput. Biol.* **2015,** *11* (5).

van Wolferen, M.; Ajon, M.; Driessen, A. J. M.; Albers, S. V. How Hyperthermophiles Adapt to Change their Lives: DNA Exchange in Extreme Conditions. *Extremophiles* **2013,** *17* (4), 545–563.

Vásquez-Ponce, F.; Higuera-Llantén, S.; Pavlov, M. S.; Ramírez-Orellana, R.; Marshall, S. H.; Olivares-Pacheco, J. Alginate Overproduction and Biofilm Formation by Psychrotolerant Pseudomonas Mandelii Depend on Temperature in Antarctic Marine Sediments. *Electron. J. Biotechnol.* **2017,** *28*, 27–34.

Vettone, A.; Perugino, G.; Rossi, M.; Valenti, A.; Ciaramella, M. Genome Stability: Recent Insights in the Topoisomerase Reverse Gyrase and Thermophilic DNA Alkyltransferase. *Extremophiles* **2014,** *18* (5), 895–904.

Voets, I. K. From Ice-Binding Proteins to Bio-Inspired Antifreeze Materials. *Soft Matter.* **2017,** *13* (28), 4808–4823.

Wang, Q.; Cen, Z.; Zhao, J. The Survival Mechanisms of Thermophiles at High Temperatures: An Angle of Omics. *Physiology (Bethesda)*. **2015,** *30* (2), 97–106.

Wan, X.; Peng, Y. F.; Zhou, X. R.; Gong, Y. M.; Huang, F. H.; Moncalián, G. Effect of Cerulenin on Fatty Acid Composition and Gene Expression Pattern of DHA-Producing Strain Colwellia Psychrerythraea Strain 34H. *Microb. Cell Fact.* **2016,** *15* (1), 1–13.

Webb, K. M.; Diruggiero, J. Role of Mn2+ and Compatible Solutes in the Radiation Resistance of Thermophilic Bacteria and Archaea. *Archaea* **2012,** 11.

Yousefi-Nejad, M.; Manesh, H. N.; Khajeh, K. Proteomics of Early and Late Cold Shock Stress on Thermophilic Bacterium, Thermus sp. GH5. *J. Proteom.* **2011,** *74* (10), 2100–2111.

CHAPTER 4

Potential Applications of Bioactive Compounds from Arid Zone Microorganisms

NATALIA QUINTERO-PACHECO[1], ANDREA VÁSQUEZ-GARCÍA[1], ANNA MARÍA POLANÍA[2], and LILIANA LONDOÑO-HERNANDEZ[1]

[1]*Biotics Group, School of Basic Sciences, Technology and Engineering, Universidad Nacional Abierta y a Distancia–UNAD, Colombia*

[2]*MIBIA Group, Biology Department, Faculty of Natural and Exact Sciences, Universidad del Valle, Colombia*

ABSTRACT

Arid regions are the interface between semi-arid, often vegetated areas and hyper-arid, biologically unproductive deserts. They are characterized by low relative humidity, high levels of solar radiation, lack of oxygen, and a large thermal oscillation. Under these conditions, many plant species grow and develop a wide variety of bioactive compounds as a result of their adaptation to environmental stress, including the biosynthesis of secondary metabolites with pharmacological activities of great interest. Some microorganisms harboring these zones are xerophilic, thermophilic, halophilic, and alkaliphilic *Actinobacteria*, members of the genus *Streptomyces* such as *Streptomyces deserti, Streptomyces bullii,* strains belonging to *Micromonospora, Saccharothrix,* and *Streptosporangium*, among others. These species have been little investigated but have been reported as producers of new bioactive compounds. It has been shown that *Actinobacteria* are among the three most abundant phyla in arid zones; in addition, they present relatively high thermal stability, bioavailability, and solubility. According to the reported research,

Exploration and Valorization of Natural Resources from Arid Zones. Cristobal Noé Aguilar, Norma Paola Meléndez-Renteria, & Karen Nathiely Ramirez-Guzman (Eds.)

some of the compounds produced by these microorganisms are: ansamycins and macrolactones, which have antibacterial and antitumor activity, as well as aminobenzoquinones that have inhibitory activity against bacteria and dermatophyte fungi. In this sense, there is a need to report the potential of these bioactive compounds obtained from these microorganisms in order to seek their potential use.

Arid regions are the interface between semi-arid, often vegetated areas and hyper-arid, biologically unproductive deserts. They are characterized by low relative humidity, high levels of solar radiation, lack of oxygen, and a large thermal oscillation. Under these conditions, many plant species and microorganisms grow and develop a wide variety of bioactive compounds because of their adaptation to environmental stress, including the biosynthesis of secondary metabolites with different activities of great interest. Some microorganisms that harbor these zones are Actinobacteria, Cyanobacteria, and Microalgae, which can be xerophilic, thermophilic, halophilic, and alkaliphilic. These species have been reported to be producers of new bioactive compounds with potential use as antibiotics, antioxidants, antitumor agents, biosurfactants, bioactive peptides, and natural pigments. This chapter describes the characteristics, recovery, properties, and applications of bioactive compounds from microorganisms with extreme growth conditions, such as those of arid zones.

4.1 INTRODUCTION

Arid and semi-arid regions cover an area of about 66.7 million km^2 of the planet and are home to 2 billion people (Pinheiro Junior et al., 2019), although this number is likely to increase in the coming years due to current climate change (León-Sobrino et al., 2019). These regions are characterized by low vegetation cover, low nutrient content, high salinity, extreme temperatures, reduced water, and instability caused by high winds and UV radiation (Alori et al., 2020; Ye et al., 2019). However, these environments can harbor a phylogenetically diverse microbial community such as Actinobacteria, Cyanobacteria, and Microalgae, which can be found in the soil and associated with plants. In recent years, the search for new natural products has intensified worldwide and has focused on extreme environments, because the adaptation of microorganisms to climatic conditions has allowed them to produce bioactive compounds with interesting characteristics and properties that can be applied in many industries (Sayed et al., 2020). According to a preliminary consensus by Guaadaoui et al. (2014), bioactive compounds are naturally occurring substances with the capacity and ability to interact with one or more components of living tissue that present a wide range of probable effects on the organism, especially on human health.

The recovery of bioactive compounds from microorganisms in arid areas is generally carried out by bioprocesses, either by submerged fermentation (SmF) or by solid state fermentation (SSF). These bioprocesses can be based on the cultivation of microorganisms in a liquid medium containing nutrients (SmF), or in solid particles in the absence of water (SSF), or almost in the absence of water that allow the growth and metabolism of microorganisms (Martins et al., 2011). The fermentation process should be chosen according to the needs of the microorganism and the characteristics of the bioactive compound to be obtained. Several studies have focused on the isolation of microorganisms from arid areas and the optimization of growth conditions to improve the production of bioactive compounds, where the substrate is the most important characteristic for fermentation outcome (Gera et al., 2014; Ratnaweera et al., 2015). The substrate is specific for each microorganism and can vary according to the type of nutrient and its respective concentration. Nutrients such as glucose, mineral salts, yeast extract, peptone, among others, have been reported in the isolation of actinomycetes (Korayem et al., 2015). Fermentation is considered a green technology for the extraction or bioaccumulation of bioactive compounds because of its economic and environmental advantages. Several bioactive compounds, such as antibiotics, pigments, enzymes, antioxidants, antitumor agents, biosurfactants, and bioactive peptides, have been extracted by fermentation (Subramaniyam and Vimala, 2012).

Bioactive compounds have potential applications that can contribute to the development of new products in industry. The pharmaceutical industry is particularly interested in the search for compounds with antibacterial and antifungal activity to combat antibiotic resistance; and in the search for compounds with cytotoxic activity to be used as anticancer agents. These activities have been identified in Actinobacteria such as *Streptomyces*, where bioactive compounds such as ansamycins, chaxapeptins, angucyclines, and actinomycins have been reported (Bao et al., 2018; Elsayed et al., 2015; Rateb et al., 2011). In agribusiness, there is a growing interest in microbial surfactants (lipopeptides and lipopolysaccharides) for their environmentally friendly properties, mainly because they are biodegradable and for various functional properties, such as emulsification, wetting, foaming, corrosion inhibition dispersion, cleaning, surface activity, etc., which can be used to improve the quality of contaminated soils, to control plant pathogens, or to promote plant–microbe interactions (Adetunji and Olaniran, 2021). The cosmetic industry is interested in the potential of natural bioactive compounds such as mycosporine-like amino acids (MAA) and scytonemin for their photoprotective properties against UV radiation (Nowruzi et al., 2020). Natural pigments are also of great interest to the cosmetic industry, as well as to the food, nutraceutical, and pharmaceutical industries. Some natural functional pigments, such as phycobiliproteins and carotenoids, have been used for their antioxidant properties (Maoka, 2020; Pagels et al., 2019). This chapter focuses on characteristics, recovery, properties, and applications of bioactive compounds from microorganisms with extreme growth conditions, such as those of arid zones.

4.2 ARID REGIONS

According to the United Nations Committee on the Combat of Desertification (UNCCD), about 40% of the total land area is considered "drylands soils." Within this classification, arid and semiarid regions are the driest; these regions are characterized by a very low ratio between mean annual precipitation (MAP) and potential evapotranspiration; this ratio is lower than 0.05 for arid areas and lower than 0.002 for extreme hyperarid areas. Likewise, these regions present high levels of solar radiation and little vegetation cover (Bull, 2011; Couradeau et al., 2016). Extreme desiccation conditions in hyper-arid deserts are characterized by high temperatures, nanomolar scale nutrient concentration, low water activity, and intense radiation; however, in other ecosystems, low temperature, high salinity, pH, or concentrations of metals may occur in the arid area (Koeberl et al., 2011).

4.2.1 CHARACTERISTICS OF ARID REGIONS

The development of microorganisms in these areas is affected by abiotic stresses, which negatively influence them, for example factors such as salinity, drought, or extreme temperatures (high or low), and other environmental extremes. In deserts, for example, during the day, the temperature can reach 40 or 50°C, and during the night it can reach 0°C (Ramawat, 2010). In arid areas, solar radiation can reach 840 GJ km^2/year. One of the areas receiving the highest solar radiation in the world corresponds to the region from the Arabian Peninsula to the Sahara. These high levels of radiation are attributed to the proximity of the deserts to the equator and low cloud cover, which causes these high levels of radiation to reach the earth's surface (Goudie, 2010; Greenville, 2018).

Another important abiotic stress factor is the scarcity of water; the main sources of water are generated by precipitation, through rainfall, groundwater, surface flows, fog, or atmospheric vapor. The combination of high evaporation rates due to high temperatures causes a very rapid depletion of water reservoirs in these areas (Wang et al., 2017). There are some factors that determine the arid characteristic of deserts, such as sedimentation, temperature variations, and lack of water. In much of the desert areas, the soil is often referred to as "aridisols," i.e., dry with low concentrations of N_2 and organic matter, large amounts of phosphate, Ca_2CO_3 and Mg_2CO_3 in all or some parts of the soil, and slightly alkaline pH (McHugh et al., 2017). Another characteristic of dry soils with little vegetation cover is the formation of biological soil crusts (BSC), which are composed of mixtures of cyanobacteria, lichens, fungi, and mosses that cover the surface of the soil and preserve soil stability and fertility. These crusts nourish the soil by capturing dust containing essential nutrients such as N_2, K, and P, which improve water retention in these soils (Weber et al., 2015; (Faist et al., 2017).

4.2.2 MICROORGANISMS IN ARID ZONES

Bacteria are part of the soil microbiome and are abundant and diverse microorganisms that significantly influence plant growth. Because arid areas have very little vegetation, the presence and abundance of bacteria are vital for proper nutrient cycling and carbon storage (De Menezes et al., 2017; Fierer, 2017). Bacteria that grow under conditions of low water activity present a higher energy expenditure since they must accumulate larger amounts of water; in fact, some more resistant bacteria reach a state of hydrobiosis when the water activity (a_w) is less than 0.88, and under these conditions the cells do not metabolize, but continue to be viable. When bacteria develop in arid environments, they manage to adjust their access to the water they need for their physiological development. Most of these bacteria are found near soils containing halite, quartz, or gypsum and, through dispersion, they manage to acquire water trapped in these minerals to achieve their growth (Azua-Bustos et al., 2012; Mohammadipanah and Wink, 2016a).

4.2.2.1 ACTINOBACTERIAS

Arid regions are at an intermediate stage between semiarid zones (those with vegetation) and hyper-arid deserts (biologically unproductive) and harbor many actinobacteria (Neilson et al., 2012). These microorganisms are Gram positive, filamentous or not, and some genera can produce spores (Ludwig et al., 2012). To date, 5 subclasses, 10 orders, 56 families, and 286 genera have been found (Euzeby, 2015). Actinobacteria can be found in different environments, including extreme terrestrial and aquatic environments and different special habitats (Dhakal et al., 2017; Goodfellow et al., 2018). Filamentous actinobacteria have been isolated mainly in desert regions such as the Sahara, Mongolia, Mojave, Taklimakan, Tataouine, and Thar, the Atacama Desert in Chile being the most studied (Meklat et al., 2011; Kurapova et al., 2012; Harwani, 2013; Yu et al., 2015; Bull et al., 2016). Actinobacteria have been reported to produce complex structure natural bioactive compounds such as antibiotics, enzymes, immunoregulatory, and antitumor agents. Metabolites derived from these microorganism exhibit antifungal, antimicrobial, antiparasitic, anticancer, and other properties (Barka et al., 2015; Lee et al., 2020).

Actinobacteria present in extreme environments or that develop in special environments present biological activities, structures, and adaptation mechanisms very different from those of temperate habitats and have been little studied and reported. However, it has been reported to present great potential as a source of new natural products due to their properties and functionality for applications in agriculture, medicine, and industry (Qin et al., 2019).

Bioactive molecules obtained from Actinobacteria from arid regions have shown bioavailability, high solubility, and thermal stability (Mohammadipanah and Wink, 2016a). These bacteria, specifically those of the phylum of the same name, are characterized by the production of specialized secondary metabolites; *Streptomyces* strains in particular account for two-thirds of the antibiotics known to date, the vast majority being of clinical use (Be´rdy, 2012). Because of this, there has been widespread interest in investigating these bacteria, since it has been shown that they can synthesize up to ten times more specialized metabolites than those known in laboratory experiments (Baltz, 2017; Katz and Baltz, 2016).

Two new *Streptomyces* species were found in the Atacama Desert (Santhanam et al., 2012, 2013) that produce compounds such as *ansamycin* (antibiotic) and 22-membered macrolactones that present antitumor and antibacterial activities (Rateb, Houssen, Harrison, et al., 2011b). Also, another *Streptomyces* strain was isolated in this desert and found to produce aminobenzoquinones, which are inhibitory to bacteria and dermatophytic fungi (Schulz et al., 2011b). The species *Streptomyces leeuwenhoekii C33T* was isolated in the Atacama Desert and has been called a privileged actinobacterium because it has a genome with 35 biosynthetic gene clusters that can produce metabolites of great interest (Busarakam, 2014). It has been evidenced that this species can produce compounds such as chaxamycins A-D, which present anticancer activity by inhibiting tumor formation; Chaxamycins A, B, and C have antitumor effects because they inhibit the intrinsic ATPase activity of the human heat shock protein Hsp 90. Chaxamycin D shows antibacterial activity against Gram-positive bacteria, and selectively against *Staphylococcus aureus* ATCC 25923, bacteria responsible for several cutaneous and mucosal infections (Gomez-Escribano et al., 2015). It also shows activity against clinical isolates of methicillin-resistant *Staphylococcus aureus*.

4.2.2.2 CYANOBACTERIA

Cyanobacteria are part of the Gran-negative photoautotrophic prokaryotes. Most of the species can perform oxygenic photosynthesis, which is why they have the function of supplying oxygen to the atmosphere (Lyons et al., 2014). They are very important at the environmental level for oxygen production, carbon fixation and the role they play in biogeochemical cycles, as they produce about 25% of carbon-based compounds in the oceans (Flombaum et al., 2013). These bacteria have a great capacity for physiological adaptation, which allows them to inhabit aquatic environments, deserts, or even the Arctic (Bullerjahn and Post, 2014; Oren et al., 2019).

For cyanobacteria in arid or semiarid environments, the hydroscopic properties of these microorganisms are conferred by extracellular polymeric substances (EPS), which represent a great advantage at the physiological level, since it allows them to resist abiotic factors such as drought and UV radiation. This protection conferred by EPS on this type of microorganisms is related to the absorption of water caused by the highly hydrophilic chemical structure of EPS (Rossi and de Philippis, 2015; Raja et al., 2015).

In recent years, research with cyanobacteria has increased due to the diversity and versatility of their species and the simplicity in their scalability, from their cultures it is possible to obtain different high value-added products (Wijffels et al., 2013; Bellini et al., 2018). EPS obtained from cyanobacteria are of great biotechnological interest since they can have applications in bioremediation to reduce the presence of heavy metals in solution (Mota et al., 2016), in obtaining coagulants (Bhatnagar et al., 2014), as a physical support for stem cell proliferation (Bellini et al., 2018), to inhibit growth against Gram-negative and Gram-positive bacteria, and they also present antifungal activity (Najdenski et al., 2013). It can also have applications in medicine, since EPS has high antioxidant activity that is influenced by the presence of glucuronic acid, which has purifying activity (Tiwari et al., 2020).

Some examples of cyanobacteria are: *Arthospira sp., Microcystis sp., Anabaena sp., Nostoc sp.*, and *Oscillatoria sp.* which produce large amounts of secondary metabolites. This may be attributed to their tolerance to environmental stress due to their ability to accumulate bioactive compounds such as phycobiliproteins, fatty acids, exopolysaccharides (EPS), cyanotoxins, and carotenoids (Kumar et al., 2019; Morone et al., 2019). The latter change due to adverse conditions such as excessive light, oxidative or nutritional stress, and UV radiation (Zakar et al., 2016). The content and composition of carotenoids is also influenced by the depth of the desert layers.

In the work carried out by Ruiz-Domnguez et al. (2021a), they reported a new cyanobacterium of the genus *Geitlerinema sp.* which is a source of bioactive compounds of interest and was isolated from the Chilean coastal desert. The main bioactive compounds identified by conventional and pressurized liquid extraction were total proteins (~52 wt%), phycobiliproteins (especially C-phycocyanin), and methyl palmitate (C16:0). In general, high temperatures improved extraction yield; however, low temperatures favored extraction of phycobiliproteins. These results indicate that *Geitlerinema sp.* could be a cyanobacterium with functional properties for nutritional, pharmaceutical, or industrial use as sustainable bioactive compounds.

Spirulina (*Arthrospira*) is one of the most frequently described cyanobacteria for use in human and animal nutrition. In addition, it is a strain that has certification of Generally Recognized as Safe (GRAS) by the Food and Drug Administration (FDA) since the metabolites it produces do not affect the health of the consumer (Costa et al., 2019; Morais et al., 2014). This strain also has a high nutritional value due to its high protein content (around 60% in dry weight); it also stands out for its easy digestibility (Lupatini et al., 2017; Trivedi et al., 2015). These metabolites produced by these cyanobacteria help desert microorganisms survive, improve soil characteristics, and produce compounds of industrial interest (Poong et al., 2018) (TABLE 4.1).

4.2.2.3 ACTINOMYCETES

Aerobic actinomycetes are Gram-positive bacteria that play an important role in the soil ecosystem and produce relevant bioactive compounds; their

TABLE 4.1 Bioactive Compounds Produced by Desert Cyanobacteria.

Location	Cyanobacteria strain	Bioactive compound	Application	References
Chilean coastal desert	*Geitlerinema sp.*	C-phycocyanin, phycoerythrin, and allophycocyanin	Could be used in biorefinery processes as potential source of functional ingredients for the food or pharmacy industry as well as for biofuel production.	(Ruiz-Domínguez et al., 2021b)
Arid and exposed region in Chetimari, Niger	*Pseudanabaena sp. CCNU1*	Palythine-serine and carotenoids	May be useful as a producer of UV photoprotectants.	(Boucar et al., 2021c)
Salar Grande, Atacama Desert, Chile	*Halothece sp.*	Scytonemin (extracellular-located pigment)	Functions as a photoprotective compound against harmful UV and provides protection against deleterious short-wavelength radiation. This compound is being evaluated as a source of potent sunscreens.	(Orellana et al., 2020b)
Dalate Banner region (located on the eastern edge of the Qubqi Desert	*Scytonema javanicum*	Exopolysaccharides	Antioxidant and antitumor activities.	(Wu et al., 2018b)
Cold desert in Himachal Pradesh, India	*Nodularia sphaerocarpa* PUPCCC 420.1	Phycobiliproteins (PBP) (water soluble fluorescent pigments)	Are widely used as natural colorant in the food, nutraceutical and cosmetic industries.	(Kaushal et al., 2017)
Volcanic area in the Atacama Desert, Chile	*Chroococcidiopsis sp.*	Carotenoids	Are compounds more effective in the direct quenching of singlet oxygen and playing a role in the nonphotochemical quenching mechanism of cyanobacteria.	(Vítek et al., 2017)
Kakuma Campus of Kanazawa University, Japan	*Nostoc commune*	Glycosylated mycosporine (UV-absorbing pigments)	Could have multifunctional roles as sunscreens and antioxidants.	(Nazifi et al., 2015b)

TABLE 4.1 *(Continued)*

Location	Cyanobacteria strain	Bioactive compound	Application	References
Gurbantunggut Desert, China	*Microcoleus vaginatus*	Polysaccharides and other organic matter	Increase nitrogenase activity of bacteria in soil, enhance the microalgal activity, and increase invertase activity.	(Xu et al., 2013)
Field sites in Xuan'en County, China	*Nostoc commune*	Extracellular polymeric substances (EPS)	They have great potential to be used as humectants in cosmetic moisturizers without the need to be combined with an occlusive agent.	(H. Li et al., 2011b)
Canyonlands National Park, USA	*Microcoleus vaginatus*	Exopolysaccharides	Protection against DNA breakages and peroxidation of lipids through blocking reactive oxygen species caused by UV-B radiations.	(L.-Z. Chen et al., 2009)

presence is influenced by environmental conditions such as temperature, pH, humidity, and plant diversity (Yu et al., 2015). These microorganisms represent between 10 and 33% of the total soil microbes (Uddin, 2013). Among the genera that stand out for their production of antimicrobial secondary metabolites are *Streptomyces, Nocardia, Micromonospora,* and *Saccharothrix*. Some of these metabolites are vancomycin (*Streptomyces orientalis*), daptomycin (*Streptomyces roseosporus*), streptomycin (*Streptomyces griseus*), lincomycin (*Streptomyces lincolnensis*), and gentamicin (*Micromonospora purpurea*), substances used to counteract the effect of pathogenic microorganisms (Bérdy, 2005; Dubourg et al., 2017).

Currently, there is an increase in the resistance of drugs against pathogenic bacteria, so it is necessary to find antimicrobial compounds to meet this demand; in this sense, studies have been conducted in unusual habitats such as deserts, forests, caves, wetlands, among others, to find actinomycetes strains capable of producing these bioactive compounds (Harwani, 2013). In the research conducted by Tiwari et al. (2015), the antimicrobial potential of some actinomycetes from the Thar desert in India has been analyzed. Thirteen actinomycetes were tested, which exhibited increased antimicrobial activity against various test microorganisms, including multidrug-resistant bacteria (MRSA). According to the 16S rRNA sequencing method, five genera of *Streptomyces* (8 strains) were obtained, followed by *Actinomadura* (2 strains), *Nocardia* (1 strain), *Nonomuraea* (1 strain), and *Spirillospora* (1 strain). These strains showed antimicrobial activity against the panel of multidrug-resistant pathogens evaluated. Similarly, they showed greater antimicrobial activity against Gram-positive bacteria than against Gram-negative bacteria. This agrees with a study conducted by Ding et al., (2013) who isolated 53 actinomycetes strains from desert ecosystems located in northeast of Qinghai-Tibet Plateau. The results obtained in this work are promising and, therefore, further studies on the purification, characterization, and identification of these secondary metabolites are needed.

In the study conducted by Fatahi-Bafghi et al. (2019), they isolated actinomycetes strains from the Lut desert in Iran and evaluated their antimicrobial activity. Their results showed that 8 isolates showed antimicrobial activity, (*M. citrea*) the one with the best antibacterial activity against *S. aureus* (HA-MRSA). Four *Micromonospora* isolates identified as *M. aurantiaca, M. chaiyaphumensis, M. coriariae*, and *M. saelicesensis* were also identified as producing antibacterial compounds capable of acting against microorganisms such as *E. faecium, A. baumannii, S. aureus*, and *K. pneumonia*. These results show that the Lut desert is a potential actinomycetes ecosystem that could produce compounds with antimicrobial activity.

4.2.2.4 MICROALGAE

Microalgae are characterized by their ability to adapt to specific environments because they can withstand high temperatures, low radiation, and water potential. They are also good colonizers of desert environments, as they have the ability to grow heterotrophically, autotrophically, or mixotrophically (Subashchandrabose et al., 2013). In the past, some researchers mentioned that desert crusts contained a limited number of microalgae, because they relied on their observations through light microscopes. However, the investigations of Lewis and Flechtner (2002) and Lewis and Flechtner (2004) allowed glimpses of different stages of zoospore and gamete and crustal structures and functions in a wide variety of desert algae.

Some of the most frequently identified algal species in the deserts of Mexico and Baja California have been *Schizothrix calcicola* (*cyanobacteria*) and *Nostoc commune*, *Myrmecia astigmatica* (chlorophyte) and diatoms such as *H. amphioxys f. capitata*, *Hantzschia amphioxys*, *L. mutica*, *Luticola cohnii*, and *Pinnularia borealis var. scalaris*, and novel strains of *Cylindrocystis brebissonii var. deserti var. nov*, *Fasckulochloris mexicana sp. nov.* and *Elakatothrix obtusata sp. nov.* which belong to 32 genera. To date, approximately 400 species of microalgae have been identified in the deserts of the world (Perera et al., 2018).

Rasheed et al. (2020) investigated a species of microalgae (*Chlorella*, QUCCCM3) isolated from the Qatar desert. The aim was to establish their growth characteristics and composition of bioactive compounds to determine their possible application as raw materials in food processing. The results showed the highest growth rate at 30°C, with 0.64 ± 0.04 day^{-1}, high protein content ($43 \pm 2.3\%$), and fatty acids such as linoleic acid, docosahexaenoic acid, and alpha-linolenic acid. A relevant finding was the fact that this microalga showed high free radical scavenging capacity with an antiproliferative effect against K562 chronic myeloid leukemia cancer cells. This species can be produced all year round in the Qatar desert, which is an advantage to be used for commercial purposes in animal feed or as nutraceutical supplements.

4.3 BIOACTIVE COMPOUND RECOVERY

The recovery of bioactive compounds from arid zone microorganisms is generally done through fermentation processes. Fermentation can be defined as a biological process that occurs under anaerobic conditions within the cell. For biotechnology, fermentation is a process that uses microorganisms or their parts to produce metabolites of interest (Bansod et al., 2021). Taking into account the latter definition, fermentation can be divided into two (a) liquid or submerged fermentation (Smf) where microorganisms grow in liquid media and (b) solid fermentation (SSf), in which the growth of microorganisms occurs in an environment with no or low free water content. Comparatively, Smf has some advantages in terms of instrumentation and control of pH, temperature, dissolved oxygen, concentration of compounds and other variables, separation of biomass after fermentation and ease of process scale-up, while the advantages of SSf are lower energy demand during sterilization, higher metabolite productivity, and the process is less susceptible to contamination (Soccol et al., 2017). In general, for fermentation processes, the variables that play a fundamental role in the yields in obtaining metabolites are mainly the concentration of the culture medium or substrate composition, pH, temperature, and inoculum size.

Studies on the processes to obtain compounds of industrial interest using microorganisms from arid zones have focused on two aspects: (1) the isolation conditions of the microorganisms and (2) the optimization of the growth conditions of these isolated microorganisms to improve the production of metabolites. Some microorganisms have specific growth conditions and therefore commercial culture media do not meet their nutritional needs for growth and development, so determining the appropriate concentration of nutrients is vital for the isolation of microorganisms from arid zones and the determination of bioactive compounds (Ratnaweera et al., 2015).

For the isolation of actinomycetes, commercial culture media and mixtures prepared in the laboratory with different concentrations of glucose and mineral salts have been tested, among others. Studies developed by Hozzein et al. (2008) aimed to find a suitable medium for the isolation of actinomycetes from the Egyptian desert found that the best composition for this medium was glucosa 0.5 g/L, yeast extract 0.5 g/L, K_2HPO_4 1 g/L, $MgSO_4.7H_2O$ 0.5 g/L, NaCl 0.5 g/L y 1 mL de 1 mL of microelements solution. The above results may be since this medium composition does not favor the growth of other microorganisms since it lacks several nutrients and therefore becomes a selective medium for actinomycetes. As can be seen in Table 4.2, for most of the isolated microorganisms, commercial media have been used with some variation, either addition of mineral salts or antimicrobial agents to inhibit the growth of other undesirable microorganisms.

After defining the conditions for the isolation of the microorganisms, it is necessary to establish the variables of the fermentation process that will allow the growth of the microorganisms in the laboratory and the subsequent generation of the bioactive compounds of interest. In liquid fermentations,

TABLE 4.2 Culture Media Used for the Isolation and Characterization of Microorganisms in Arid Zones.

Microorganisms	Source	Medium	Bioactivity	Reference
Actinobacteria	Cotton fields in semiarid zones of Haryana	KenKnight Medium (Glucose 1.0 g/L, potassium dihydrogen phosphate 0.1 g/L, Sodium nitrate 0.1, potassium chloride 0.1 g/L, magnesium sulfate 0.1 g/L) with ampicillin 25 g/mL	Antibacterial activity	(Gera et al., 2014)
Endophytic fungi: *Aspergiullus niger, Fusarium sp.*	*Opuntia dilleni* of Sri Lanka	Commercial mediums: Potato dextrose agar (PDA) yeast peptone starch agar (SYP), yeast peptone dextrose agar (YPD)	Antibacterial activity	(Ratnaweera et al., 2015)
Acidophilic actinomycetes	Acidic soil samples from each site Phatthalung Botanical Garden, Phatthalung Province, Thailand	Starch Casein Agar (SCA), Soil Extract Agar (SEA) and Humic Vitamin Agar (HVA)	Antimicrobial Activity	(Niyasom et al., 2015)
Endophytic fungi: *Acremonium sp., Gliocladiopsis sp., Fusarium sp., Col-letotrichum sp., Aspergillus sp., Phlebia sp., Earliella sp.*, and *Pseudolagarobasidium sp.*	Ginger (*Zingiber officinale*) from Thrissur, Kerala	Commercial medium: Potato dextrose agar (PDA)	Antibacterial activity	(Anisha and Radhakrishnan, 2017)
Arthrinium phaeospermum, Colletotrichum siamense, C. tropicale, Fusarium oxysporum, F. chlamydosporum, F. proliferatum, F. solani, Lasiodiplodia Theobromae, Nodulisporium sp., Paecilomyces formosus, Penicillium citrinum, and *Pestalotiopsis microspora.*	Roots of different species of plants mangroves from Luzon Island, Philippines	Potato Dextrose Agar medium supplemented with 33 g/L marine Salts, and 500 mg/L streptomycin	Antimicrobial Activity	(Moron et al., 2018)

TABLE 4.2 *(Continued)*

Microorganisms	Source	Medium	Bioactivity	Reference
Endophytic fungi: *Chaetomium madrasense, Chaetomium sp., Chaetomium globosum, Aspergillus hiratsukae, Aspergillus ochraceus, Alternatia tenuissima* and	Leaves of Avicennia marina from Yanbu Industrial City, Saudi Arabia	Commercial medium: Malt Extract Agar	Antimicrobial and antioxidant activities	(Khalil et al., 2020)
Actinomycetes: *Streptomyces, Nocardiopsis, Saccharomonospora* and *Actinoalloteichus*	Soil samples from arid zone of Barmer and Jaisalmer districts of Rajasthan (Indian Thar Desert), India.	Commercial medium: actinomycetes isolation agar	Antibacterial activity	(S. Kumar et al., 2021)

commercial media have generally been used at a temperature of 22–28 °C, for 6–21 days, with agitation between 0 and 220 rpm. Among the studies conducted, González-Menéndez et al. (2018) isolated fungal strains from the arid areas of Andalusia to evaluate their antifungal and antitumor activity. During the study they found some fungi of the genera *Neocamarosporium*, *Preussia*, *Alternaria*, *Ascochyta*, *Phoma*, *Comoclathris*, *Neomicrosphaeropsis*, *Aureobasidium*, and *Pleospora*. Once these microorganisms were obtained, liquid fermentations were carried out to obtain the extracts and evaluate the biological activity of the obtained metabolites. The process was carried out using commercial culture media with different sources of carbon and nitrogen, and with the addition of polymeric resins that improve yields, for 14 days at 22°C, 220 rpm, and 70% relative humidity. This process yielded 3 novel bioactive compounds $C_{17}H_{26}O_4$ (1; MDN-0209), produced both with and without the resin addition, and $C_{17}H_{24}O_4$ (2; MDN-0210) and $C_{10}H_{12}O_5$ (3; MDN-0211), induced only in the presence of the XAD-16 resin. Among other investigations, Gacem et al. (2020) evaluated the antimicrobial and antioxidant potential of a Streptomyces strain isolated from El-Ogbane Forest soils in Algeria. The fermentations were carried out using commercial media with the addition of mineral salt mixtures at 30°C for 15 days, obtaining compounds such as spectinabilin, undecylprodigiosin, and metacycloprodigiosin that exhibited the studied bioactivities.

Depending on the microorganism and metabolites of interest, it is necessary to study other variables within the process. This is the case for cyanobacteria, whose growth is affected by factors such as light intensity. Montero-Lobato et al., (2020) isolated and characterized cyanobacteria from arid zones of the Atacama Desert, Chile, and evaluated the production of phycobiliproteins by varying, among other factors, light intensity. The isolated microorganism *Chroococcidiopsis sp* achieved its best incident light performance of 10 μmol photons-m-2-s-1. These studies show the importance of determining the optimal growth conditions to scale up the processes and obtain bioactive compounds industrially.

4.4 BIOACTIVE COMPOUNDS PROPERTIES

The search for bioactive compounds has intensified worldwide because each compound has different properties that can be applied according to the characteristics required by each industry. For example, natural antimicrobial products are receiving special attention for their clinical and pharmaceutical applications, mainly to combat the emergence of multidrug-resistant pathogens, often referred to as antibiotic resistance. One type of these compounds is found in microorganisms such as *Streptomyces*, which can produce secondary metabolites such as ansamycins (Rateb, Houssen, Arnold, et al., 2011), which are an important class of macrolactam natural products that have been shown a broad range of bioactivities, such as anti-microbial, anticancer, and antiviral activities (Escalante et al., 2021; Park et al., 2019). The biosynthesis of these macrolactams involves an initial 3-amino-5-hydroxybenzoic acid (AHBA) unit and the modular synthesis of type I polyketides, and they are classified into naphthalenes and benzenes according to the function of the aromatic moiety. Each group can be further classified according to the number and type of extensor units recruited. From the group of naphthalene ansamycins, rifamycin stands out for its strong antimicrobial activity against many Gram-positive and Gram-negative bacteria by inhibiting RNA polymerase (Adams et al., 2021; Z. Li et al., 2018; Tang et al., 2018).

Bioactive compounds are important in the field of medicine because they can be used as anticancer agents, due to their cytotoxic activities that prevent cancer cells from dividing and growing. Different studies have demonstrated high levels of cytotoxicity of the *Streptomyces* strain genus isolated from arid zones, against hepatocellular carcinoma cell lines (HepG2 ACC 180) and mouse fibroblasts (NIH-3T3) (Albayay et al., 2019); against human breast adenocarcinoma cells (MCF-7); and different leukemia cells (HL-60, K562 and THP-1) (Almuhayawi et al., 2021). This feature attracts the attention of the pharmaceutical industries because it can be used as a drug in the treatment of various types of cancer. Angucyclines and actinomycins are two bioactive compounds that exhibited significant cytotoxicity against a panel of human cancer cell lines A549, H157, MCF7, MDA-MB-231, and HepG2 (Bao et al., 2018). The group of natural products of angucyclines is the largest group of polycyclic aromatic polyketides, rich in chemical scaffolds, which are characterized by the nucleus of tetracyclic benz [a] anthracene and most of them exist as O or C glycosides in nature (Kharel et al., 2012). Actinomycins belong to a family of chromopeptide lactones and are composed of a chromophore group and two pentapeptide chains with a variable composition of amino acids. The best-known member of this family, actinomycin D, has been widely used in the clinic as an antitumor drug (Machushynets et al., 2021).

Chaxapeptin is another compound with an important pharmaceutical and clinical applications because it exhibits several potentially therapeutic characteristics, such as antimicrobial activity, receptor antagonism, and enzyme inhibition (Bull et al., 2016). This bioactive compound showed inhibitory activity in a cell invasion assay with the human lung cancer cell line A549 in *Streptomyces leeuwenhoekii* strain C58, recovered from the Atacama Desert. Chaxapeptin was identified as a lasso peptide (LP) using bioinformatic tools such as genome mining (Elsayed et al., 2015a). Lasso peptides belong to a specific family of natural products synthesized ribosomally and post-translationally modified peptides (RiPP) with a unique 3D structure, in which the C-terminus threads through an N-terminal macrolactam ring in a right-handed conformation (Martin-Gómez and Tulla-Puche, 2018). They are classified into three subclasses based on their N-terminal residues and the number of disulfide bridges. Class I LPs have an N-terminal Cys and contain two disulfide bridges; class II LPs have an N-terminal Gly and are devoid of disulfide bridges, and class III LPs have an N-terminal Gly and contain one disulfide bridge. Owing to their exceptional stability to enzymatic degradation and thermal or chemical denaturation, LPs are highly attractive as potential scaffolds for epitope grafting (Rateb et al., 2018).

The discovery of new antibiotics is essential for pharmaceutical industries, not only for clinical but also for agricultural applications, as bioactive compounds can be used as antibiosis and act as plant pathogen control agents to replace synthetic chemical fungicides. For this, researchers have paid increasing attention to cyclic lipopeptide compounds produced by microorganisms (Fira et al., 2018). Bacillomycin and fengycins are cyclic lipopeptides with an antibiotic that is potentially active against pathogenic fungi, identified from *Bacillus subtilis* recovered from the arid zone of Matmata, Tunisia (Nasfi et al., 2018). Lipopeptides from Bacillus are low-molecular-weight cyclic amphiphilic oligopeptides synthesized by multi-enzyme complexes called non-ribosomal peptide synthetases (NRPSs) and are grouped into the surfactin, iturin, and fengycin families, according to their peptide scaffold (Kaspar et al., 2019). Furthermore, surfactin is another lipopeptide that has shown important applications in the phytosanitary, pharmaceutical, food, and cosmetic industries, mainly due to its characteristics as a biosurfactant (Khem Raj and Shamsher S, 2015). Biosurfactants (BS) belong to the surfactant family and are amphipathic compounds that act by reducing the interfacial and surface tensions between the particles of two liquids or between a liquid and a solid. They can be low-molecular weight (biosurfactants), generally glycolipids or lipopeptides, phospholipids, neutral lipids or fatty acids etc., or high-molecular weight compounds (bioemulsifiers), which are mainly lipopolysaccharides, lipoproteins, or a combination of both (Adetunji and Olaniran, 2021). Biosurfactants have received special attention because of their functional capabilities and ecological properties. These bioactive compounds can be used as stabilizing agents of emulsions in the food, cosmetic, textile, chemical and pharmaceutical industries (Raddadi et al., 2018). Biosurfactant activity of *Streptomyces sp.* 5S isolated from arid Egyptian soil has been reported, with an emulsification index of 31.74% (Korayem et al., 2015).

Extracellular polymeric substances (EPS) are a bioactive compound with bioemulsifying properties produced by cyanobacteria. This compound has important ecological applications such as restoration of degraded areas and remediation of heavy metal-contaminated soil and water (Sabarinathan et al., 2021). EPS are biopolymers that possess an anionic character mainly due to the high content of uronic acids and are composed of polysaccharides, lipids, proteins, and DNA in the form of heteropolymers, like lipopolysaccharides or glycoproteins. Different monosaccharide units, such as hexoses and pentoses numbered 40–50 linked by glycosidic bonds, constitute polysaccharides (Singh et al., 2018). EPSs are essential in the formation of biological soil crust (BSC) in arid and semiarid environments because they help to bind soil particles together and contribute to improving soil moisture, allowing other microorganisms to colonize and form the BSC (Dahech et al., 2021). Furthermore, EPSs have the characteristic of having negative charges in the external cell structure, making them promising immobilizing agents for the removal of positively charged heavy metal ions from effluents and wastewater, because metal cations can be taken up by negatively charged groups present in the cell wall and those present in the polysaccharide layers surrounding the cell wall (Philippis and Micheletti, 2018).

The cosmetic industry has evolved over time, focusing on the search for new compounds that can protect against UV radiation and thus improve the characteristics of topical skin products. Numerous species of cyanobacteria have been used for this purpose, as they can produce bioactive compounds such as mycosporine-like amino acids (MAAs) and scytonemin, which have been reported for their potential use as natural UV blockers (Nowruzi et al., 2020). MAAs act as sunscreen agents, thanks to their ability to absorb light in both the UV-A (315–400 nm) and UV-B (280–315 nm) range without producing free radicals (Rosic, 2019). All MAAs molecules contain a central cyclohexanone or cyclohexenimine ring that is responsible for UV absorption. Their biosynthesis in cyanobacteria occurs through specific pathways such as the shikimate pathway and the pentose phosphate pathway (Derikvand et al., 2017). Two novel MAAs were recently found in the crusts of biological soils of extreme habitats and were named klebsormidin A and klebsormidin B. These compounds showed absorption maxima of 324 nm, supporting their role as a natural UV−sunscreen (Hartmann et al., 2020).

Scytonemin is a small, yellow-brown-to-dark-red, or mahogany-colored hydrophobic pigment that serves as a protectant against short-wavelength solar UV radiation because it has a maximum absorption at 384 nm. There are two forms of scytonemin based on the redox and acid–base conditions during the extraction process: an oxidized form (green) named fuscochlorin and a reduced form (red) known as fuscorhodin. This pigment works not only as a powerful UV protector but also as a powerful antioxidant and anti-inflammatory (Rastogi et al., 2015).

Phycobiliproteins are a family of colored proteins that are responsible for the uptake of light mainly in the green and yellow areas of the light spectrum. This characteristic makes them compounds of great interest as natural pigments in the food, nutraceutical, cosmetic, and pharmaceutical industries. Phycobiliproteins are classified into four groups based on their absorption of photosynthetic active radiation (PAR) light. Phycoerythrin (PE, red pigment) shows maximal absorbance in the wavelength range 490–570 nm, phycocyanin (PC, blue pigment) at 610–625 nm, phycoerythrocyanin (PEC) at 560-600, and allophycocyanin (APC, bluish pigment) at 650–660 nm (Pagels et al., 2019). C-phycocyanin (PC), which functions as a blue photosynthetic pigment, has been also used in the health area, as a biomarker, for its fluorescence properties, and as a possible therapeutic agent for the treatment of diseases associated with oxidative stress due to its antioxidant, anti-inflammatory, and immunomodulatory properties (Alfaro-Alfaro et al., 2020; Díaz Domínguez et al., 2016). This bioactive compound is commonly found in cyanobacteria such as Spirulina (*Arthrospira*) platensis and represents more than 20% of the dry weight of Spirulina (Ferreira-Santos et al., 2020), but it has also been identified in *Chroococcidiopsis sp*., an endolithic cyanobacterium, isolated from gypsum rocks from the Atacama Desert (Chile) (Montero-Lobato et al., 2020a).

Carotenoids are another interesting group of lipid-soluble pigments with nutraceutical, cosmetic, and food value because they present yellow, orange, red, and purple colors and have antioxidant properties. They are divided into two groups: carotenes (are hydrocarbons) and xanthophylls (containing oxygen atoms) (Maoka, 2020). Carotenes, such as β-carotene, and xanthophylls, such as echinenone, canthaxanthin, zeaxanthin, synechoxanthin, and myxoxanthophyll have been found in cyanobacteria from arid areas (Abed et al., 2018; Boucar et al., 2021a). β-carotene is a precursor of vitamin A and is highly insoluble in the aqueous system and is prone to degradation under various environmental conditions. For this reason, some studies have sought mechanisms to protect β -carotene for use in food systems (Wei et al., 2018). This bioactive compound has been used in the treatment of cardiovascular and thyroid diseases by reducing oxidative stress (Miller et al., 2020; Rotondo Dottore et al., 2018). Canthaxanthin is a ketocarotenoid that is used as a food coloring, giving a red-orange color to egg yolks, chicken skin, koi carp skin, and the flesh of salmon and trout when included in dietary supplements. This occurs because canthaxanthins are a lipophilic compound that can accumulate in the fatty tissues of foods. In humans, these carotenoids are released from the food matrix and incorporated into mixed micelles made of bile acid, fatty acids, and lipids (Rebelo et al., 2020).

4.5 APPLICATIONS

4.5.1 HEALTH APPLICATIONS

The *Saccharothrix* SA198 strain was isolated from a sample of Saharan soil from Algeria, due to climatic conditions this soil contains many actinomycetes, among which the *Saccharothrix* genus stands out, of which strains have been found that produce new antibiotics such as dithiolopyrrolones (Merrouche et al., 2011) and anthracyclines (Zitouni et al., 2004). In *in vitro* studies, this strain showed antimicrobial activity against Gram-positive and Gram-negative bacteria and filamentous fungi. It exhibited strong to very strong activity against bacteria such as *Enterococcus faecalis* and *Bacillus subtilis* and against fungi such as *Penicillum expansum, Mucor ramannianus,* and *Aspergillus carbonarius*. Two new antibiotics A4 and A5 were isolated, showing strong activity against fungi (Boubetra et al., 2013).

Streptomyces leeuwenhoekii was identified as a producer of two new bioactive compounds, the first chaxamycins, which present antagonism against the strain of *Staphylococcus aureus* ATCC 25923 and the second compound, chaxalactins, a bacterial agent against Gram positive (Rateb, Houssen, Harrison et al., 2011a). Isolation of Streptomyces sp. strain C34, from hyper-arid soil of the Chilean Atacama Desert allowed the identification of three new compounds of the rare class of macrolactone polyketides, called chaxalactins A–C (1-3). Also, known compounds of ferroxamine E (4), hygromycin A (5), and 500-dihydrohygromycin A (6) were detected. Compounds showed strong activity against Gram-positive strains, but weak activity against Gram-negative strains (Rateb, Houssen, Harrison, et al., 2011a).

Two other bioactive compounds, atacamycins and chaxapeptin, are produced by *Streptomyces* sp., C38 (Nachtigall et al., 2011) and C58 (Elsayed et al., 2015b), respectively. These strains were isolated from hyperarid soil of the Chaxa Lagoon of the Salar de Atacama located in northern Chile.

These two compounds are ribosome-synthesized Lasso peptides with post-translational modification so that cell invasion assays demonstrated cell line inhibition against human lung cancer.

Another new group of bioactive secondary metabolites are the A-D abenkines, produced by the *Streptomyces* sp. DB634, isolated from the Chilean Altiplano (Schulz et al., 2011a). Abenquinas show inhibitory activity against bacteria and dermatophyte fungi and inhibitor of phosphodiesterase type 4b (Mohammadipanah and Wink, 2016b). An isolate identified *Streptomyces* sp. DA3-7 from Saudi Arabian desert soil samples, which had broad-spectrum antimicrobial activity against Gram-positive and Gram-negative bacteria, as well as fungi. This strain had minimum inhibitory concentrations of 78 and 156 µg/mL against *Enterococcus faecalis* and *Salmonella typhimurium*, respectively. The active compound was identified as pyridine-2,5-diacetamide, which showed antimicrobial activity against *Cryptococcus neoformans* and *Escherichia coli* with a lower minimum inhibitory concentration (31.25 µg/mL). The authors report that the antimicrobial activities of this new compound are reported for the first time in their study (Nithya et al., 2018). Strain *Streptomyces* sp. TK-VL_333 isolated from southwestern Algeria produces two compounds 4-(4-hydroxyphenoxy)butan-2-one Acetic acid-2-hydroxy-6-(3-oxo-butyl)-phenyl ester with antibacterial and antifungal bioactivity (Kavitha et al., 2010).

4.5.2 AGRICULTURAL APPLICATIONS

Streptomyces may be involved in plant growth because they are able to colonize the rhizosphere, compartments in the rhizoplane of the host plant. This process is achieved through some specific properties, such as the production of antibiotics, enzymes, and phytohormones, and chemotaxis. Within enzyme production, chitinase, cellulase, lipase, and β-1,3-glucanase stand out. This microorganism can also be used in the agricultural sector, as biofertilizers or biological control agent to promote plant growth. Biocontrol is related to the production of some metabolites such as nigericin and geldanamycin (*S. violaceusniger* YCED-9), chitinase (*S. violaceusniger* YH27A), and siderophores (*S. coelicolor*). The production of enzymes, antibiotics, and phytohormones supports its ability to biocontrol pathogens, against which *Streptomyces parvulus* protects different plants by using a symbiotic association. Bacteria and fungi such as *Fusarium graminearum, Aspergiullus carbonarius, Aspergillus westerdijkiae, Colletrichum musae,*

Candida albicans, Staphyloccus aureus ATCC 25923, *Staphyloccus aureus* ATCC 6538, and *Staphyloccus aureus* ATCC 33591 (Khushboo et al., 2022).

Plant growth-promoting properties displayed by actinobacteria include the production of compounds such as 1-aminocyclopropane-1-carboxylic acid (ACC), deaminase, auxin, gibberellic acid (GA), indole-3-acetic acid (IAA), siderophores, nitrogen fixation, and solubilization of phosphorus and potassium (Xie and Pathom-aree, 2021). For example, a strain with high IAA production is *Kocuria turfanensis* 2M4 isolated in India, with results such as a significant increase in total length and fresh peanut biomass after 15 days of germination by a study carried out in a pot (*Arachis hypogaea* L.) (Goswami et al., 2014). Likewise, *Nocardiopsis dassonvillei* MB22 was isolated from the soil of the Algerian Sahara, which stimulated the growth of wheat seedlings, in terms of shoot and root length and dry weight (Allali et al., 2019).

Phormidium tenue produces a polysaccharide that has an effect on plant growth, specifically the bush *Caragana korshinskii* was studied. The addition of the polysaccharide significantly increased seed germination and bush metabolic activity. The presence of polysaccharide increased the rate of seed germination, ion absorption, and photosynthetic efficiency by affecting the electron transport chain and decreasing oxidative damage by moving oxygen, thus promoting shrub yield in desert areas (Xu et al., 2013).

4.5.3 FOOD INDUSTRY APPLICATIONS

The *Streptomyces* genus produces a wide range of secondary metabolites, such as the production of extracellular enzymes, among which are cellulases, amylases, nucleases, and lipases. Various strains of this genus such as *S. clavuligerus, S. griseus*, and *S. halstedii* produce lipases. These enzymes are used in detergents and for other applications. However, more studies are needed to explore the activity and production of lipases from this microorganism (Khushboo et al., 2022).

Phycobiliproteins from cyanobacteria are natural pigments that are used as colorants in the food industry; for example, aqueous extracts of unpurified blue phycocyanin from spirulina have been added to products such as sports drinks, candies, ice cream, jellies, chewing gum, and dairy products. Specifically, brilliant blue phycocyanin has been selected over less brilliant natural colorants such as gardenia blue and indigo in candy production (Pandey et al., 2013).

Galetovic et al. (2020) formulated a new food based on skim milk fortified with cyanobacterial phycobiliproteins (PBP), phycoerythrin (PE), and phycocyanin (PC). In this study, antioxidant activity and absence of toxicity were verified, in addition to sensory tests. The antioxidant results indicate that phycobiliproteins from cyanobacteria isolated from Atacama, Chile, have an antioxidant power comparable to that of fruits such as pineapple and passion fruit. The work also showed that cyanobacterial phycobiliproteins were not toxic to *Caenorhabditis elegans*; nematode survival was 100% at all concentrations used. In the sensory tests, the highest score was obtained by the PE-fortified skim milk. Therefore, a functional food based on skim milk containing PBP can be considered an innovative beverage for the food industry, considering that PBPs were stable at ultrahigh temperature.

4.5.4 COSMETIC INDUSTRY APPLICATIONS

Pseudanabaena sp. CCNU1 in the arid and exposed region of Chetimari, Niger. This strain isolated as UV-B treatment significantly improved the production of active photoprotectants, including carotenoids (β-carotene, echinenone, canthaxanthin, zeaxanthin, sinecoxanthin and UV-absorbing mycosporin-like amino acids (MAA). UV radiation B mainly induced the production of MAA, a compound identified as palythine-serine; the content of this compound was high after UV-B treatment for 4 days (Boucar et al., 2021b). Cyanobacteria of the genus Halothece, isolated from the Atacama Desert, synthesize scytonemin (absorption maximum at 370 nm), a lipid-soluble heterocyclic dimer and UV-filtering pigment. In this study, UV rays were shown to positively modulate scytonemin biosynthesis in this genus (Orellana et al., 2020a). Two genotypes (A and B) of *Nostoc commune* colonies with different UV absorption spectra were examined, which produced structurally distinct MAAs in each genotype. The aqueous extract of genotype A showed a UV absorption spectrum with an absorption maximum at 335 nm, while genotype B presented an absorption maximum of 310 nm. MAA produced in this study had potent radical scavenging activities *in vitro* and the results confirmed that MAA have multiple functions as UV protector and relevant antioxidant (Nazifi et al., 2015a).

The oxidative damage induced by UV-B rays and the protective effect of exopolysaccharides (EPS) produced by *Microcoleus vaginatus*, a cyano-bacterium isolated from the desert, were investigated. After being irradiated with UV-B radiation, EPS, photosynthetic activity, total carbohydrates, and sucrose production of irradiated cells decreased. The results suggested that EPS exhibits a significant protective effect on DNA strand breaks and lipid peroxidation by effectively scavenging UV-B-induced reactive oxygen species in *M. vaginatus* (L. Chen et al., 2009).

4.5.5 OTHER APPLICATIONS

The cyanobacterium *Scytonema javanicum* isolated from the Qubqi Desert was cultured in artificial synthetic wastewater to explore the feasibility of nutrient transfer from wastewater to the desert. After inoculation of the strain in the wastewater, phosphorus and nitrogen were gradually removed from the waters. According to the results obtained in the study, this cyanobacterium is a promising species for the transfer of nutrients from wastewater to the desert, which has a growth rate. The maximum biomass yield was reached at 3.91 mg Chl-a L-1 and the maximum biomass productivity 0.15 mg Chl-a L^{-1} d^{-1} (Wu et al., 2018a).

The biomedical potential of the polysaccharide produced by Nostoc commune, regarding the antioxidant capacity and moisture retention of the polymer, was studied using the animal model Caenorhabditis elegans; it was shown that the polysaccharide can increase the activity of the antioxidant enzyme, decrease the level of peroxidation of lipids, and reduce oxidative damage. Polysaccharide was also found to have strong in vitro moisture retention and absorption capabilities compared to compounds such as chitosan and urea (H. Li et al., 2011a).

In different dry salt lakes in northeastern Algeria, the diversity of halo-archaea was investigated with respect to its potential to produce hydrolytic enzymes. A total of 68 aerobic halophilic archaea were isolated from saline sediments. Of the isolates, 89.7% were capable of producing at least two types of extracellular halophilic enzymes. Among the enzymes produced are cellulases, esterase, gelatinase, and inulinase. The production of this type of compounds shows a promising source of molecules for various biotechnological applications (L. Chen et al., 2009).

4.6 PERSPECTIVES

The growing consumer demand for natural products has led different industries to focus on the search for alternative products that can be obtained in a sustainable way. This search has focused on novel compounds generated by microorganisms living in extreme environments, such as microorganisms isolated in arid areas, which have demonstrated their ability to produce different bioactive compounds with many desirable characteristics for the fields of food, cosmetics, environment, medicine, and pharmacy. Bioactive compounds have generated great impact due to their multiple characteristics that allow them to replace synthetic additives, so this market has promising growth potential.

KEYWORDS

- **emerging technology**
- **biotechnology**
- **additives**
- **metabolites**
- **bioactive compounds**

REFERENCES

Abed, R. M. M.; Palinska, K. A.; Köster, J. Characterization of Microbial Mats from a Desert Wadi Ecosystem in the Sultanate of Oman. *Geomicrobiol. J.* **2018,** *35* (7), 601–611.

Adams, R. A.; Leon, G.; Miller, N. M.; Reyes, S. P.; Thantrong, C. H.; Thokkadam, A. M.; Lemma, A. S.; Sivaloganathan, D. M.; Wan, X.; Brynildsen, M. P. Rifamycin Antibiotics and the Mechanisms of their Failure. *J. Antibiot.* **2021,** *74* (11), 786–798.

Adetunji, A. I.; Olaniran, A. O. Production and Potential Biotechnological Applications of Microbial Surfactants: An Overview. *Saudi J. Biol. Sci.* **2021,** *28* (1), 669–679.Albayay, C.; Silber, J.; Imholf, J.; Asenjo, J.; Andrews, B.; Nouioui, I.; Dorador, C. The Polyextreme Ecosystem, Salar de Huasco at the Chilean Altiplano of the Atacama Desert Houses Diverse Streptomyces spp. with Promising Pharmaceutical Potentials. *Diversity* **2019,** *11* (5), 69.

Alfaro-Alfaro, Á. E.; Alpízar-Cambronero, V.; Duarte-Rodríguez, A. I.; Feng-Feng, J.; Rosales-Leiva, C.; Mora-Román, J. J. C-Ficocianinas: Modulación del Sistema Inmune y su Posible Aplicación como Terapia Contra el cáncer. *Rev. Tecnol. Marcha* **2020,** *33,* 125–139.

Allali, K.; Goudjal, Y.; Zamoum, M.; Bouznada, K.; Sabaou, N.; Zitouni, A. Nocardiopsis Dassonvillei Strain MB22 from the Algerian Sahara Promotes Wheat Seedlings Growth and Potentially Controls the Common Root Rot Pathogen Bipolaris Sorokiniana. *J. Plant Pathol.* **2019,** *101* (4), 1115–1125.

Almuhayawi, M. S.; Mohamed, M. S. M.; Abdel-Mawgoud, M.; Selim, S.; Al Jaouni, S. K.; Abdelgawad, H. Bioactive Potential of Several Actinobacteria Isolated from Microbiologically Barely Explored Desert Habitat, Saudi Arabia. *Biology* **2021,** *10* (3).

Alori, E. T.; Emmanuel, O. C.; Glick, B. R.; Babalola, O. O. Plant–Archaea Relationships: A Potential Means to Improve Crop Production in Arid and Semi-Arid Regions. *World J. Microbiol. Biotechnol.* **2020,** *36* (9), 133.

Anisha, C.; Radhakrishnan, E. K. Metabolite Analysis of Endophytic Fungi from Cultivars of Zingiber Officinale Rosc. Identifies Myriad of Bioactive Compounds Including Tyrosol. *3 Biotech* **2017,** *7* (2).

Azua-Bustos, A.; Urrejola, C.; Vicuña, R. Life at the Dry Edge: Microorganisms of the Atacama Desert. *FEBS Lett.* **2012,** *586* (18), 2939–2945.

Baltz, R. H. Gifted Microbes for Genome Mining and Natural Product Discovery. *J. Indus. Microbiol. Biotechnol.* **2017,** *44,* 573–588.

Bao, J.; He, F.; Li, Y.; Fang, L.; Wang, K.; Song, J.; Zhou, J.; Li, Q.; Zhang, H. Cytotoxic Antibiotic Angucyclines and Actinomycins from the Streptomyces sp. XZHG99T. *J. Antibiot.* **2018,** *71* (12), 1018–1024.

Barka, E. A.; Vatsa, P.; Sanchez, L.; Gaveau-Vaillant, N.; Jacquard, C.; Meier-Kolthoff, J. P.; Klenk, H.-P.; Clément, C.; Ouhdouch, Y.; van Wezel, G. P. Taxonomy, Physiology, Andnatural Products of Actinobacteria. *Microbiol. Mol. Biol. Rev.* **2015,** *80* (1), 1–43.

Bérdy, J. Thoughts and Facts About Antibiotics: Where We are Now and Where we are Heading. *J. Antibiot.* **2012,** *65,* 385–395.

Bellini, E.; Ciocci, M.; Savio, S.; Antonaroli, S.; Seliktar, D.; Melino, S.; Congestri, R. Trichormus Variabilis (Cyanobacteria) Biomass: From the Nutraceutical Products to Novel EPS-Cell/Protein Carrier Systems. *Mar. Drugs* **2018,** *16* (9), 298.

Bérdy, J. Bioactive Microbial Metabolites. *J. Antibiot.* **2005,** *58* (1), 1–26.

Bhatnagar, M.; Parwani, L.; Sharma, V.; Ganguly, J.; Bhatnagar, A. Exopolymers from Tolypothrix Tenuis and Three Anabaena sp. (Cyanobacteriaceae) as Novel Blood Clotting Agents for Wound Management. *Carbohydr. Polym.* **2014,** *99,* 692–699.

Boubetra, D.; Sabaou, N.; Zitouni, A.; Bijani, C.; Lebrihi, A.; Mathieu, F. Taxonomy and Chemical Characterization of New Antibiotics Produced by Saccharothrix SA198 Isolated from a Saharan Soil. *Microbiol. Res.* **2013,** *168* (4), 223–230.

Boucar, M. C. M.; Shen, L. Q.; Wang, K.; Zhang, Z. C.; Qiu, B. S. UV-B Irradiation Enhances the Production of Unique Mycosporine-Like Amino Acids and Carotenoids in the Subaerial Cyanobacterium Pseudanabaena sp. CCNU1. *Eur. J. Phycol.* **2021a,** *56* (3), 316–323.

Boucar, M. C. M.; Shen, L. Q.; Wang, K.; Zhang, Z. C.; Qiu, B. S. UV-B Irradiation Enhances the Production of Unique Mycosporine-Like Amino Acids and Carotenoids in the Subaerial Cyanobacterium Pseudanabaena sp. CCNU1. *Eur. J. Phycol.* **2021b,** *56* (3), 316–323.

Boucar, M. C. M.; Shen, L.-Q.; Wang, K.; Zhang, Z.-C.; Qiu, B.-S. UV-B Irradiation Enhances the Production of Unique Mycosporine-Like Amino Acids and Carotenoids in the Subaerial Cyanobacterium *Pseudanabaena* sp. CCNU1. *Eur. J. Phycol.* **2021c,** *56* (3), 316–323.

Bull, A. T. Actinobacteria of the Extremobiosphere. In *Extremophiles Handbook*, Horikoshi, K. Ed.;)Springer, 2011; pp 1203–1240.

Bull, A. T.; Asenjo, J. A.; Goodfellow, M.; Gómez-Silva, B. The Atacama Desert: Technical Resources and the Growing Importance of Novel Microbial Diversity. *Annu. Rev. Microbiol.* **2016,** *70,* 215–234.

Bullerjahn, G. S.; Post, A. F. Physiology and Molecular Biology of Aquatic Cyanobacteria. *Front. Microbiol.* **2014,** *5.*

Busarakam, K. *Novel Actinobacterial Diversity in Arid Atacama Desert Soils as a Source of New Drug Leads*; 2014.

Chen, L.; Wang, G.; Hong, S.; Liu, A.; Li, C.; Liu, Y. UV-B-Induced Oxidative Damage and Protective Role of Exopolysaccharides in Desert Cyanobacterium Microcoleus Vaginatus. *J. Integr. Plant Biol.* **2009,** *51* (2), 194–200.

Chen, L.-Z.; Wang, G.-H.; Hong, S.; Liu, A.; Li, C.; Liu, Y.-D. UV-B-induced Oxidative Damage and Protective Role of Exopolysaccharides in Desert Cyanobacterium *Microcoleus vaginatus*. *J. Integr. Plant Biol.* **2009,** *51* (2), 194–200.

Costa, J. A. V.; Freitas, B. C. B.; Rosa, G. M.; Moraes, L.; Morais, M. G.; Mitchell, B. G. Operational and Economic Aspects of Spirulina-Based Biorefinery. *Bioresour. Technol.* **2019,** *292,* 121946.

Couradeau, E.; Karaoz, U.; Lim, H. C.; da Rocha, U. N., Northen, T.; Brodie, E.; Garcia-Pichel, F. Bacteria Increase Arid-Land Soil Surface Temperature Through the Production of Sunscreens. *Nat. Commun.* **2016,** *7,* 1–7.

Dahech, P.; Schlömann, M.; Ortiz, C. Light Intensity Stimulates the Production of Extracellular Polymeric Substances (EPS) in a Culture of the Desert Cyanobacterium Trichormus sp. *J. Appl. Phycol.* **2021,** *33* (5), 2795–2804.

De Menezes, A. B.; Richardson, A. E.; Thrall, P. H. Linking Fungal–Bacterial Co-occurrences to Soil Ecosystem Function. *Curr. Opin. Microbiol.* **2017,** *37,* 135–141.

Derikvand, P.; Llewellyn, C. A.; Purton, S. Cyanobacterial Metabolites as a Source of Sunscreens and Moisturizers: A Comparison with Current Synthetic Compounds. *Eur. J. Phycol.* **2017,** *52* (1), 43–56.

Dhakal, D.; Pokhrel, A. R.; Shrestha, B.; Sohng, J. K. Marinerare Actinobacteria: Isolation, Characterization, and Strategies for Harnessingbioactive Compounds. *Front. Microbiol.* **2017,** *8,* 1106.

Díaz Domínguez, G.; Marsán Suárez, V.; del Valle Pérez, L. O. Principales Propiedades Inmunomoduladoras y Antinflamatorias de la Ficobiliproteína C-Ficocianina. *Rev. Cub. Hematol. Inmunol. Hemoter.* **2016,** *32* (4), 447–454.

Ding, D.; Chen, G.; Wang, B.; Wang, Q.; Liu, D.; Peng, M.; Shi, P. Culturable Actinomycetes from Desert Ecosystem in Northeast of Qinghai-Tibet Plateau. *Ann. Microbiol.* **2013,** *63* (1), 259–266.

Dubourg, G.; Abat, C.; Raoult, D. Why New Antibiotics are Not Obviously Useful Now. *Int. J. Antimicrob. Agents* **2017,** *49* (5), 549–553.

Elsayed, S. S.; Trusch, F.; Deng, H.; Raab, A.; Prokes, I.; Busarakam, K.; Asenjo, J. A.; Andrews, B. A.; Van West, P.; Bull, A. T.; Goodfellow, M.; Yi, Y.; Ebel, R.; Jaspars, M.; Rateb, M. E. Chaxapeptin, a Lasso Peptide from Extremotolerant Streptomyces Leeuwenhoekii Strain C58 from the Hyperarid Atacama Desert. *J. Organ. Chem.* **2015a,** *80* (20), 10252–10260.

Elsayed, S. S.; Trusch, F.; Deng, H.; Raab, A.; Prokes, I.; Busarakam, K.; Asenjo, J. A.; Andrews, B. A.; Van West, P.; Bull, A. T.; Goodfellow, M.; Yi, Y.; Ebel, R.; Jaspars, M.; Rateb, M. E. Chaxapeptin, a Lasso Peptide from Extremotolerant Streptomyces leeuwenhoekii Strain C58 from the Hyperarid Atacama Desert. *J. Org. Chem.* **2015b,** *80* (20), 10252–10260.

Escalante, A.; Mendoza-Flores, R.; Gosset, G.; Bolívar, F. The Aminoshikimic Acid Pathway in Bacteria as Source of Precursors for the Synthesis of Antibacterial and Antiviral Compounds. *J. Indus. Microbiol. Biotechnol.* **2021,** *48* (9–10).

Euzeby, J. P. *List of Prokaryotic Names with Standing in Nomenclature*; 2015.

Faist, A. M.; Herrick, J. E.; Belnap, J.; Van Zee, J. W.; Barger, N. N. Biological Soil Crust and Disturbance Controls on Surface Hydrology in A Semi-Arid Ecosystem. *Ecosphere* **2017,** *8* (3), e01691.

Fatahi-Bafghi, M.; Rasouli-nasab, M.; Yasliani-Fard, S.; Habibnia, S.; Gharehbaghi, F.; Eshraghi, S. S.; Kabir, K.; Heidarieh, P. Diversity and Antimicrobial Activity of Actinomycetes Isolated from Lut Desert: The Extremely Arid Climatic Zones of Iran. *Int. J. Pept. Res. Ther.* **2019,** *25* (3), 1201–1207.

Ferreira-Santos, P.; Nunes, R.; De Biasio, F.; Spigno, G.; Gorgoglione, D.; Teixeira, J. A.; Rocha, C. M. R. Influence of Thermal and Electrical Effects of Ohmic Heating on C-Phycocyanin Properties and Biocompounds Recovery from Spirulina Platensis. *LWT* **2020,** *128* (January).

Fierer, N. Embracing the Unknown: Disentangling the Complexities of the Soil Microbiome. *Nat. Rev. Microbiol.* **2017,** *15*, 579–590.

Fira, D.; Dimkić, I.; Berić, T.; Lozo, J.; Stanković, S. Biological Control of Plant Pathogens by Bacillus Species. *J. Biotechnol.* **2018,** *285*, 44–55.

Flombaum, P.; Gallegos, J. L.; Gordillo, R. A.; Rincon, J.; Zabala, L. L.; Jiao, N.; Karl, D. M.; Li, W. K. W.; Lomas, M. W.; Veneziano, D.; Vera, C. S.; Vrugt, J. A.; Martiny, A. C. Present and Future Global Distributions of the Marine Cyanobacteria Prochlorococcus and Synechococcus. *Proc. Natl Acad. Sci.* **2013,** *110* (24), 9824–9829.

Gacem, M. A.; Ould-El-Hadj-Khelil, A.; Boudjemaa, B.; Wink, J. Antimicrobial and Antioxidant Effects of a Forest Actinobacterium V002 as New Producer of Spectinabilin, Undecylprodigiosin and Metacycloprodigiosin. *Curr. Microb.* **2020,** *77* (10), 2575–2583.

Galetovic, A.; Seura, F.; Gallardo, V.; Graves, R.; Cortés, J.; Valdivia, C.; Nuñez, J.; Tapia, C.; Neira, I.; Sanzana, S.; Gómez-Silva, B. Use of Phycobiliproteins from Atacama Cyanobacteria as Food Colorants in a Dairy Beverage Prototype. *Foods* **2020,** *9* (2), 1–13.

Gera, R.; Bhatia, R.; Kumar, V.; Kayasth, M.; Walia, M.; Kaur, H.; Goyal, S. Diversity and Antibacterial Activity of Actinobacteria Isolated from Cotton Fields in Semi Arid Zones of Haryana. *J. Cotton Res. Dev.* **2014,** *28* (1), 129–134.

Gomez-Escribano, J. P.; Castro, J. F.; Razmilic, V.; Chandra, G.; Andrews, B.; Asenjo, J. A.; Bibb, M. J. The Streptomyces Leeuwenhoekii Genome: de Novosequencing and Assembly in Single Contigs of the Chromosome, Circular Plasmid pSLE1 and Linear Plasmid pSLE2. *BMC Genom.* **2015,** *16* (1), 1–11.

González-Menéndez, V.; Crespo, G.; de Pedro, N.; Diaz, C.; Martín, J.; Serrano, R.; Mackenzie, T. A.; Justicia, C.; González-Tejero, M. R.; Casares, M.; Vicente, F.; Reyes, F.; Tormo, J. R.; Genilloud, O. Fungal Endophytes from Arid Areas of Andalusia: High Potential Sources for Antifungal and Antitumoral Agents. *Sci. Rep.* **2018,** *8* (1).

Goodfellow, M.; Nouioui, I.; Sanderson, R.; Xie, F.; Bull, A. T. Rare Taxa and Dark Microbial Matter: Novel Bioactive Actinobacteriaabound in Atacama Desert Soils. *Anton. Leeuw.* **2018,** *111*, 1315–1332.

Goswami, D.; Pithwa, S.; Dhandhukia, P.; Thakker, J. N. Delineating Kocuria Turfanensis 2M4 as a Credible PGPR: A Novel IAA-Producing Bacteria Isolated from Saline Desert. *J. Plant Interact.* **2014,** *9* (1), 566–576.

Goudie, A. Holocene Book Review: Deserts and Desert Environments Julie Laity, Oxford and Chichester. *Holocene* **2010,** *20,* 147–148.

Greenville, A. The Biology of Deserts, Ward, David, 2nd edition. Oxford University Press, Oxford, 2016. xv + 370 pp. Price AUD $66 (paperback, also available as hardback and Ebook). ISBN: 9780198732761. *Austral Ecol.* **2018,** *43* (8), e20.

Guaadaoui, A.; Benaicha, S.; Elmajdoub, N.; Bellaoui, M.; Hamal, A. What is a Bioactive Compound? A Combined Definition for a Preliminary Consensus. *Int. J. Food Sci. Nutr.* **2014,** *3* (3), 17–179.

Hartmann, A.; Glaser, K.; Holzinger, A.; Ganzera, M.; Karsten, U. Klebsormidin A and B, Two New UV-Sunscreen Compounds in Green Microalgal Interfilum and Klebsormidium Species (Streptophyta) From Terrestrial Habitats. *Front. Microbiol.* **2020,** *11* (March), 1–12.

Harwani, D. Biodiversity of Rare Thermophilic Actinomycetes in the Great Indian Thar Desert: An Overview. *Indo Am. J. Pharm. Res.* **2013a,** *3,* 9349–9356.

Harwani, D. Biodiversity of Thermophilic Actinomycetes on the Great Indian Thar Desert: An Overview. *Indo Am. J. Pharm. Res.* **(2013b,** *3,* 934–939.

Hozzein, W. N.; Ali, M. I.; Rabie, W. A New Preferential Medium for Enumeration and Isolation of Desert Actinomycetes. *World J. Microbiol. Biotechnol.* **2008,** *24,* 1547–1552.

Kaspar, F.; Neubauer, P.; Gimpel, M. Bioactive Secondary Metabolites from Bacillus subtilis: A Comprehensive Review. *J. Nat. Prod.* **2019,** *82* (7), 2038–2053.

Katz, M. R.; Baltz, R. H. Natural Product Discovery: Past, Present and Future. *J. Indus. Microbiol. Biotechnol.* **2016,** *43,* 155–176.

Kaushal, S.; Singh, Y.; Khattar, J. I. S.; Singh, D. P. Phycobiliprotein Production by a Novel Cold Desert Cyanobacterium Nodularia Sphaerocarpa PUPCCC 420.1. *J. Appl. Phycol.* **2017,** *29* (4), 1819–1827.

Kavitha, A.; Prabhakar, P.; Vijayalakshmi, M.; Venkateswarlu, Y. Purification and Biological Evaluation of the Metabolites Produced by Streptomyces sp. TK-VL_333. *Res. Microbiol.* **2010,** *161* (5), 335–345.

Khalil, A. M. A.; Abdelazis, A. M.; Khaleil, M. M.; Hashem, A. H. Fungal Endophytes From Leaves of Avicennia Marina Growing in Semi-Arid Environment as a Promising Source for Bioactive Compunds. *Lett. Appl. Microbiol.* **2020,** *72,* 263–274.

Kharel, M. K.; Pahari, P.; Shepherd, M. D.; Tibrewal, N.; Nybo, S. E.; Shaaban, K. A.; Rohr, J. Angucyclines: Biosynthesis, Mode-of-Action, New Natural Products, and Synthesis. *Nat. Prod. Rep.* **2012,** *29* (2), 264–325.

Khem Raj, M.; Shamsher S, K. Lipopeptides as the Antifungal and Antibacterial Agents.pdf. *Biomed. Res. Int.* **2015,** *15,* 1–9.

Khushboo, K. P.; Dubey, K. K.; Usmani, Z.; Sharma, M.; Gupta, V. K. Biotechnological and Industrial Applications of Streptomyces Metabolites. *Biofuel. Bioprod. Biorefin.* **2022,** *16* (1), 244–264.

Koeberl, M.; Müller, H.; Ramadan, E. M.; Berg, G. Desert Farming Benefits from Microbial Potential in Arid Soils and Promotes Diversity and Plant Health. *PLoS One* **2011,** *6,* e24452.

Korayem, A. S.; Abdelhafez, A. A.; Zaki, M. M.; Saleh, E. A. Optimization of Biosurfactant Production by Streptomyces Isolated from Egyptian Arid Soil Using Plackett–Burman Design. *Ann. Agric. Sci.* **2015,** *60* (2), 209–217.

Kumar, J.; Singh, D.; Tyagi, M. B.; Kumar, A. Cyanobacteria: Applications in Biotechnology. In *Cyanobacteria*; Elsevier, 2019; pp 327–346.

Kumar, S.; Solanki, D. S.; Parihar, K.; Tak, A.; Gehlot, P.; Pathak, R.; Singh, S. K. Actinomycetes Isolates of Arid Zone of Indian Thar Desert and Efficacy of their Bioactive Compounds Against Human Pathogenic Bacteria. *Biol. Futura* **2021,** *72* (4), 431–440.

Kurapova, A. I.; Zenova, G. M.; Studnitsyn, I. I.; Kizilova, A. K.; Man-ucharova, N. A.; Norovsuren, Z.; Zvyagintsev, D. J. Thermotolerant and Thermophilic Actinomycetes from Soilsof Mongolia Desert Steppe Zone. *Microbiology* **2012,** *81,* 98–108.

Lee, L.-H.; Go, B.-H.; Chan, K. G. Editorial: Actinobacteria: Profificproducers of Bioactive Metabolites. *Front. Microbiol.* **2020,** *11* (1612), 1–3.

León-Sobrino, C.; Ramond, J.-B.; Maggs-Kölling, G.; Cowan, D. A. Nutrient Acquisition, Rather Than Stress Response Over Diel Cycles, Drives Microbial Transcription in a Hyper-Arid Namib Desert Soil. *Front. Microbiol.* **2019,** *10.*

Lewis, L. A.; Flechtner, V. R. Green Algae (Chlorophyta) of Desert Microbiotic Crusts: Diversity of North American Taxa. *Taxon* **2002,** *51* (3), 443.

Lewis, L. A.; Flechtner, V. R. Cryptic Species of *Scenedesmus* (Chlorophyta) from Desert Soil Communities of Western North America. *J. Phycol.* **2004,** *40* (6), 1127–1137.

Li, H.; Xu, J.; Liu, Y.; Ai, S.; Qin, F.; Li, Z.; Zhang, H.; Huang, Z. Antioxidant and Moisture-Retention Activities of the Polysaccharide from Nostoc Commune. *Carbohydr. Polym.* **2011a,** *83* (4), 1821–1827.

Li, H.; Xu, J.; Liu, Y.; Ai, S.; Qin, F.; Li, Z.; Zhang, H.; Huang, Z. Antioxidant and Moisture-Retention Activities of the Polysaccharide from Nostoc Commune. *Carbohydr. Polym.* **2011b,** *83* (4), 1821–1827.

Li, Z.; Zhu, D.; Shen, Y. Discovery of Novel Bioactive Natural Products Driven by Genome Mining. *Drug Discov. Therapeut.* **2018,** *12* (6), 318–328.

Ludwig, W.; Euzeby, J.; Schumann, P.; Busse, H.-J.; Trujillo, M. E.; Kämpfer, P. Road Map of the Phylum Actinobacteria. In *Bergey's Manual of Systematic Bacteriology*, 2nd ed.; Whitman, W. B.; Goodfellow, M.; Kämpfer, P.; Busse, H.-J.; Trujillo, M. E.; Suzuki, K.-I.; Ludwig, W. Eds.; Springer, 2012; pp 1–28.

Lupatini, A. L.; Colla, L. M.; Canan, C.; Colla, E. Potential Application of Microalga *Spirulina Platensis* as a Protein Source. *J. Sci. Food Agric.* **2017,** *97* (3), 724–732.

Lyons, T.; Reinhard, C.; Planavsky, N. The Rise of Oxygen in Earth's Early Ocean and Atmosphere. *Nature* **2014,** *506,* 307–315.

Machushynets, N.; Elsayed, S. S.; Du, C.; Siegler, M. A.; de la Cruz, M.; Genilloud, O.; Hankemeier, T.; van Wezel, G. P. New from Old: Discovery of the Novel Antibiotic Actinomycin L in Streptomyces sp. MBT27. *BioRxiv*; 2021. DOI: 2021.10.12.464064.

Maoka, T. Carotenoids as Natural Functional Pigments. *J. Nat. Med.* **2020,** *74* (1), 1–16.

Martin-Gómez, H.; Tulla-Puche, J. Lasso Peptides: Chemical Approaches and Structural Elucidation. *Org. Biomol. Chem.* **2018,** *16* (28), 5065–5080.

Martins, S.; Mussatto, S. I.; Martínez-Avila, G.; Montañez-Saenz, J.; Aguilar, C. N.; Teixeira, J. A. Bioactive Phenolic Compounds: Production and Extraction by Solid-State Fermentation. A Review. *Biotechnol. Adv.* **2011,** *29* (3), 365–373.

McHugh, T. A.; Compson, Z.; van Gestel, N., Hayer, M.; Ballard, L.; Haverty, M.; Hines, J.; Irvine, N.; Krassner, D.; Lyons, T.; Musta, E. J.; Schiff, M.; Zint, P.; Schwartz, E. Climate Controls Prokaryotic Community Composition in Desert Soils of the Southwestern United States. *FEMS Microbiol. Ecol.* **2017,** *93* (10), 1–10.

Meklat, A.; Sabaou, N.; Zitouni, A.; Mathieu, F.; Lebrini, A. Isolation, Taxonomy and Antagonistic Properties of Halophilic Actinomycetes in Saharan Soils of Algeria. *Appl. Environ. Microbiol.* **2011,** *77,* 6710–6714.

Merrouche, R.; Bouras, N.; Coppel, Y.; Mathieu, F.; Sabaou, N.; Lebrihi, A. New Dithiolo-pyrrolone Antibiotics Induced by Adding Sorbic Acid to the Culture Medium of Saccharo-thrix Algeriensis NRRL B-24137. *FEMS Microbiol. Lett.* **2011,** *318* (1), 41–46.

Miller, A. P.; Coronel, J.; Amengual, J. The Role of β-Carotene and Vitamin A in Atherogenesis: Evidences from Preclinical and Clinical Studies. *Biochim. Biophys. Acta Mol. Cell Biol. Lipids.* **2020,** *1865* (11), 1–32.

Mohammadipanah, F.; Wink, J. Actinobacteria from Arid and Desert Habitats: Diversity and Biological Activity. *Front. Microbiol.* **2016a,** *6* (1541), 1–10.

Mohammadipanah, F.; Wink, J. Actinobacteria from Arid and Desert Habitats: Diversity and Biological Activity. *Front. Microbiol.* **2016b,** *6* (JAN), 1–10.

Montero-Lobato, Z.; Fuentes, J. L.; Garbayo, I.; Ascaso, C.; Wierzchos, J.; Vega, J. M.; Vílchez, C. Identification, Biochemical Composition and Phycobiliproteins Production of Chroococcidiopsis sp. from Arid Environment. *Proc. Biochem.* **2020a,** *97* (July), 112–120.

Montero-Lobato, Z.; Fuentes, J. L.; Garbayo, I.; Ascaso, C.; Wierzchos, J.; Vega, J. M.; Vílchez, C. Identification, Biochemical Composition and Phycobiliproteins Production of Chroococcidiopsis sp. from Arid Environment. *Proc. Biochem.* **2020b,** *97,* 112–120.

de Morais, M. G.; da Vaz, B. S.; de Morais, E. G.; Costa, J. A. V. Biological Effects of *Spirulina* (*Arthrospira*) Biopolymers and Biomass in the Development of Nanostructured Scaffolds. *Biomed. Res. Int.* **2014,** *2014,* 1–9.

Moron, L. S.; Lim, Y.-W.; dela Cruz, T. E. E. Antimicrobial Activities of Crude Culture Extracts from Mangrove Fungal Endophytes Collected in Luzon Island, Philippines. *Philipp. Sci. Lett.* **2018,** *11.*

Morone, J.; Alfeus, A.; Vasconcelos, V.; Martins, R. Revealing the Potential of Cyanobacteria in Cosmetics and Cosmeceuticals—A New Bioactive Approach. *Algal Res.* **2019,** *41,* 101541.

Mota, R.; Rossi, F.; Andrenelli, L.; Pereira, S. B.; de Philippis, R.; Tamagnini, P. Released Polysaccharides (RPS) from Cyanothece sp. CCY 0110 as Biosorbent for Heavy Metals Bioremediation: Interactions Between Metals and RPS Binding Sites. *Appl. Microbiol. Biotechnol.* **2016,** *100* (17), 7765–7775.

Nachtigall, J.; Kulik, A.; Helaly, S.; Bull, A. T.; Goodfellow, M.; Asenjo, J. A.; Maier, A.; Wiese, J.; Imhoff, J. F.; Süssmuth, R. D.; Fiedler, H. P. Atacamycins A-C, 22-Membered Antitumor Macrolactones Produced by Streptomyces sp. C38. *J. Antibiot.* **2011,** *64* (12), 775–780.

Najdenski, H. M.; Gigova, L. G.; Iliev, I. I.; Pilarski, P. S.; Lukavský, J.; Tsvetkova, I. V.; Ninova, M. S.; Kussovski, V. K. Antibacterial and Antifungal Activities of Selected Microalgae and Cyanobacteria. *Int. J. Food Sci. Technol.* **2013,** *48* (7), 1533–1540.

Nasfi, Z.; Busch, H.; Kehraus, S.; Linares-Otoya, L.; König, G. M.; Schäberle, T. F.; Bachoual, R. Soil Bacteria Isolated from Tunisian Arid Areas Show Promising Antimicrobial Activities Against Gram-Negatives. *Front. Microbiol.* **2018,** *9* (NOV), 1–13.

Nazifi, E.; Wada, N.; Asano, T.; Nishiuchi, T.; Iwamuro, Y.; Chinaka, S.; Matsugo, S.; Sakamoto, T. Characterization of the Chemical Diversity of Glycosylated Mycosporine-Like Amino Acids in the Terrestrial Cyanobacterium Nostoc Commune. *J. Photochem. Photobiol. B: Biol.* **2015a,** *142,* 154–168.

Nazifi, E.; Wada, N.; Asano, T.; Nishiuchi, T.; Iwamuro, Y.; Chinaka, S.; Matsugo, S.; Sakamoto, T. Characterization of the Chemical Diversity of Glycosylated Mycosporine-Like Amino Acids in the Terrestrial Cyanobacterium Nostoc Commune. *J. Photochem. Photobiol. B: Biol.* **2015b,** *142,* 154–168.

Neilson, J. W.; Quade, J.; Ortiz, M.; Nelson, W. M.; Legatzki, A.; Tian, F.; LaComb, M.; Betancourt, J. L.; Wing, R. A.; Soderlund, C. A.; Maier, R. M. Life at the Hyperarid Margin: Novel Bacterial Diversity in Arid Soils of the Atacama Desert, Chile. *Extremophiles* **2012,** *16,* 553–566.

Nithya, K.; Muthukumar, C.; Biswas, B.; Alharbi, N. S.; Kadaikunnan, S.; Khaled, J. M.; Dhanasekaran, D. Desert Actinobacteria as a Source of Bioactive Compounds Production with a Special Emphases on Pyridine-2,5-Diacetamide a New Pyridine Alkaloid Produced by Streptomyces sp. DA3-7. *Microbiol. Res.* **2018,** *207* (November 2017), 116–133.

Niyasom, C.; Boonmak, S.; Meesri, N. Antimicrobial Activity of Acidophilic Actinomycetes Isolated from Acidic Soil. *KMITL Sci. Tech. J.* **2015,** *15* (2), 62–69.

Nowruzi, B.; Sarvari, G.; Blanco, S. The Cosmetic Application of Cyanobacterial Secondary Metabolites. *Algal Res.* **2020,** *49.*

Orellana, G.; Gómez-Silva, B.; Urrutia, M.; Galetović, A. Uv-A Irradiation Increases Scytonemin Biosynthesis in Cyanobacteria Inhabiting Halites at Salar Grande, Atacama Desert. *Microorganisms* **2020a,** *8* (11), 1–12.

Orellana, G.; Gómez-Silva, B.; Urrutia, M.; Galetović, A. UV-A Irradiation Increases Scytonemin Biosynthesis in Cyanobacteria Inhabiting Halites at Salar Grande, Atacama Desert. *Microorganisms* **2020b,** *8* (11), 1690.

Oren, N.; Raanan, H.; Kedem, I.; Turjeman, A.; Bronstein, M.; Kaplan, A.; Murik, O. Desert Cyanobacteria Prepare in Advance for Dehydration and Rewetting: The Role of Light and Temperature Sensing. *Mol. Ecol.* **2019,** *28* (9), 2305–2320.

Pagels, F.; Guedes, A. C.; Amaro, H. M.; Kijjoa, A.; Vasconcelos, V. Phycobiliproteins from Cyanobacteria: Chemistry and Biotechnological Applications. *Biotechnol. Adv.* **2019,** *37* (3), 422–443.

Pandey, V. D.; Pandey, A.; Sharma, V. Review Article Biotechnological Applications of Cyanobacterial Phycobiliproteins. *Int. J. Curr. Microb. Appl. Sci.* **2013,** *2* (9), 89–97.

Park, S.; Park, J. A.; Jeon, J. H.; Lee, Y. Traditional and Novel Mechanisms of Heat Shock Protein 90 (HSP90) Inhibition in Cancer Chemotherapy Including HSP90 Cleavage. *Biomol. Therapeut.* **2019,** *27* (5), 423–434.

Perera, I.; Subashchandrabose, S. R.; Venkateswarlu, K.; Naidu, R.; Megharaj, M. Consortia of Cyanobacteria/Microalgae and Bacteria in Desert Soils: An Underexplored Microbiota. *Appl. Microbiol. Biotechnol.* **2018,** *102* (17), 7351–7363.

Philippis, R.; Micheletti, E. Heavy Metal Removal with Exopolysaccharide-Producing Cyanobacteria. In *Handbook of Advanced Industrial and Hazardous Wastes Management*, 2018; p. 931.

Pinheiro Junior, C. R.; Pereira, M. G.; de Souza O. Filho, J.; Beutler, S. J. Can Topography Affect the Restoration of Soil Properties After Deforestation in a Semiarid Ecosystem? *J. Arid Environ.* **2019,** *162,* 45–52.

Poong, S.-W.; Lim, P.-E.; Phang, S.-M.; Wong, C.-Y.; Pai, T.-W.; Chen, C.-M.; Yang, C.-H.; Liu, C.-C. Transcriptome Sequencing of an Antarctic Microalga, Chlorella sp. (Trebouxiophyceae, Chlorophyta) Subjected to Short-Term Ultraviolet Radiation Stress. *J. Appl. Phycol.* **2018,** *30* (1), 87–99.

Qin, S.; Li, W.-J.; Klenk, H.-P.; Hozzein, W. N.; Ahmed, I. Editorial: Actinobacteria in Special and Extreme Habitats: Diversity, Function Roles and Environmental Adaptations, Second Edition. *Front. Microbiol.* **2019,** *10* (944), 1–3.

Raddadi, N.; Giacomucci, L.; Marasco, R.; Daffonchio, D.; Cherif, A.; Fava, F. Bacterial Polyextremotolerant Bioemulsifiers from Arid Soils Improve Water Retention Capacity and Humidity Uptake in Sandy Soil. *Microb. Cell Factories* **2018,** *17* (1), 1–12.

Raja, R.; Hemaiswarya, S.; Ganesan, V.; Carvalho, I. S. Recent Developments in Therapeutic Applications of Cyanobacteria. *Critic. Rev. Microbiol.* **2015,** 1–12.

Ramawat, K. G. *Desert Plants: Biology and Biotechnology*; Springer, 2010.

Rasheed, R.; Saadaoui, I.; Bounnit, T.; Cherif, M.; al Ghazal, G.; al Jabri, H. Sustainable Food Production and Nutraceutical Applications from Qatar Desert Chlorella sp. (Chlorophyceae). *Animals* **2020,** *10* (8), 1413.

Rastogi, R. P.; Sonani, R. R.; Madamwar, D. Cyanobacterial Sunscreen Scytonemin: Role in Photoprotection and Biomedical Research. *Appl. Biochem. Biotechnol.* **2015,** *176* (6), 1551–1563.

Rateb, M. E.; Ebel, R.; Jaspars, M. Natural Product Diversity of Actinobacteria in the Atacama Desert. *Antonie Van Leeuwenhoek* **2018,** *111* (8), 1467–1477.

Rateb, M. E.; Houssen, W. E.; Arnold, M.; Abdelrahman, M. H.; Deng, H.; Harrison, W. T. A.; Okoro, C. K.; Asenjo, J. A.; Andrews, B. A.; Ferguson, G.; Bull, A. T.; Goodfellow, M.; Ebel, R.; Jaspars, M. Chaxamycins A-D, Bioactive Ansamycins from a Hyper-Arid Desert Streptomyces sp. *J. Nat. Prod.* **2011,** *74* (6), 1491–1499.

Rateb, M. E.; Houssen, W. E.; Harrison, W. T. A.; Deng, H.; Okoro, C. K.; Asenjo, J. A.; Andrews, B. A.; Bull, A. T.; Goodfellow, M.; Ebel, R.; Jaspars, M. Diverse Metabolic Profiles of a Streptomyces Strain Isolated from a Hyper-Arid Environment. *J. Nat. Prod.* **2011a,** *74* (9), 1965–1971.

Rateb, M. E.; Houssen, W. E.; Harrison, W. T. A.; Deng, H.; Okoro, C. K.; Asenjo, J. A.; Andrews, B. A.; Bull, A. T.; Goodfellow, M.; Ebel, R.; Jaspars, M. Diverse Metabolic Profiles of a *Streptomyces* Strain Isolated from a Hyper-arid Environment. *J. Nat. Prod.* **2011b,** *74* (9), 1965–1971.

Ratnaweera, P. B.; de Silva, E. D.; Williams, D. E.; Andersen, R. Antimicrobial Activities of Endophytic Fungi Obtained from the Arid Zone Invasive Plant Opuntia Dillenii and Isolation of Equisetin, from Endophytic Fusarium sp. *BMC Complement. Altern. Med.* **2015,** *15* (220), 1–7.

Rebelo, B. A.; Farrona, S.; Rita Ventura, M.; Abranches, R. Canthaxanthin, a red-hot carotenoid: Applications, synthesis, and biosynthetic evolution. *Plants*, **2020,** *9*(8), 1–18. https://doi.org/10.3390/plants9081039

Rosic, N. N. Mycosporine-like amino acids: Making the foundation for organic personalised sunscreens. *Marine Drugs*, **2019,** *17*(11), 1–17. https://doi.org/10.3390/md17110638

Rossi, F.; de Philippis, R. Role of Cyanobacterial Exopolysaccharides in Phototrophic Biofilms and in Complex Microbial Mats. *Life* **2015,** *5* (2), 1218–1238.

Rotondo Dottore, G.; Ionni, I.; Menconi, F.; Casini, G.; Sellari-Franceschini, S.; Nardi, M.; Vitti, P.; Marcocci, C.; Marinò, M. Antioxidant Effects of β-Carotene, but Not of Retinol and Vitamin E, in Orbital Fibroblasts from Patients with Graves' Orbitopathy (GO). *J. Endocrinol. Invest.* **2018,** *41* (7), 815–820.

Ruiz-Domínguez, M. C.; Fuentes, J. L.; Mendiola, J. A.; Cerezal-Mezquita, P.; Morales, J.; Vílchez, C.; Ibáñez, E. Bioprospecting of Cyanobacterium in Chilean Coastal Desert, Geitlerinema sp. Molecular Identification and Pressurized Liquid Extraction of Bioactive Compounds. *Food Bioprod. Process.* **2021a,** *128,* 227–239.

Ruiz-Domínguez, M. C.; Fuentes, J. L.; Mendiola, J. A.; Cerezal-Mezquita, P.; Morales, J.; Vílchez, C.; Ibáñez, E. Bioprospecting of Cyanobacterium in Chilean Coastal Desert, Geitlerinema sp. Molecular Identification and Pressurized Liquid Extraction of Bioactive Compounds. *Food Bioprod. Process.* **2021b,** *128,* 227–239.

Bansod, T. R.; Hingane, L. D. At. Post Shiv Nagar, Bramhapuri Tah Bramhapuri Dist Chandrapur. Fermentation Technology. *World J. Pharmaceut. Res.* **2021,** *10* (3), 260–299.

Sabarinathan, K.; Gomathy, M.; Kumar, A.; Kannan, R.; Arutkani, E. Cyanobacteria-Mediated Bioremediation of Problem Soils. In *Microorganisms for Sustainability*, Vol. 1; 2021; p 142.

Santhanam, R.; Okoro, C. K.; Rong, X.; Huang, Y.; Bull, A. T.; Weon, H.-Y.; Andrews, B. A.; Asenjo, J. A.; Goodfellow, M. Streptomyces Atacamensis sp. nov., Isolated from an Extreme Hyper-Arid Soil of the Atacama Desert, Chile. *Int. J. Syst. Evol. Microbiol.* **2012,** *62* (Pt 11), 2680–2684.

Santhanam, R.; Rong, X.; Huang, Y.; Andrews, B. A.; Asenjo, J. A.; Goodfellow, M. Streptomyces bullii sp. nov.; isolated from a hyper-arid Atacama Desert soil. *Antonie van Leeuwenhoek*, **2013,** *103*(2), 367–373. https://doi.org/10.1007/s10482-012-9816-x

Sayed, A. M.; Hassan, M. H. A.; Alhadrami, H. A.; Hassan, H. M.; Goodfellow, M.; Rateb, M. E. Extreme Environments: Microbiology Leading to Specialized Metabolites. *J. Appl. Microbiol.* **2020,** *128* (3), 630–657.

Schulz, D.; Beese, P.; Ohlendorf, B.; Erhard, A.; Zinecker, H.; Dorador, C.; Imhoff, J. F. Abenquines A-D: Aminoquinone Derivatives Produced by Streptomyces sp. Strain DB634. *J. Antibiot.* **2011a,** *64* (12), 763–768.

Schulz, D.; Beese, P.; Ohlendorf, B.; Erhard, A.; Zinecker, H.; Dorador, C.; Imhoff, J. F. (2011b). Abenquines A–D: Aminoquinone Derivatives Produced by Streptomyces sp. Strain DB634. *J. Antibiot.* **2011b,** *64* (12), 763–768.

Singh, S.; Kant, C.; Yadav, R. K.; Reddy, Y. P.; Abraham, G. Cyanobacterial Exopolysaccharides: Composition, Biosynthesis, and Biotechnological Applications. In *Cyanobacteria: From Basic Science to Applications*; Elsevier Inc., 2018; pp 347–358.

Soccol, C. R.; da Costa, E. S. F.; Letti, L. A. J.; Karp, S. G.; Woiciechowski, A. L.; de Souza Vandenberghe, L. P. Recent Developments and Innovations in Solid State Fermentation. *Biotechnol. Res. Innov*; 2017.

Subashchandrabose, S. R.; Ramakrishnan, B.; Megharaj, M.; Venkateswarlu, K.; Naidu, R. Mixotrophic Cyanobacteria and Microalgae as Distinctive Biological Agents for Organic Pollutant Degradation. *Environ. Int.* **2013,** *51,* 59–72.

Subramaniyam, R.; Vimala, R. Solid State and Submerged Fermentation for the Production of Bioactive Substances : A Comparative Study. *Int. J. Sci. Nat.* **2012,** *3* (3), 480–486.

Tang, D.; Liu, L. L.; He, Q. R.; Yan, W.; Li, D.; Gao, J. M. Ansamycins with Antiproliferative and Antineuroinflammatory Activity from Moss-Soil-Derived Streptomyces Cacaoi Subsp. Asoensis H2S5 [Review-article]. *J. Nat. Prod.* **2018,** *81* (9), 1984–1991.

Tiwari, K.; Upadhyay, D. J.; Mösker, E.; Süssmuth, R.; Gupta, R. K. Culturable Bioactive Actinomycetes from the Great Indian Thar Desert. *Ann. Microbiol.* **2015,** *65* (4), 1901–1914.

Tiwari, O. N.; Muthuraj, M.; Bhunia, B.; Bandyopadhyay, T. K.; Annapurna, K.; Sahu, M.; Indrama, T. Biosynthesis, Purification and Structure-Property Relationships of New Cyanobacterial Exopolysaccharides. *Polym. Test.* **2020,** *89,* 106592.

Trivedi, J.; Aila, M.; Bangwal, D. P.; Kaul, S.; Garg, M. O. Algae based Biorefinery—How to Make Sense? *Renew. Sustain. Ener. Rev* **2015,** *47,* 295–307.

Tuan Hamzah, T. N.; Lee, S. Y.; Hidayat, A.; Terhem, R.; Faridah-Hanum, I.; Mohamed, R. Diversity and Characterization of Endophytic Fungi Isolated From the Tropical Mangrove Species, Rhizophora mucronata, and Identification of Potential Antagonists Against the Soil-Borne Fungus, Fusarium Solani. *Front. Microbiol.* **2018,** *1707,* 1–17.

Uddin, M. Bioactive Metabolite Production by Streptomyces Albolongus in Favourable Environment. *Jo. Microbiol. Infect. Dis.* **2013,** *3* (2), 75–82.

Vítek, P.; Ascaso, C.; Artieda, O.; Casero, M. C.; Wierzchos, J. Discovery of Carotenoid Red-Shift in Endolithic Cyanobacteria from the Atacama Desert. *Sci. Rep.* **2017,** *7* (1), 11116.

Wang, L.; Kaseke, K. F.; Seely, M. K. Effects of Non-Rainfall Water Inputs on Ecosystem Functions. *WIREs WATER* **2017,** *4* (1), e1179.

Weber, B.; Wu, D.; Tamm, A.; Ruckteschler, N.; Rodríguez-Caballero, E.; Steinkamp, J.; Meusel, H.; Elbert, W.; Behrendt, T.; Sörgel, M.; Cheng, Y.; Crutzen, P. J.; Su, H.; Pöschl, U. Biological Soil Crusts Accelerate the Nitrogen Cycle Through Large NO and HONO Emissions in Drylands. *Proc. Natl Acad. Sci. USAm.* **2015,** *112* (50), 15384–15389.

Wei, Y.; Sun, C.; Dai, L.; Zhan, X.; Gao, Y. Structure, Physicochemical Stability and in Vitro Simulated Gastrointestinal Digestion Properties of β-Carotene Loaded Zein-Propylene

Glycol Alginate Composite Nanoparticles Fabricated by Emulsification-Evaporation Method. *Food Hydrocoll.* **2018,** *81,* 149–158.

Wijffels, R.; Kruse, O.; Hellingwerf, K. Potential of Industrial Biotechnology with Cyanobacteria and Eukaryotic Microalgae. *Curr. Opin. Biotechnol.* **2013,** *24,* 405–413.

Wu, L.; Zhu, Q.; Yang, L.; Li, B.; Hu, C.; Lan, S. Nutrient Transferring from Wastewater to Desert Through Artificial Cultivation of Desert Cyanobacteria. *Bioresour. Technol.* **2018a,** *247,* 947–953.

Wu, L.; Zhu, Q.; Yang, L.; Li, B.; Hu, C.; Lan, S. Nutrient Transferring from Wastewater to Desert Through Artificial Cultivation of Desert Cyanobacteria. *Bioresour. Technol.* **2018b,** *247,* 947–953.

Xie, F.; Pathom-aree, W. Actinobacteria From Desert: Diversity and Biotechnological Applications. *Front. Microbiol.* **2021,** *12* (December).

Xu, Y.; Rossi, F.; Colica, G.; Deng, S.; De Philippis, R.; Chen, L. Use of Cyanobacterial Polysaccharides to Promote Shrub Performances in Desert Soils: A Potential Approach for the Restoration of Desertified Areas. *Biol. Fertil. Soils* **2013,** *49* (2), 143–152.

Ye, J.; Delgado-Baquerizo, M.; Soliveres, S.; Maestre, F. T. Multifunctionality Debt in Global Drylands Linked to Past Biome and Climate. *Glob. Change Biol.* **2019,** *25* (6), 2152–2161.

Yu, J.; Zhang, L.; Liu, Q.; Qi, X.; Ji, Y.; Kim, B. S. Isolation and Characterization of Actinobacteria from Yalujiang Coastal Wetland, North China. *Asian Pac. J. Trop. Biomed.* **2015a,** *5* (7), 555–560.

Yu, L.-H.; Luo, X.-S.; Huang, Q. Diversity of Inonizing Radiation-Resistant Bacteria Obtained from the Taklimakan Desert. *J. Basic Microbiol.* **2015b,** *55,* 135–140.

Zakar, T.; Laczko-Dobos, H.; Toth, T. N.; Gombos, Z. Carotenoids Assist in Cyanobacterial Photosystem II Assembly and Function. *Front. Plant Sci.* **2016,** *7.* DOI: 10.3389/fpls.2016.00295

Zitouni, A.; Mathieu, F.; Coppel, Y.; Pont, F.; Sabaou, N.; Lebrihi, A. Mutactimycin PR, a New Anthracycline Antibiotic from Saccharothrix sp. SA 103 II. Physico-chemical Properties and Structure Elucidation. *J. Antibiot.* **2004,** *57* (6), 373–378.

CHAPTER 5

Plant Resources with Medicinal Applications

KARLA FABIOLA CHACÓN-VARGAS[1], JAIME RAÚL ADAME-GALLEGOS[1], DAYANIRA MORALES-CORRAL[1], and GUADALUPE VIRGINIA NEVÁREZ-MOORILLÓN[1]

[1]*Facultad de Ciencias Químicas, Universidad Autónoma de Chihuahua, Chihuahua, Mexico*

ABSTRACT

The use of medicinal plants to meet different requirements of the human being has a long history since it has helped man to survive, the use of herbalism has been a tool used by many people that to this day continue to exploit it mainly for treatment for diseases or conditions. This is due to the capacity of plants to produce a large number of metabolites also known as biomolecules, with different characteristics and properties, which in the plant have the purpose of its survival, for its adaptation, as a mechanism of defense against pathogens of different natures and mainly adverse conditions, which is the case of plants belonging to arid zones since they are in more adverse conditions regarding the intensity of sunlight, extreme temperatures and mainly the low percentage of humidity to mention a few, their capacity to produce phytochemicals is greater and diverse in their biological activities.

Among the molecules produced by these plants from arid areas are alkaloids, terpenes and a great diversity of polyphenols, which have biological activities of industrial interest in different areas such as food, pharmaceutical, cosmetics, agro-industrial, among others. Among the most notable activities is the capacity, antibacterial, antiparasitic, antifungal, hypoglycemic activity, antioxidant, etc. Although these attributes have been demonstrated, it is

Exploration and Valorization of Natural Resources from Arid Zones. Cristobal Noé Aguilar, Norma Paola Meléndez-Renteria, & Karen Nathiely Ramirez-Guzman (Eds.)

important to delve into the study of these natural resources since it is important to create strategies to not overexploit their use and lose sight of a sustainable system that does not harm the environment in addition to reviewing the contradictions of use of these plants as well as toxicity, dosage and their interaction with other drugs.

5.1 INTRODUCTION

The use of plant extracts for medicinal purposes has a long history, almost as long as the rise of civilization. The knowledge of the healing nature of plants has been a powerful tool for human survival; those with this wisdom often had privileged positions or were considered powerful enemies. Even today, most societies use plant resources to control or cure a particular disease. The characterization of plant metabolites with medical applications has been derived in many modern allopathic drugs. The plants produce many metabolites as part of their defense mechanisms against harsh environmental conditions, plagues, or microbial pathogens. In arid environments, the conditions that the plants must overcome to survive are even more drastic than in temperate or humid ecosystems. Therefore, plants usually produce more active metabolites with different biological activities. This chapter aims to revise the current research on the uses of plants from arid areas for medicinal applications, particularly for three applications: antidiabetic, antiparasitic, and antimicrobial activities. The emphasis will be placed on the characterization of the active plant components and proposed mechanisms of action. Research gaps will also be revised to propose the areas of future research.

5.2 MEDICINAL PLANTS AND THEIR IMPORTANCE

The term "medicinal plants" refers to a wide variety of plants to which beneficial health properties are attributed; medicinal plants contain chemical substances that have a physiological effect on the body to prevent or cure diseases (Jamshidi-Kia et al., 2018; Sofowora et al., 2013). These substances are known as active principles, phytochemicals, or secondary metabolites; generally, they are products of the secondary metabolism of plants. Phytochemicals vary according to the properties of plants, and they can have different biological effects such as antimicrobial, antiseptic, sedative, analgesic, stimulant of the nervous system, hypoglycemic, among others. The use of medicinal plants refers to the use of herbal material, herbal preparations, or herbal products, which contain the active principles of one or several parts of plants such as seeds, stem, root, leaf, fruit, skin, and flowers (Jamshidi-Kia et al., 2018; WHO, 2013).

The WHO classifies the use of medicinal plants as "traditional and complementary medicine," along with acupuncture and manual therapies; the primary purpose is to prevent or treat diseases to improve the quality of life (WHO, 2018). Traditional medicine includes health practices that are not part of conventional medicine and are not fully integrated into the healthcare system. In Africa, Asia, and Latin America, about 80% of the population has made use of medicinal plants to treat primary health needs; for millions of people, it is the primary source of health care, and sometimes the only one, it is an accessible and affordable alternative (Akinyemi et al., 2018; WHO, 2013). In developed countries, the use of medicinal plants is increasing, and it is considered that at least 30% have made use of traditional medicine (WHO, 2003). Currently, medicinal plants are also used to treat mental health and spiritual activities (Akinyemi et al., 2018). The use of medicinal plants is related to the evolution of humankind; records have been found from almost 5000 years ago of their use in India, Greece, and Egypt. Possibly, the beginning of the use of plants was intuitive, to become an empirical use and later was limited by facts and evidence (Boy et al., 2018; Petrovska, 2011; Süntar, 2020).

The distribution of plants worldwide is not uniform, and it is estimated that about 70 000 (10%) species are used as medicinal alternatives and are mainly collected from natural resources. Of the total number of medicinal plants used worldwide, only 15% have been scientifically investigated to confirm their properties (Boy et al., 2018; Jamshidi-Kia et al., 2018; Süntar, 2020). Although it is still necessary to increase research projects on medicinal plants, advances have been fundamental for the development of modern medicine; drugs such as morphine, acetylsalicylic acid, digoxin, among others, have their origin in the study of secondary metabolites of plants. About 11% of conventional medicines considered basic and essential by the WHO come from plants (Akinyemi et al., 2018).

Despite the wide use of medicinal plants, many questions about the safety and efficacy of most plants with traditional knowledge still need to be resolved. Effectiveness, mechanism of action, exact dosage, duration of treatment, side effects, acute and chronic toxicity, among others, need to be identified (Jamshidi-Kia et al., 2018). WHO is currently implementing a program of strategies to achieve greater recognition of traditional medicine and support its integration into national health systems. The objective is to promote quality, safe, and effective traditional medicine; on the other hand, it seeks to conserve and protect the reserves of medicinal plants and knowledge of traditional medicine from being used sustainably and rationally (WHO, 2013).

5.3 SECONDARY METABOLITES

Plants, in general, have a similar biochemical profile, generating primary metabolites from amino acids, carbohydrates, lipids, and nucleic acids necessary for plant growth and development. Secondary metabolites, or phytochemicals, are produced from the surplus of primary metabolites; these phytochemicals are not involved in plant development; however, they are related to plant interactions with other organisms and the surrounding environment (Awuchi, 2019). Phytochemicals are intended to defend plants from biotic (insects, parasites, bacteria, fungi, herbivores) and abiotic (drought, ultraviolet light, temperature, etc.) adverse factors (Pagare et al., 2015). Secondary metabolites are synthesized in small amounts and not in a generalized form; some metabolites are specific to a particular plant's genus or family; in other cases, the concentrations of secondary metabolites may vary between plant parts and according to environmental conditions. Several studies have identified some phytochemicals as the active principles or bioactive molecules of medicinal plants; the most important are: alkaloids, terpenes, and phenolic compounds (Mahajan et al., 2020; Süntar, 2020; Verma and Shukla, 2015).

5.3.1 ALKALOIDS

Alkaloids are heterocyclic secondary metabolites with at least one nitrogen atom in their structure and derived from amino acids such as tryptophan, lysine, and tyrosine. In plants, the alkaloids have a defensive function against herbivores and pathogens. Around 12,000 molecules have been identified in this group of secondary metabolites. Alkaloids have a long history in medicine due to their biological activity in humans; they generate physiological and psychological responses, primarily due to interaction with neurotransmitters; it has also been postulated that alkaloids intercalate with DNA. The alkaloids have a high value and therapeutic effect as muscle relaxants, tranquilizers, antitussives, or analgesics in low doses; however, almost all alkaloids are toxic in high doses (Awuchi, 2019; Chitra et al., 2019).

5.3.2 TERPENES

Terpenes or terpenoids comprise the largest group of secondary metabolites (more than 40,000 different molecules); they have a common biosynthetic pathway from acetyl-CoA. The terpenes constitute the majority of essential oils obtained from plants, and they are intensely aromatic and of interest in the food and cosmetic industry. The terpene's odor helps plants repel insects and herbivores, among other functions. Terpenes are divided into monoterpenes, sesquiterpenes, diterpenes, triterpenes, and polyterpenes. Several terpenoid compounds have medicinal importance due to their biological properties, such as antimicrobial, anticarcinogenic, and others (Awuchi, 2019; Pagare et al., 2015; Tetali, 2019). The mode of action of terpenes has not been fully elucidated, but it is postulated that they may cause membrane breakdown through lipophilic compounds (Andrade-Ochoa et al., 2021).

5.3.3 PHENOLIC DERIVATIVES

Phenolic secondary metabolites are characterized by a phenyl ring with one or more hydroxyl groups. They are highly heterogeneous compounds, and some are soluble in water and others in organic acids; about 10,000 derivatives have been identified, including tannins, coumarins, flavonoids, lignins, stilbenes, arylpyrrhones, among others. In plants, phenolic secondary metabolites are derived from the shikimic acid pathway (Ashraf et al., 2018; Mahajan et al., 2020; Pagare et al., 2015). Their mechanisms of action are variable; for example, the toxicity of phenols in microorganisms is attributed to enzymatic inhibition by oxidation reactions.

5.4 SECONDARY METABOLITES IN PLANTS FROM ARID ECOSYSTEMS

Arid zone plants have an excellent capacity to adapt to dry and hostile environments where other plants cannot grow. In conditions of heat and lack of water, these plants have developed several forms of adaptation: spines provide shade; waxes and hairs regulate temperature; succulents have a characteristic morphology that allows them to retain water. Additionally, plants in these zones release toxins into the soil to prevent the growth of other plants and avoid competition for water.

Water is a key molecule in the physiological activities of plants, and it functions as a medium for the transport of metabolites and nutrients to all parts of the plant. When water supply becomes limiting or transpiration rate increases in plants, water stress occurs (Ashraf et al., 2018; Verma and Shukla, 2015). To develop resistance and adapt to water stress, plants induce changes in gene expression that activate metabolic processes, which promote drought resistance and may include biosynthesis and accumulation of secondary metabolites. These adaptations in metabolism involve the production of phenolic, flavonoid, terpenoid, and nitrogen compounds that are species and genotype-specific (Zahedi et al., 2021).

Consequently, due to selection pressures and resource availability, plants in arid zones have developed a wide diversity of secondary metabolites (Vilela et al., 2011). In most medicinal plants from arid zones, the proportion of functional metabolites increases, causing natural populations of medicinal plants with high quantity and quality of specialized metabolites. The essential oils obtained from these plants are ecotypes from arid and mountainous regions with low seasonal rainfall (Ahmad et al., 2004; Zahedi et al., 2021). Generally, plants growing in desert environments develop additional defenses to mitigate the effects of high temperatures, ultraviolet rays, drought, and salinity (Rioja Soto, 2020).

5.5 PROPERTIES OF MEDICINAL PLANTS FROM ARID ZONES

There is a wide diversity of plants with medicinal properties in arid zones, and they are traditionally used to treat different health problems. Few plants have been scientifically investigated to evaluate biological activity, identify the active metabolites, and determine their mechanism of action. This chapter describes plants from arid zones with medicinal properties: antidiabetic, antiparasitic, and antimicrobial activities. There are plants that may have two or more properties, or even properties different from those mentioned above, such as: anticancer, anti-inflammatory, antioxidant, antihypertensive, among others; however, they will not be discussed in this chapter.

Currently, diabetes and antimicrobial/antiparasitic resistance are public health problems. Diabetes is a chronic disease with possible complications in other organs and high prevalence in developing countries (WHO, 2020). On the other hand, antimicrobial and antiparasitic resistance is increasing and is a threat to public health (WHO, 2019). Therefore, it is necessary to search for and develop new pharmaceutical strategies; natural resources of arid zones are a great alternative accessible, cheap, low toxicity, and with extensive traditional knowledge.

5.5.1 ARTEMISIA LUDOVICIANA

Artemisina ludovicina is a plant of the *Asteraceae* family. The plant is widely distributed; it can be found from the United States to Guatemala. *A. ludoviciana* grows in regions with warm, semi-warm, dry, semi-dry, and temperate climates. In Mexico, it is commonly known as "estafiate" and grows in the northeast part of the country. It is a plant that can measure up to 1m in height. The branches are grayish. The leaves are divided into three and are elongated. The flowers are yellow and the plant gives off a strong odor. In traditional medicine, infusions of estafiate are used to treat intestinal disorders (diarrhea, pain, dysentery, parasitosis, etc.) (APMTM, 2009a).

Antiparasitic activity. Several investigations have evidenced the antiparasitic effect of *A. ludoviciana*. Aqueous and solvent extracts of the mature plant were found to be active *in vitro* against the intestinal parasitic protozoa *Entamoeba histolytica* and *Giardia lamblia* (Fernández et al., 2005). The ethanolic extract of *A. ludoviciana* showed an antimalarial effect in a murine model of malaria with *Plasmodium yoelii* (Malagón. et al. 1997). The essential oil of *A. ludovicina* showed activity against *Acanthamoeba castellanii, Leishmania infantum*, and *Trichomonas vaginalis*; additionally, camphor, 1,8-cineole, and camphene were identified as its main components (Baldemir et al., 2018). Likewise, the methanolic extract showed activity against *Fasciola hepatica* (Ezeta-Miranda et al., 2016).

Hypoglycemic activity. The infusion of *A. ludoviciana* showed hypoglycemic and antihyperglycemic effect in laboratory animals with induced diabetes. In addition, two components of estafiate were identified, 1) arglanine, which promotes insulin release; 2) salvinin, which inhibits the α-glucosidase inhibitor that prevents carbohydrate absorption. It is suggested that both mechanisms act in synergy to reduce glucose levels (Anaya-Eugenio et al., 2014).

Antimicrobial activity. The effect of *A. ludoviciana* extract on severe gastrointestinal bacteria was evaluated. Estafiate can inhibit the growth and enterotoxin production of *Vibrio choloreae* (Garcia et al., 2006), *Campylobacter jejuni* (Castillo et al., 2011), and *Helicobacter pylori* (Castillo-Juárez et al., 2009). The identified compounds are cineol, thujone, tannins, and santonin.

5.5.2 FLOURENSIA CERNUA

Fluorensia cernua is a shrubby plant that grows wild in arid and semi-arid regions of the southwestern United States and northern Mexico. It is commonly known as "hojasen" or "tarbush." The plant grows from 1 to 2 meters. The leaves are thick and oval, up to 2–3 cm long. The flowers are yellow, hanging, and contain various florets. The fruit is an achene up to 1 cm long. Most plant parts are very resinous and have a bitter taste (CIJE, 2022). In traditional medicine, *F. cernua* is used in infusions to treat various gastrointestinal diseases such as stomach pain, diarrhea, and dysentery. It is also used as a purgative, expectorant, and antirheumatic.

Antimicrobial activity. Scientific reports show that *F. vernua* is especially active against *Mycobacterium tuberculosis,* even in multi-resistant bacteria (Molina-Salinas et al., 2006; Molina-Salinas et al., 2011). The biological activity has been correlated with benzofurans in the plant extracts (Jasso de Rodríguez et al., 2012). Additionally, antioxidant and antifungal properties have been reported in phytopathogenic microorganisms (Alvarez-Pérez et al., 2020).

5.5.3 IBERVILLEA SONORAE

Ibervillea sonorae is a guided plant of the family *Cucurbitaceae.* The species is native to northwestern Mexico, locally known as "wareque, wereke, or guareque." It is a large root with a vine that climbs trees, the flower is yellow, the fruit is a ball, it is green striped when tender, and red stripes when ripe (Naturalista, 2022). In traditional Mexican medicine, the root of *Ibervillea sonorae* has been widely used to treat diabetes. In addition, the leaves are used for the treatment of skin diseases, stomach ulcers, and other diseases (Herbal Safety, 2022; Naturalista, 2022).

Hypoglycemic activity. Several pharmacological studies have shown that extracts from the roots of *Ibervillea sonorae* have hypoglycemic activity in healthy and diabetes-induced laboratory animals (Alarcon-Aguilar et al., 2002, 2005; Rivera-Ramírez et al., 2011). Interestingly, it was found to have a better hypoglycemic effect when the root extract is administered peritoneally compared to oral administration (Alarcon-Aguilar et al., 2002). The phytochemicals involved and the mechanisms of action are not known with certainty, but it has been determined that the extract of *I. sonorae* can inhibit α-glucosidase to decrease carbohydrate absorption and stimulate insulin secretion from the β-cells of the pancreas (Semotiuk et al., 2020).

The chemical composition reported for this plant includes alcohols, ketones, alkanes and amides, palmitic acid, and methyl palmitate associated with hypoglycemic activities (Morales et al., 2012; Sinagawa-García et al, 2015).

5.5.4 JATROPHA DIOICA

Jatropha dioica is a plant native to Mexico that habits sites of dry and semi-dry climates, belongs to the *Euphorbiaceae* family. Its common name is "blood of drago" because it has a colorless juice that darkens in contact with air. The plant is a bush from 50cm to 1.50m in height. Its branches are dark red. The leaves are longer than wide. The flowers are pink, small, and in groups. The fruits are globose, between 1 and 2 cm long, and have a seed (APMTM, 2009b).

Traditionally used to prevent hair loss, the root is boiled or macerated to extract a liquid to be applied to the hair (Razo and Alvarado, 2014). The root is the most used part of the plant and has many therapeutic properties. There are reports of obtaining root extracts with different solvents, suggesting hexane as the most effective (Ramírez-Moreno et al., 2020). Various components such as essential oil, resin, saponins, alkaloids, and oxalic acid are obtained from the roots. Three terpenoids have been identified (citlalitrone, jatrofone, and riolosatrione), and an esteral, R-sitosterol, to which antimicrobial properties have been attributed (Ramírez-Moreno et al., 2020).

Antimicrobial activity. *J. dioica* extracts have biological activity against different pathogenic microorganisms in humans such as *Bacillus cereus, Escherichia coli, Salmonella typhi, Staphylococcus aureus, Enterobacter aerogenes, Enterobacter cloacae, Salmonella typhimurium, Cryptococcus neoformans, Candida albicans, Candida parapsilosis*, and *Sorothrix schenckii* (Ramírez-Moreno et al., 2020; Silva-Belmares et al., 2014). In toxicity evaluations, *J. doica* extracts do not show a genotoxic or cytotoxic effect in mouse peripheral blood at high doses (Araujo-Espino et al., 2017).

5.5.5 LARREA TRIDENTATA

Larrea tridentata, belongs to the *Zygophyllaceae* family; it is commonly known as "governor, chaparral, or creosote bush." It is a bushy plant from 60cm to 3m high, branched; the leaves are divided into leaflets and covered with resin. The flowers are yellow and solitary. The fruits are capsules with a hairy cover. The plant dominates some areas of the desert of the southwestern

United States and northern Mexico where the climate is dry, very dry, or temperate; it inhibits the growth of other plants around it to store water (APMTM, 2009c).

Multiple secondary metabolites have been identified in *L. tridentate* extracts, and the lignans are the most important constituents due to their promising biological activity. Flavonoid aglycones, glycosylated flavonoids, and essential oils have been found in the plant resin. The essential oils contain more than 300 volatile and 67 nonvolatile compounds. In addition, saponins, sapogenins, sterols, tannins, monoterpenes, and sesquiterpenes have also been reported (Gnabre et al., 2015). This plant has several medicinal applications; traditionally, infusions of the leaves and branches are often used to treat kidney stones and urinary tract affections.

Antimicrobial activity. Research with extracts of *L. tridentata* has identified antimycobacterial activity attributed to lignans and flavonoids from the plant (Favela-Hernández et al., 2012). Ethanolic extracts of the leaves also have activity against pathogenic gram-positive bacteria *Staphylococcus aureus* and *Streptococcus pyogenes* (Turner et al., 2021). The lignin nordihydroguaiaretic acid (NDGA) is the most studied secondary metabolite of *L. tridentata.* NDGA possesses remarkable biological activity, it acts through sequestration of reactive oxygen species, inhibition of lipoxygenases (LOX), and activation of the endogenous antioxidant response mediated by nuclear factor erythroid 2-related factor 2 (NRF2) (Reyes-Melo et al., 2021).

Antiparasitic activity. Additionally, research has also observed antiparasitic properties of *L. tridentata* attributed to lignans compounds, nordihydroguairetic acid, and demethoxyisoguaiacin. An effect on the protozoa *Entamoeba histolytica, Giardia lamblia, Naegleria fowleri, Trypanosoma brucei rhodesiense, Trypanosoma cruzi, Leishmania donovani,* and *Plasmodium falciparum* has been reported. NDGA was the most active compound against parasites (Bashyal et al., 2017; Schmidt et al., 2012). In the veterinary area, *L. tridentata* have activity against ruminant intestinal helminths (García et al., 2018). NDGA is a potential and highly controversial oxidant. It was widely used during the 1950s as a food preservative and to preserve natural fibers. In the early 1960s, it was banned after reports of toxicity, chronic use of creosote bush and NDGA causes kidney toxicity and hepatotoxicity (Arteaga et al., 2005).

5.5.6 *LIPPIA GRAVEOLENS OR LIPPIA BERLANDIERI*

Oregano is a plant with worldwide distribution, and several botanical families have been identified; the genus *Lippia sp* is native to America and belongs to the *Verbenacea* family. *Lippia graveolens* HBK or *Lippia berlandieri* Schauer, also known as Mexican oregano, is the most important in the country, and it has wide use in gastronomy. *Lippia graveolens* grows in arid and semiarid areas of the country; although it grows in the wild, there are cultivation areas to produce and market oregano and its derivatives. The plant develops bushes up to 2.5 m tall with 1.20 m foliage, has branched stems with a large number of oval leaves with serrated edges that are 1–3 cm long, dark green, rough texture, with hairs and small white flowers that form inflorescences in clusters. The smell and taste of this plant are robust (Cid-Pérez et al., 2016).

Mainly from the leaves, essential oil (EO) is obtained by distillation, and its major components are thymol and carvacrol, both terpenes that are isomers, with only a difference in the position of the hydroxyl group. The concentrations of thymol and carvacrol can vary according to climate, altitude, time of harvest, etc. Generally, *L. berlandieri* has a higher concentration of carvacrol than thymol; however, chemotypes with a higher amount of thymol can be found (Levario-Gómez et al. 2020). Mexican oregano EO has been widely studied as an antimicrobial agent. The microbicidal properties are mainly attributed to thymol and carvacrol. EO's effectiveness depends on the concentrations present of its major components and the structure of the microorganism (Adame-Gallegos et al., 2016; Andrade-Ochoa et al., 2021).

Antimicrobial activity. A broad antibacterial activity of the essential oil of *L. berlandieri* has been demonstrated. Gram-positive bacteria such as *Bacillus cereus*, *Staphylococcus aureus*, and *Listeria monocytogenes;* and gram-negative bacteria cause gastrointestinal infections such as *Escherichia coli, Salmonella typhimurim, Shigella sonnei* (Andrade-Ochoa et al., 2021), and even on the formation of their biofilms (Reyes-Jurado et al., 2020). The effect of EO on clinical isolates of *Enterococcus faecalis* and *E. coli* causing urinary tract infections has been confirmed (Zapién-Chavarría et al., 2019). In eukaryotic microorganisms, antifungal activity was found on filamentous fungi such as *Penicillium expansum, Aspergillus ochraceus, Aspergillus niger,* and *Alternaria alternata* (Levario-Gómez et al., 2020).

Antiparasitic activity. There are reports of the effect on protozoa such as *Giardia lamblia* trophozoites and trypanosomatids: epimastigotes of *Trypanosoma cruzi*, the causative agent of Chagas disease, and promastigotes of *Leishmania mexicana,* the causative agent of cutaneous leishmaniasis (Andrade-Ochoa et al., 2021). Additionally, oregano EO and carvacrol showed relevant larvicidal activity against insects such as *Culex quinquefasciatus,* a vector of lymphatic filariasis (Andrade-Ochoa et al., 2018). According to the above, oregano essential oil *L. berlandieri* has a high antimicrobial potential that should be exploited.

5.5.7 OPUNTIA FICUS-INDICA

This plant is commonly known as "nopal" in Mexico and belongs to the *Cactaceae* family. It is present in arid and semiarid areas of the country; it can be found in the wild or controlled crop plantations. The plant is up to 5m tall, and the stem is branched, dull green in color. The stems are flattened; the leaves are rounded areoles with a few yellow spines. The flowers are yellow or orange, 7 to 10 cm wide. Its fruits are green, red, or purple, with a sweet flavor, and are known as prickly pears (APMTM, 2009d; Hikal et al., 2021). The fruit of *O. ficus-indica* contains the alkaloids indole, betanin, isomers iso, and neobetanin, and indicaxanthin and opuntia xanthin. The leave and stems contain the alkaloids mescaline and tyramine. The flowers contain the flavonoid isorhamnetin and the sterol beta-sitosterol (APMTM, 2009d).

Nopal is used as a food in Mexico and as an alternative treatment for burns, wounds, edema, and indigestion. It has been traditionally used to treat diabetic patients. Different studies have shown that its alcoholic extracts have anti-inflammatory, antibacterial, hypoglycemic, and hypolipemic properties (Kaur et al., 2012; Tilahun and Welegerima, 2018).

Hypoglycemic activity. In developing countries, the plant has a dual function, medicinal and nutritional. The consumption of nopal favors the prevention of diabetes in healthy patients and avoids long-term complications in diabetic patients (Mtat et al., 2021). Several pharmacological studies have shown that the consumption of Opuntia extracts improves hyperglycemia in diabetic experimental animals and hypercholesterolemia and renal function (Gomaa et al., 2012; Hwang et al., 2017; Salahet al., 2020). The hypoglycemic effect of *Opuntia ficus-indica* has been attributed to two processes. (1) Inhibition of the enzyme alpha-glucosidase, which decreases postprandial glucose absorption at the intestinal level, (2) increased glucose uptake by activating the AMPK/p38MAPK pathway; the active metabolite is still unknown (Leem et al., 2016).

Antimicrobial activity. Regarding antibacterial properties, there are reports of different alcoholic extracts of the cladodes or fruits and their activity against Gram-positive and Gram-negative bacteria (Hikal et al., 2021; Welegerima and Zemene, 2017).

5.5.8 *POLIOMINTHA LONGIFLORA*

Poliomintha longiflora is the second most important species of Mexican oregano, belonging to the *Lamiaceae* family. The plant grows wild in north-eastern Mexico; it is a low shrub, the leaves have smooth edges, the stems are light green oval, and the flowers are tubular and purple-pink. It is widely used as a spice for food flavoring, mainly used as a substitute for European oregano because it has a milder flavor concerning *L. graveolens* (Cid-Pérez et al., 2016). The essential oil of *P. longiflora* has also been attributed medicinal properties, being a variant of oregano, the major components of its essential oil are thymol and carvacrol. Generally, *P. longiflora* has a higher concentration of thymol than carvacrol; hence, its flavor and odor are milder than *L. berlandieri* (Cid-Pérez et al., 2019; Díaz-de León et al., 2020).

Antimicrobial activity. *P. longiflora* shows fungicidal activity against filamentous fungi (Levario-Gómez et al., 2020) and activity against Gram-negative and Gram-positive bacteria such as *E. coli* and *Enterococcus faecalis*, respectively (Zapién-Chavarría et al., 2019).

5.5.9 *PROSOPIS GLANDULOSA*

Prosopis glandulosa belongs to the *Fabaceae* family and is commonly known as "honey mesquite," found in dry, arid regions of the southwestern United States and northern Mexico; it is classified as an invasive plant. *Prosopis glandulosa* has an arboreal growth, reaches 6 to 9 meters, has drooping, rounded, large, and flexible branches with abundant foliage. This tree has elongated, light yellow spikes and straight yellow pods; the flowers are arranged in a dense inflorescence that is cylindrical in shape and yellow or white. Traditionally, infusions of dried and ground pods are used to stabilize blood glucose and blood pressure (Felger and Moser, 1971; Ffolliott and Thames, 1982).

Hypoglycemic activity. The efficacy of a local commercial product produced from *P. glandulosa* pods was evaluated as a hypoglycemic agent in rats with induced diabetes. *P. glandulosa* was found to modestly reduce blood glucose levels. The effect was related to the stimulation of insulin secretion and the induction of β-cell formation in type 1 diabetes. The active component has not yet been identified (George et al., 2011).

Antiparasitic activity. The indolizidine alkaloids *P. glandulosa:* Δ1,6-juliprosopine, juliprosopine, and juliprosine were isolated from *P. glandulosa* leaf extract and found to be effective against the tissue protozoa *Plasmodium falciparum* and *Leishmania donovani* (Rahman et al., 2011).

Antimicrobial activity. The same indolizidine alkaloids mentioned above showed a broad antimicrobial effect on *Cryptococcus neoformans, Candida albicans, Candida krusei, Candida glabrata, Cryptococcus neoformans, Aspergillus fumigatus*, and the bacteria *Staphylococcus aureus*, methicillin-resistant *S aureus*, and *Mycobacterium intracellulare* (Rahman et al., 2011).

5.5.10 RHUS VIRENS

Rhus virens, a plant of the *Anacardiaceae* family, is distributed from the southern United States to central Mexico It is commonly known: as lambrisco or lantrisco *Rhus virens* is a 1–6 m shrub with oblong-ovate pale leaves of 2–5 cm, and the fruit is a red or orange drupe of 6–8 cm. The infusion of this plant has a hypoglycemic effect in healthy people (CIIDZA, 2022.).

Hypoglycemic activity. To evaluate the antidiabetic effect of *Rhus virens*, the whole plant infusion or powder infusion was administered to healthy volunteers, in both cases a hypoglycemic effect was observed. In the analysis of the functional groups, aromatic compounds, amides, aliphatic compounds, phthalates, and sugar diversities were found. Shikimic acid, galloylquinic acid, gallic acid, and vanillic acid O-hexoside were identified in *Rhus virens*. The proposed mechanism suggests that the active compounds inhibit α-glucosidase and prevent the complete absorption of carbohydrates (Aguilar-Lozano, 2012; Juárez-Aragón et al., 2019).

5.5.11 TECOMA STANS

This plant has several common names: "thundering, yellow elderberry, golden rain, yellow bugle, yellow bells, bow stick." *T. stans* belongs to the family *Bignoniaceae*. It is a shrub, 1 to 8 m tall. The leaves are compound and serrated. The small, yellow, and bell-shaped flowers are grouped in clusters at the tips of the branches. The fruits are elongated capsules, green and brown, with flattened seeds when ripe. It is a native plant of North America and high regions of South America and occurs in warm climates; semi-warm, dry, very dry, and temperate; it is a widely distributed wild plant that grows along roadsides (Anand and Basavaraju, 2021; APMTM, 2009e).

The infusion of *Tecoma stans* is frequently used for various ailments. About 120 compounds have been isolated and identified from the plant;

monoterpene alkaloids, phenolic acids, flavonoids, and fatty acids are the main bioactive components that contribute to its therapeutic benefits (Anand and Basavaraju, 2021).

Antimicrobial activity. The potential effect of *T. stans* against several microorganisms has been reported. Gram-negative bacteria of the *Enterobacteriaceae* family, main causative agents of gastrointestinal and urinary tract infections: *Klebsiella pneumoniae, Serrratia marcescens, Proteus vulgaris, Salmonella tiphy*, and *E. coli.* In Gram-positive bacteria: *Bacillus subtilis, Staphylococcus aureus, Streptococcus pneumoniae*, methicillin-resistant *Staphylococcus aureus* (Bakr et al., 2019; Gonçalves et al., 2022; Mohamed et al., 2013). In fungi such as *Aspergillus fumigatus* and *Candida albicans* Flavonoids from *T. stans* are responsible for antimicrobial activity (Sunita Verma, 2016).

Hypoglycemic activity. Pharmacological studies of the aqueous extract of *T. stans* have reported hypoglycemic properties in laboratory animals. Additionally, it is postprandial antihyperglycemic, hypocholesterolemic, and hypotriglyceridemic. The suggested mechanism of action is by inhibition of intestinal α-glucosidase (Aguilar-Santamaría et al., 2009). The leaves contain the alkaloid tecomine, this component has been associated with the ability to reduce blood glucose and hypolipidemic properties to reduce plasma cholesterol levels in diabetic mice (Costantino et al., 2003).

5.5.12 OTHER PLANTS WITH MEDICINAL PROPERTIES IN ARID ZONES

In Table 5.1, other arid zone plants with reported hypoglycemic, antimicrobial, and/or antiparasitic properties are described.

5.6 CONCLUSION

Plants in arid zones have developed a great diversity of secondary metabolites due to selection pressures and the availability of resources. These medicinal plants have a high potential as natural resources to be used rationally and sustainably to contribute to in research and development of new pharmaceutical alternatives.

However, from a scientific context, it is necessary to broaden the knowledge of medicinal plants, identify their composition, active principles, mechanisms of action, toxicity, side effects, as well as their interactions with

TABLE 5.1 Properties of Medicinal Plants from Arid Zones.

Botanical name	Family	Comun name	Part of plant	Bioactivities	References
Acacia farnesiana	*Fabaceae*	Huizache	Leaves	Antimicrobial	Hernández-García et al. (2019); Ramli et al. (2011)
Acacia rigidula	*Fabaceae*	Blackbrush acacia Chaparro prieto	Leaves	Antibacterial	Cavazos et al. (2021)
Agave lechuguilla	*Agavaceae*	Lechuguilla	Whole plant Guishe	Hypoglycemic	Esquivel-Gutiérrez et al. (2021)
Artemisia ludoviciana	*Asteraceae*	Estafiate	Leaves Whole plant	Hypoglycemic Antiparasitic Antibacterial	Anaya-Eugenio et al. (2014); Castillo-Juárez et al. (2009); Fernández et al. (2005)
Castela tortuosa	*Simaroubaceae*	Chaparro amargo	Branches, leaves	Antiparasitic Antibacterial	Zamilpa et al. (2019)
Ferocactus latispinus ***Ferocactus histrix***	*Cactaceae*	Biznaga burra Biznaga Devil's Tongue Barrel	Fresh fruits	Hypoglycemic	Medicinales Aromáticas et al. (2010)
Flourensia cernua	*Asteraceae*	Hojasen	Leaves	Antimicrobial	Molina-Salinas et al. (2006)
Galphimia glauca	*Malpighiaceae*	Shower of gold	Aerial parts	Antimicrobial	Sharma et al. (2018)
Ibervillea sonorae	*Cucurbitaceae.*	Wereque	Root	Hypoglycemic	Rivera-Ramírez et al. (2011)
Jatropha dioica	*Euphorbiaceae*	Blood of drago	Root	Antimicrobial	Silva-Belmares et al. (2014)
Krameria ramosissima	*Krameriaceae*	Calderona	Root	Antimicrobial	Villarreal-García et al. (2014)
Larrea tridentata	*Zygophyllaceae*	Governor	Leaves	Antimicrobial Antiparasitic	Favela-Hernández et al. (2012); Turner et al. (2021)
Lippia berlandieri ***Lippia graveolens***	*Verbenacea*	Mexican oregano	Leaves	Antimicrobial Antiparasitic	Adame-Gallegos et al. (2016); Andrade-Ochoa et al. (2021)
Marrubium vulgare	*Lamiaceae*	White horehound,	Root Stem	Hypoglycemic Antimicrobial Antiparasitic	Aćimović et al. (2020); Molina-Garza et al. (2014); Vergara-galicia et al. (2012); Villanueva et al. (2017)

TABLE 5.1 *(Continued)*

Botanical name	Family	Comun name	Part of plant	Bioactivities	References
Opuntia ficus-indica	*Cactaceae*	Nopal	Leaves	Hypoglycemic Antimicrobial	Hwang et al. (2017); Salah et al. (2020)
Poliomintha longiflora	*Lamiaceae*	Mexican oregano	Leaves	Antimicrobial	Levario-Gómez et al. (2020); Zapién-Chavarría et al. (2019)
Prosopis glandulosa	*Fabaceae*	Honey mesquite	Leaves	Hypoglycemic Antimicrobial Antiparasitic	George et al. (2011); Rahman et al. (2011)
Rhus virens	*Anacardiacea*	Lambrisco	Whole plant	Hypoglycemic	Juárez-Aragón et al. (2019)
Salvia connivens	*Lamiaceae*	Charahuesca	Leaves	Antiparasitic	Bautista et al. (2020)
Tecoma stans	*Bignoniacea*	Yellow bells Tronadora	Whole plants	Antimicrobial Hypoglycemic	Aguilar-Santamaría et al. (2009); Bakr et al. (2019)
Thymophylla pentachaeta	*Asteraceae*	Limoncillo Parralena	Root	Antimicrobial	Cordisco et al. (2021)
Turnera diffusa	*Turneraceae*	Damiana	Leaves	Hypoglycemic Antimicrobial	Baéz-Parra et al. (2019); Parra-Naranjo et al. (2017)

other medicines. This requires multidisciplinary studies with ethnobotanists, chemists, pharmacologists, agronomists, among others. Arid zone plants are valuable natural resources for new pharmaceutical alternatives.

KEYWORDS

- **medicinal plants**
- **hypoglycemic**
- **antimicrobial**
- **antiparasitic**

REFERENCES

Aćimović, M.; Jeremić, K.; Salaj, N.; Gavarić, N.; Kiprovski, B.; Sikora, V.; Zeremski, T. Marrubium Vulgare L.: A Phytochemical and Pharmacological Overview. *Molecules* **2020,** *25* (12). https://doi.org/10.3390/molecules25122898

Adame-Gallegos, J. R.; Andrade-Ochoa, S.; Nevarez-Moorillon, G. V. Potential Use of Mexican Oregano Essential Oil against Parasite, Fungal and Bacterial Pathogens. *J. Essential Oil-Bearing Plants* **2016,** *19* (3), 553–567. https://doi.org/10.1080/0972060X.2015.1116413

Aguilar-Lozano, C. A. *Evaluación del efecto hipoglicemiante in vitro de diferentes extracciones de Lambrisco (rhus virens Lindh) y caracterización de su composición química.* Universidad Autónoma de Chihuahua.

Aguilar-Santamaría, L.; Ramírez, G.; Nicasio, P.; Alegría-Reyes, C.; Herrera-Arellano, A. Antidiabetic Activities of *Tecoma stans* (L.) Juss. ex Kunth. *J. Ethnopharmacol.* **2009,** *124* (2), 284–288. https://doi.org/10.1016/J.JEP.2009.04.033

Ahmad, H.; Bhatti, G. R.; Latif, A. Medicinal Flora of the Thar Desert: An Overview of Problems and Their Feasible Solutions. *Zonas Áridas* **2004,** *4*, 73–83. http://revistas.lamolina.edu.pe/index.php/rza/article/viewFile/715/697

Akinyemi, O.; Oyewole, S.; Jimoh, K. Medicinal Plants and Sustainable Human Health: A Review. *Hortic. Int. J.* **2018,** *2* (4), 2–4. https://doi.org/10.15406/hij.2018.02.00051

Alarcon-Aguilar, F. J.; Calzada-Bermejo, F.; Hernandez-Galicia, E.; Ruiz-Angeles, C.; Roman-Ramos, R. Acute and Chronic Hypoglycemic Effect of Ibervillea Sonorae Root Extracts-II. *J. Ethnopharmacol.* **2005,** *97* (3), 447–452. https://doi.org/10.1016/J.JEP.2004.11.035

Alarcon-Aguilar, F. J.; Campos-Sepulveda, A. E.; Xolalpa-Molina, S.; Hernandez-Galicia, E.; Roman-Ramos, R. Hypoglycaemic Activity of *Ibervillea sonorae* Roots in Healthy and Diabetic Mice and Rats. *Pharm. Biol.* **2002,** *40* (8), 570–575. https://doi.org/10.1076/phbi.40.8.570.14652

Alvarez-Pérez, O. B.; Ventura-Sobrevilla, J. M.; Ascacio-Valdés, J. A.; Rojas, R.; Verma, D. K.; Aguilar, C. N. Valorization of Flourensia Cernua DC as Source of Antioxidants and

Antifungal Bioactives. *Ind. Crops Products* **2020,** *152*, 112422. https://doi.org/10.1016/J.INDCROP.2020.112422

Anand, M.; Basavaraju, R. A Review on Phytochemistry and Pharmacological Uses of *Tecoma stans* (L.) Juss. ex Kunth. *J. Ethnopharmacol.* **2021,** *265* (July 2018). https://doi.org/10.1016/j.jep.2020.113270

Anaya-Eugenio, G. D.; Rivero-Cruz, I.; Rivera-Chávez, J.; Mata, R. Hypoglycemic Properties of Some Preparations and Compounds from *Artemisia ludoviciana* Nutt. *J. Ethnopharmacol.* **2014,** *155* (1), 416–425. https://doi.org/10.1016/J.JEP.2014.05.051

Andrade-Ochoa, S.; Chacón-Vargas, K. F.; Sánchez-Torres, L. E.; Rivera-Chavira, B. E.; Nogueda-Torres, B.; Nevárez-Moorillón, G. V. Differential Antimicrobial Effect of Essential Oils and Their Main Components: Insights Based on the Cell Membrane and External Structure. *Membranes* **2021,** *11* (6), 1–17. https://doi.org/10.3390/membranes11060405

Andrade-Ochoa, S.; Sánchez-Aldana, D.; Chacón-Vargas, K. F.; Rivera-Chavira, B. E.; Sánchez-Torres, L. E.; Camacho, A. D. et a;. Oviposition Deterrent and Larvicidal and Pupaecidal Activity of Seven Essential Oils and Their Major Components Against Culex Quinquefasciatus Say (Diptera: Culicidae): Synergism–Antagonism Effects. *Insects* **2018,** *9* (1). https://doi.org/10.3390/insects9010025

APMTM. *Artemisia ludoviciana*, 2009a. http://www.medicinatradicionalmexicana.unam.mx/apmtm/termino.php?l=3&t=artemisia-ludoviciana (accessed 23 Jan 2022).

APMTM. *Jatropha dioica*, 2009b. http://www.medicinatradicionalmexicana.unam.mx/apmtm/termino.php?l=3&t=jatropha-dioica (accessed 19 Jan 2022).

APMTM. *Larreaa tridentata*, 2009c. http://www.medicinatradicionalmexicana.unam.mx/apmtm/termino.php?l=3&t=larrea-tridentata (accessed 16 Jan 2022).

APMTM. *Opuntia ficus*, 2009d. http://www.medicinatradicionalmexicana.unam.mx/apmtm/termino.php?l=3&t=opuntia-ficus-indica (accessed 19 Jan 2022).

APMTM. *Tecoma stans*, 2009e. http://www.medicinatradicionalmexicana.unam.mx/apmtm/termino.php?l=3&t=tecoma-stans (accessed 15 Jan 2022).

Araujo-Espino, D. I.; Zamora-Perez, A. L.; Zúñiga-González, G. M.; Gutiérrez-Hernández, R.; Morales-Velazquez, G.; Lazalde-Ramos, B. P. Genotoxic and Cytotoxic Evaluation of *Jatropha dioica* Sessé ex Cerv. by the Micronucleus Test in Mouse Peripheral Blood. *Regul. Toxicol. Pharmacol.* **2017,** *86*, 260–264. https://doi.org/10.1016/J.YRTPH.2017.03.017

Arteaga, S.; Andrade-Cetto, A.; Cárdenas, R. Larrea Tridentata (Creosote Bush), an Abundant Plant of Mexican and US-American Deserts and Its Metabolite Nordihydroguaiaretic Acid. *J. Ethnopharmacol.* **2005,** *98* (3), 231–239. https://doi.org/10.1016/J.JEP.2005.02.002

Ashraf, M. A.; Iqbal, M.; Rasheed, R.; Hussain, I.; Riaz, M.; Arif, M. S. *Environmental Stress and Secondary Metabolites in Plants: An Overview—Plant Metabolites and Regulation under Environmental Stress*; Elsevier Inc, 2018. https://doi.org/10.1016/B978-0-12-812689-9.00008-X

Awuchi, C. G. Medicinal Plants: The Medical, Food, and Nutritional Biochemistry and Uses. *Int. J. Adv. Acad. Res.* **2019,** *5* (11), 2488–9849. https://www.researchgate.net/profile/Chinaza-Awuchi/post/What_is_the_difference_between_bioactive_compound_and_secondary_metabolite2/attachment/5df4dd103843b09383962da1/AS%3A835996325261314%401576328464677/download/ijaar-ste-v5n11-nov19-p27.pdf

Baéz-Parra, K.; Soto-Beltrán, M.; López-Cuevas, O.; Basilio Heredia, J.; Alcaraz-Meléndez, L.; Angulo-Escalante, M. A. et al. In Vitro Antimicrobial Activity of Methanolic and Hexanic Extracts of *Turnera diffusa* Against Common Urinary Pathogens. *Revista Bio Ciencias* **2019,** *6* (SPE), 670. https://doi.org/10.15741/REVBIO.06.NESP.E670

Bakr, R.; Fayed, M.; Salem, M.; Hussein, A. *Tecoma stans*: Alkaloid Profile and Antimicrobial Activity. *J. Pharm. Bioallied Sci.* **2019,** *11* (4), 341. https://doi.org/10.4103/JPBS.JPBS_79_19

Baldemir, A.; Karaman, Ü.; İlgün, S.; Kaçmaz, G.; Demirci, B. Antiparasitic Efficacy of *Artemisia ludoviciana* Nutt. (Asteraceae) Essential Oil for *Acanthamoeba castellanii, Leishmania infantum* and *Trichomonas vaginalis*. *Indian J. Pharm. Educ. Res.* **2018,** *52*. https://doi.org/10.5530/ijper.52.3.48

Bashyal, B.; Li, L.; Bains, T.; Debnath, A.; LaBarbera, D. V. *Larrea tridentata*: A Novel Source for Anti-Parasitic Agents Active Against *Entamoeba histolytica, Giardia lamblia* and *Naegleria fowleri*, 2017. https://doi.org/10.1371/journal.pntd.0005832

Bautista, E.; Calzada, F.; Yépez-Mulia, L.; Bedolla-García, B. Y.; Fragoso-Serrano, M.; Pastor-Palacios, G.; González-Juárez, D. E. Salvia connivens, a Source of Bioactive Flavones with Amoebicidal and Giardicidal Activity. *Revista Brasileira de Farmacognosia* **2020,** *30* (5), 729–732. https://doi.org/10.1007/S43450-020-00103-8/TABLES/2

Boy, H. I. A.; Rutilla, A. J. H.; Santos, K. A.; Ty, A. M. T.; Yu, A. I.; Mahboob, T. et al. Recommended Medicinal Plants as Source of Natural Products: A Review. *Digit. Chin. Med.* **2018,** *1* (2), 131–142. https://doi.org/10.1016/s2589-3777 (19)30018-7

Castillo-Juárez, I.; González, V.; Jaime-Aguilar, H.; Martínez, G.; Linares, E.; Bye, R.; Romero, I. Anti-Helicobacter Pylori Activity of Plants Used in Mexican Traditional Medicine for Gastrointestinal Disorders. *J. Ethnopharmacol.* **2009,** *122* (2), 402–405. https://doi.org/10.1016/J.JEP.2008.12.021

Castillo, S. L.; Heredia, N.; Contreras, J. F.; García, S. Extracts of Edible and Medicinal Plants in Inhibition of Growth, Adherence, and Cytotoxin Production of *Campylobacter jejuni* and *Campylobacter coli*. *J. Food Sci.* **2011,** *76* (6), M421–M426. https://doi.org/10.1111/J.1750-3841.2011.02229.X

Cavazos, P.; Gonzalez, D.; Lanorio, J.; Ynalvez, R. Secondary Metabolites, Antibacterial and Antioxidant Properties of the Leaf Extracts of *Acacia rigidula* Benth. and *Acacia berlandieri* Benth. *SN Appl. Sci.* **2021,** *3* (552). https://doi.org/10.1007/s42452-021-04513-8

Chitra, J.; Shivani, K.; Rekha, V. Bioactivity of Secondary Metabolites of Various Plants: A Review. *Int. J. Pharm. Sci. Res.* **2019,** *10* (2), 494–504. https://doi.org/10.13040/IJPSR.0975-8232.10 (2).494-04

Cid-Pérez, T. S.; Ávila-Sosa, R.; Ochoa-Velasco, C. E.; Rivera-Chavira, B. E.; Nevárez-Moorillón, G. V. Antioxidant and Antimicrobial Activity of Mexican Oregano (*Poliomintha longiflora*) Essential Oil, Hydrosol and Extracts from Waste Solid Residues. *Plants* **2019,** *8* (1). https://doi.org/10.3390/plants8010022

Cid-Pérez, T. S.; Nevárez-Moorillón, G. V.; Torres-Muñoz, J. V.; Palou, E.; López-Malo, A. *Mexican oregano (Lippia berlandieri and Poliomintha longiflora) oils. Essential Oils in Food Preservation, Flavor and Safety*; Elsevier Inc, 2016. https://doi.org/10.1016/B978-0-12-416641-7.00063-8

CIIDZA. *Tecoma stans*, n.d. https://ciidza.ipicyt.edu.mx/catalogo/showone.php?id=24 (accessed 15 Jan 2022).

CIJE, C. de I. y J. E. Hojasen, 2022. http://www.uadec.mx/hojasen/ (accessed 19 Jan 2022).

Cordisco, E.; Petenatti, E.; Svetaz, L.; Sortino, M. Evaluation of the Antifungal Photodynamic Activity of *Thymophylla pentachaeta* Extracts Against *Candida albicans* and Its Virulence Factors. *Phytomedicine* **2021,** *90*, 153608. https://doi.org/10.1016/J.PHYMED.2021.153608

Costantino, L.; Raimondi, L.; Pirisino, R.; Brunetti, T.; Pessotto, P.; Giannessi, F. et al. Isolation and Pharmacological Activities of the *Tecoma stans* Alkaloids. *Farmaco* **2003,** *58* (9), 781–785. https://doi.org/10.1016/S0014-827X (03)00133-2

Díaz-de León, C. I.; González-Álvarez, M.; Guzmán-Lucio, M. A.; Núñez-Guzmán, G. R.; Moreno-Limón, S. El orégano de los géneros Lippia (verbenaceae) y Poliomintha (lamiaceae) en el estado de Nuevo León, Mexico. *Polibotánica* **2020,** *0* (50), 1–18. https://doi.org/10.18387/polibotanica.50.1

Esquivel-Gutiérrez, E. R.; Manzo-Avalos, S.; Peña-Montes, D. J.; Saavedra-Molina, A.; Morreeuw, Z. P.; Reyes, A. G. Hypolipidemic and Antioxidant Effects of Guishe Extract from Agave Lechuguilla, a Mexican Plant with Biotechnological Potential, on Streptozotocin-Induced Diabetic Male Rats. *Plants* **2021,** *10* (11). https://doi.org/10.3390/plants10112492

Ezeta-Miranda, A.; Vera-Montenegro, Y.; Ávila-Acevedo, G.; Manuel Álvarez-Mercado, J.; Francisco-Márquez, G. Short Communication. *RESUMEN Mexican J. Biotechnol.* **2016,** *1* (1), 177–183. www.mexjbiotechnol.com

Favela-Hernández, J. M. J.; García, A.; Garza-González, E.; Rivas-Galindo, V. M.; Camacho-Corona, M. R. Antibacterial and Antimycobacterial Lignans and Flavonoids from *Larrea tridentata. Phytother. Res.* **2012,** *26* (12), 1957–1960. https://doi.org/10.1002/ptr.4660

Felger, R. S.; Moser, M. B. Seri Use of Mesquite (Prosopis glandulosa var torreyana) Author (s): Richard S. Felger and Mary B. Moser Published by: Taylor & Francis, Ltd. On Behalf of the Arizona Archaeological and Historical Society Stable URL: https://www.jstor.org/stable. *Taylor & Francias Ltd. and ARizona ARchaeologicla Hist. Soc.* **1971,** *37* (1), 53–60.

Fernández, S. S.; Ramos Gurra, M. C.; Mata Cárdenas, B. D.; Vargas Villarreal, J.; Villarreal Treviño, L. In Vitro Antiprotozoal Activity of the Leaves of *Artemisia ludoviciana* Cite This Paper. *Fitoterapia* **2005,** *76*, 466–468. https://doi.org/10.1016/j.fitote.2005.04.009

Ffolliott, P. F.; Thames, J. L. Manual sobre taxonomía de Prosopis en Mexico, Perú y Chile. https://www.fao.org/3/q2580s/q2580s07.htm (accessed 23 Jan 2022).

García, J. E.; Gómez, L.; Mendoza-De-Gives, P.; Rivera-Corona, J. L.; Millán-Orozco, J.; Ascacio, J. A. et al. Anthelmintic Efficacy of Hydro-Methanolic Extracts of *Larrea tridentata* Against Larvae of *Haemonchus contortus*. *Trop. Anim. Health Product.* **2018,** *50*, 1099–1105. https://doi.org/10.1007/s11250-018-1535-5

García, S.; Alarcon, G.; Rodrıguez, C.; Heredia, N. Extracts of *Acacia farnesiana* and *Artemisia ludoviciana* Inhibit Growth, Enterotoxin Production and Adhesion of *Vibrio cholerae.World J. Microbiol. Biotechnol.* **2006,** *22*, 669–674. https://doi.org/10.1007/s11274-005-9087-z

George, C.; Lochner, A.; Huisamen, B. The Efficacy of *Prosopis glandulosa* as Antidiabetic Treatment in Rat Models of Diabetes and Insulin Resistance. *J. Ethnopharmacol.* **2011,** *137* (1), 298–304. https://doi.org/10.1016/J.JEP.2011.05.023

Gnabre, J.; Bates, R.; Huang, R. C. Creosote Bush Lignans for Human Disease Treatment and Prevention: Perspectives on Combination Therapy. *J. Tradit. Complement. Med.* **2015,** *5* (3), 119–126. https://doi.org/10.1016/J.JTCME.2014.11.024

Gomaa, A.; Hassan, A.; Shehab, G. G.; Abd El-Razek, F.; El-Metwally, E. Effects of Cactus Pear (*Opuntia ficus* Indica) Juice on Oxidative Stress in Diabetic Cataract Rats. *Saudi J. Health Sci.* **2012,** *1* (1), 23. https://doi.org/10.4103/2278-0521.94980

Gonçalves, T. P. R.; Parreira, A. G.; Zanuncio, V. S. dos S.; Farias, K. de S.; da Silva, D. B.; Lima, L. A. R. dos S. Antibacterial and Antioxidant Properties of Flowers from *Tecoma stans* (L.) Juss. ex Kunth (Bignoniaceae). *South Afr. J. Bot.* **2022,** *144*, 156–165. https://doi.org/10.1016/J.SAJB.2021.08.028

Herbal Safety. Wereke. https://www.utep.edu/herbal-safety/hechos-herbarios/hojas-de-datos-a-base-de-hierbas/wereke.html (accessed 23 Jan 2022).

Hernández-García, E.; García, A.; Garza-González, E.; Avalos-Alanís, F. G.; Rivas-Galindo, V. M.; Rodríguez-Rodríguez, J. et al. Chemical Composition of *Acacia farnesiana* (L) Wild

Fruits and Its Activity Against Mycobacterium Tuberculosis and Dysentery Bacteria. *J. Ethnopharmacol.* **2019,** *230*, 74–80. https://doi.org/10.1016/J.JEP.2018.10.031

Hikal, W. M.; Ahl, H. A. H. S.; Kačániová, M. A Review of Antimicrobial Activities of Cactus (*Opuntia ficus-indica*) **2021,** *3* (2), 49–56.

Hwang, S. H.; Kang, I.-J.; Lim, S. S. Antidiabetic Effect of Fresh Nopal (*Opuntia ficus-indica*) in Low-Dose Streptozotocin-Induced Diabetic Rats Fed a High-Fat Diet, 2017. https://doi.org/10.1155/2017/4380721

Jamshidi-Kia, F.; Lorigooini, Z.; Amini-Khoei, H. Medicinal Plants: Past History and Future Perspective. *J. HerbMed Pharmacol.* **2018,** *7* (1), 1–7. https://doi.org/10.15171/jhp.2018.01

Jasso de Rodrguez, D.; Hernández-Catillo, F. D.; Solís-Gaona, S.; Rodríguez Jasso, R. M. *Flourensia cernua* DC: A Plant from Mexican Semiarid Regions with a Broad. *Integrated Pest Management and Pest Control—Current and Future Tactics*, 2012. https://doi.org/10.5772/31783

Juárez-Aragón, M. C.; Del Rocio Moreno-Ramírez, Y.; Guerra-Pérez, A.; Mora-Olivo, A.; Olazarán-Santibáñez, F. E.; Torres-Castillo, J. A. Drying Effects on Phenolics and Free Radical-Scavenging Capacity of Rhus Pachyrrhachis and Rhus Virens Used in Traditional Medicine. *Molecules* **2019,** *24* (13). https://doi.org/10.3390/molecules24132438

Kaur, M.; Kaur, A.; Sharma, R. Pharmacological Actions of *Opuntia ficus indica*: A Review. *J. Appl. Pharm. Sci.* **2012,** *2* (7), 15–18. https://doi.org/10.7324/JAPS.2012.2703

Leem, K.-H.; Kim, M.-G.; Hahm, Y.-T.; Kim, H. K. Hypoglycemic Effect of *Opuntia ficus-indica* var. saboten Is Due to Enhanced Peripheral Glucose Uptake Through Activation of AMPK/p38 MAPK Pathway. *Nutrients,* **2016,** *8* (800). https://doi.org/10.3390/nu8120800

Levario-Gómez, A.; Ávila-Sosa, R.; Gutiérrez-Méndez, N.; López-Malo, A.; Nevárez-Moorillón, G. V. Modeling the Combined Effect of pH, Protein Content, and Mexican Oregano Essential Oil Against Food Spoilage Molds. *Front. Sustain. Food Syst.* **2020,** *4* (March), 1–10. https://doi.org/10.3389/fsufs.2020.00034

Mahajan, M.; Kuiry, R.; Pal, P. K. Understanding the Consequence of Environmental Stress for Accumulation of Secondary Metabolites in Medicinal and Aromatic Plants. *J. Appl. Res. Med. Aromatic Plants* **2020,** *18* (April), 100255. https://doi.org/10.1016/j.jarmap.2020.100255

Malagón, F.; Vázquez, J.; Delgado, G.; Ruiz, A. Antimalaric Effect of an Alcoholic Extract of *Artemisia ludoviciana* Mexicana in a Rodent Malaria Model. *Parassitologia* **1997,** *39* (1), 3–7. Retrieved from https://europepmc.org/article/med/9419840

Medicinales Aromáticas, P.; Martha, R.; Flores, M.; Maria, J. Boletín Latinoamericano y del Caribe de, 2010. http://www.redalyc.org/articulo.oa?id=85615688008

Mohamed, Z. M. S. et al. Antioxidant and Antibacterial Activities of Leaves and Branches Extracts of *Tecoma stans* (L.) Juss. ex Kunth Against Nine Species of Pathogenic Bacteria. *Afr. J. Microbiol. Res.* **2015,** *7* (5), 418–426. https://doi.org/10.5897/ajmr12.2274

Molina-Garza, Z. J.; Bazaldúa-Rodríguez, A. F.; Quintanilla-Licea, R.; Galaviz-Silva, L. Anti-Trypanosoma Cruzi Activity of 10 Medicinal Plants Used in Northeast Mexico. *Acta Tropica* **2014,** *136* (1), 14–18. https://doi.org/10.1016/j.actatropica.2014.04.006

Molina-Salinas, G. M.; Ramos-Guerra, M. C.; Vargas-Villarreal, J.; Mata-Cárdenas, B. D.; Becerril-Montes, P.; Said-Fernández, S. Bactericidal Activity of Organic Extracts from *Flourensia cernua* DC Against Strains of Mycobacterium Tuberculosis. *Arch. Med. Res.* **2006,** *37* (1), 45–49. https://doi.org/10.1016/J.ARCMED.2005.04.010

Molina-Salinass, G. M.; Peña-Rodríguez, L. M.; Mata-Cárdenas, D. B.; Escalante-Erosa, F.; González-Hernández, S.; Torres De La Cruz, V. M. et al. *Flourensia cernua*: Hexane Extracts a Very Active Mycobactericidal Fraction from an Inactive Leaf Decoction

Against Pansensitive and Panresistant Mycobacterium Tuberculosis, 2011. https://doi.org/10.1155/2011/782503

Morales, D. M.; Siles, S.; Mateos, A. L.; Delegación, G. A.; Madero, C. P.; 07738 Mexico, D. F. Identificación de los principales componentes de la raíz de wereque, extraíbles en medio acuoso, por GC-MS. *Avances En Ciencias e Ingeniería* **2012,** *4* (2), 15–21. http://www.exeedu.com/publishing.cl/av_cienc_ing/15

Mtat, D.; Soltane, R.; Chrouda, A.; Dhahri, A.; Pashameah, R. A.; Almulla, N. Antidiabetic Activity ofOpuntia<spp. In: *Opuntia spp.: Chemistry, Bioactivity and Industrial Applications*; Springer: Cham, 2021; pp 483–489. https://doi.org/10.1007/978-3-030-78444-7_22

Naturalista. Choya Guani (*Ibervillea sonorae*). https://www.naturalista.mx/taxa/273781-Ibervillea-sonorae (accessed 21 Jan 2022).

Pagare, S.; Bhatia, M.; Tripathi, N.; Pagare, S.; Bansal, Y. K. Secondary Metabolites of Plants and Their Role: Overview. *Curr. Trends Biotechnol. Pharm.* **2015,** *9* (3), 293–304.

Parra-Naranjo, A.; Delgado-Montemayor, C.; Fraga-López, A.; Castañeda-Corral, G.; Salazar-Aranda, R.; Acevedo-Fernández, J. J.; Waksman, N. Acute Hypoglycemic and Antidiabetic Effect of Teuhetenone a Isolated from *Turnera diffusa. Molecules* **2017,** *22* (4), 1–13. https://doi.org/10.3390/molecules22040599

Petrovska, B. B. Historical Review of Medicinal Plants' Usage. *Pharmacogn. Rev.* **2011,** *6* (11), 1–11. https://doi.org/10.4103/0973-7847.95849

Rahman, A. A.; Samoylenko, V.; Jacob, M. R.; Sahu, R.; Jain, S. K.; Khan, S. I. et al. Antiparasitic and Antimicrobial Indolizidines from the Leaves of *Prosopis glandulosa* var. Glandulosa. *Planta Medica* **2011,** *77* (14), 1639. https://doi.org/10.1055/S-0030-1270906

Ramírez-Moreno, A.; Delgadillo-Guzmán, D.; Bautista-Robles, V.; Marszalek, J. E.; Keita, H.; Kourouma, A. et al. African Journal of Pharmacy and Pharmacology Full Length Research Paper *Jatropha dioica*, an Aztec Plant with Promising Pharmacological Properties: A Systematic Review **2020,** *14* (6), 169–178. https://doi.org/10.5897/AJPP2020.5147

Ramli, S.; Harada, K. I.; Ruangrungsi, N. Antioxidant, Antimicrobial and Cytotoxicity Activities of *Acacia farnesiana* (L.) Willd. Leaves Ethanolic Extract. *Pharmacogn. J.* **2011,** *3* (23), 50–58. https://doi.org/10.5530/pj.2011.23.8

Razo, R. C.; Alvarado, B. E. CAPTOSAN: Tónico Capilar de Sangregado (*Jatropha dioica*). *Jóvenes Investigadores* **2014,** *1* (1), 493–497. http://repositorio.ugto.mx/handle/20.500.12059/2273

Reyes-Jurado, F.; Munguía-Pérez, R.; Cid-Pérez, T. S.; Hernández-Carranza, P.; Ochoa-Velasco, C. E.; Avila-Sosa, R. Inhibitory Effect of Mexican Oregano (Lippia berlandieri Schauer) Essential Oil on *Pseudomonas aeruginosa* and Salmonella Thyphimurium Biofilm Formation. *Front. Sustain. Food Syst.* **2020,** *4* (April), 1–6. https://doi.org/10.3389/fsufs.2020.00036

Reyes-Melo, K. Y.; Galván-Rodrigo, A. A.; Martínez-Olivo, I. E.; Núñez-Mojica, G.; Ávalos-Alanís, F. G.; García, A.; del Rayo Camacho-Corona, M. *Larrea tridentata* and Its Biological Activities. *Curr. Top. Med. Chem.* **2021,** *21* (26), 2352–2364. https://doi.org/10.2174/1568026621666210727170908

Rioja Soto, T. C. Los metabolitos secundarios de las plantas y potencial uso en el manejo de plagas agrícolas en agroecosistemas desérticos. *Idesia (Arica)* **2020,** *38* (1), 3–5. https://doi.org/10.4067/s0718-34292020000100003

Rivera-Ramírez, F.; Escalona-Cardoso, G. N.; Garduño-Siciliano, L.; Galaviz-Hernández, C.; Paniagua-Castro, N. Antiobesity and Hypoglycaemic Effects of Aqueous Extract of *Ibervillea sonorae* in Mice Fed a High-Fat Diet with Fructose. *J. Biomed. Biotechnol.* **2011,** *2011*. https://doi.org/10.1155/2011/968984

Salah, S.; Elshehy, H.; Abdel-Mawla, E.; Agamy, N. Protective Effect of *Opuntia ficus-indica* Against Diabetes in Alloxan-induced Diabetic Rats. *Can. J. Clin. Nutr.* **2020,** *8* (2). https://doi.org/10.14206/canad.j.clin.nutr.2020.02.03

Schmidt, T. J.; Rzeppa, S.; Kaiser, M.; Brun, R. *Larrea tridentata*—Absolute Configuration of Its Epoxylignans and Investigations on Its Antiprotozoal Activity. *Phytochem. Lett.* **2012,** *5* (3), 632–638. https://doi.org/10.1016/J.PHYTOL.2012.06.011

Semotiuk, A. J.; Hernández-Romano, E. F.; Colunga-GarcíaMarín, P.; Salazar-Olivo, L. A.; Ezcurra, E. The Mechanisms of the Myth: *Ibervillea sonorae* Inhibits α-Glucosidase and Stimulates the Secretion of Insulin In Vitro. *J. Herb. Med.* **2020,** *23*, 100354. https://doi.org/10.1016/J.HERMED.2020.100354

Sharma, A.; Angulo-Bejarano, P. I.; Madariaga-Navarrete, A.; Oza, G.; Iqbal, H. M. N.; Cardoso-Taketa, A.; Villarreal, M. L. Multidisciplinary Investigations on Galphimia Glauca: A Mexican Medicinal Plant with Pharmacological Potential. *Molecules* **2018,** *23* (11), 1–22. https://doi.org/10.3390/molecules23112985

Silva-Belmares, Y.; Rivas-Morales, C.; Viveros-Valdez, E.; de la Cruz-Galicia, M. G.; Carranza-Rosales, P. Antimicrobial and Cytotoxic Activities from *Jatropha dioica* Roots. *Pak. J. Biol. Sci.* **2014,** *17* (5), 748–750. https://doi.org/10.3923/PJBS.2014.748.750

Sinagawa-García, S.; Gutiérrez-Díez, A.; Mora-Olivo, A.; Juárez-Aragón, M.; Torres-Castillo, J. Características descriptivas de la raíz de wereke (Ibervillea sonorae Greene) y generalidades bioquímicas de su extracto acuoso Wereke Root (Ibervillea sonorae Greene) Descriptive Characteristics and Biochemical Generalities of Its Aqueous Extract. *Revista Internacional de Botánica Experimental* **2015,** *84*, 358–367.

Sofowora, A.; Ogunbodede, E.; Onayade, A. The Role and Place of Medicinal Plants in the Strategies for Disease Prevention. *Afr. J. Tradit. Complement. Altern. Med.* **2013,** *10* (5), 210. https://doi.org/10.4314/AJTCAM.V10I5.2

Sunita Verma. Phytochemical and Pharmacological Review Study on *Tecoma stans* Linn. *J. Med. Plants Stud.* **2016,** *4* (5), 162–164. https://www.plantsjournal.com/archives/?year=2016&vol=4&issue=5&part=C&ArticleId=432

Süntar, I. Importance of Ethnopharmacological Studies in Drug Discovery: Role of Medicinal Plants. *Phytochem. Rev.* **2020,** *19* (5), 1199–1209. https://doi.org/10.1007/s11101-019-09629-9

Tetali, S. D. Terpenes and Isoprenoids: A Wealth of Compounds for Global Use. *Planta* **2019,** *249* (1), 1–8. https://doi.org/10.1007/s00425-018-3056-x

Tilahun, Y.; Welegerima, G. Pharmacological Potential of Cactus Pear (*Opuntia ficus Indica*): A Review Yemane Tilahun and Gebrekidan Welegerima. *J. Pharmacogn. Phytochem.* **2018,** *7* (3), 1360–1363.

Turner, T.; Ruiz, G.; Gerstel, J.; Langland, J. Characterization of the Antibacterial Activity from Ethanolic Extracts of the Botanical, *Larrea tridentata. BMC Complement. Med. Therap.* **2021,** *21* (1), 1–12. https://doi.org/10.1186/S12906-021-03344-9/TABLES/2

Vergara-galicia, J.; Aguirre-crespo, F.; Tun-suarez, A.; Aguirre, A.; Estrada-carrillo, M.; Jaimes-huerta, I. et al. Acute Hypoglycemic Effect of Ethanolic Extracts from *Marrubium vulgare*. *Phytopharmacology* **2012,** *3* (1), 54–60.

Verma, N.; Shukla, S. Impact of Various Factors Responsible for Fluctuation in Plant Secondary Metabolites. *J. Appl. Res. Med. Aromatic Plants* **2015,** *2* (4), 105–113. https://doi.org/10.1016/j.jarmap.2015.09.002

Vilela, A. E.; González-Paleo, L.; Ravetta, D. A. Metabolismo secundario de plantas leñosas de zonas áridas: Mecanismos de producción, funciones y posibilidades de aprovechamiento. *Ecol. Austral* **2011,** *21* (3), 317–327.

Villanueva, J. R.; Esteban, J. M.; Villanueva, L. R. A Reassessment of the Marrubium Vulgare L. Herb's Potential Role in Diabetes Mellitus Type 2: First Results Guide the Investigation toward New Horizons. *Medicines* **2017,** *4* (3), 57. https://doi.org/10.3390/MEDICINES4030057

Villarreal-García, L. E.; Oranday-Cárdenas, A.; de la Garza-Ramos, M. A.; Rivas-Morales, C.; Verde-Star, M. J.; Gómez-Treviño, J. A.; Torres-de la Cruz, V. Neolignanos de Krameria ramosissima (A. Gray) S. Watson con actividad contra Porphyromonas gingivalis, evaluación citotóxica y mutagénica. *Revista Mexicana de Ciencias Farmaceuticas* **2014,** *45* (2), 69–76.

Welegerima, G.; Zemene, A. Antibacterial Activity of *Opuntia ficus-indica* Skin Fruit Extracts. *Biotechnol. Int.* **2017,** *10* (3), 74–83.

WHO. *Medicina tradicional Informe de la Secretaría. 56ª Asamblea Mundial De La Salud* (Vol. A/56), 2003. http://www.who.int/topics/traditional_medicine/definitions/es/

WHO. *Estrategia de la OMS sobre medicina tradicional 2014–2023*; World Health Organization, 2013.

WHO. *Traditional and Complementary Medicine in Primary Health Care*; World Health Organization, 2018; pp 1–11.

WHO. Resistencia a los antibióticos, 2019. https://www.who.int/es/news-room/fact-sheets/detail/resistencia-a-los-antibióticos (accessed 20 Jan 2019).

WHO. Diabetes, 2020. https://www.who.int/es/news-room/fact-sheets/detail/diabetes (accessed 24 Jan 2022).

Zahedi, S. M.; Karimi, M.; Venditti, A. Plants Adapted to Arid Areas: Specialized Metabolites. *Natural Product Res.* **2021,** *35* (19), 3314–3331. https://doi.org/10.1080/14786419.2019.1689500

Zamilpa, A.; García-Alanís, C.; López-Arellano, M. E.; Hernández-Velázquez, V. M.; Valladares-Cisneros, M. G.; Salinas-Sánchez, D. O.; Mendoza-De Gives, P. In Vitro Nematicidal Effect of *Chenopodium ambrosioides* and *Castela tortuosa* n-Hexane Extracts Against *Haemonchus contortus* (Nematoda) and Their Anthelmintic Effect in Gerbils. *J. Helminthol.* **2019,** *93* (4), 434–439. https://doi.org/10.1017/S0022149X18000433

Zapién-Chavarría, K. A.; Plascencia-Terrazas, A.; Venegas-Ortega, M. G.; Varillas-Torres, M.; Rivera-Chavira, B. E.; Adame-Gallegos, J. R. et al. Susceptibility of Multidrug-Resistant and Biofilm-Forming Uropathogens to *Mexican oregano* Essential Oil. *Antibiotics* **2019,** *8* (4). https://doi.org/10.3390/antibiotics8040186

CHAPTER 6

Plant Resources with Industrial Applications

DAYANE CORREIA GOMES and RUBI VIVEROS CONTRERAS

Centro de Investigaciones Biomédicas, Universidad Veracruzana, Xalapa-Ver, Mexico

ABSTRACT

Arid regions are of utmost importance since they have the capacity to absorb and reflect solar radiation, in addition to keeping the components of the atmosphere in balance. Arid, semi-arid and super-arid regions cover around 41% of the global terrestrial ecosystem, on which 2.5 billion people depend for their survival. 44% of the world's food harvest is produced in these regions even though they have low humidity and adverse climates, with extreme temperatures, low amounts of nutrients in the soil, eroded soils; however, they have some plant species such as the cactus Opuntia spp. which is of great interest since it contains characteristics such as a high amount of fiber, water and its low caloric intake. Another important specimen in these areas for the food sector are walnuts, oregano, other aromatic plants that, due to their ability to produce phytochemicals, they have a high range of flavors, smells and organoleptic and therapeutic qualities, in addition to a diversity of chili peppers among others. Another commercial application is the alcoholic beverage industry, which is the production of tequila, which is carried out by a process where it is extracted from an original plant. from arid areas such as blue agave, in the textile industry the resources from arid areas such as jute, hemp, agave sisal and cotton also stand out, which due to its versatility grows in both humid and arid climates, arid areas. They represent a growing interest for the industry and the economy as their qualities and resources are increasingly known.

Exploration and Valorization of Natural Resources from Arid Zones. Cristobal Noé Aguilar, Norma Paola Meléndez-Renteria, & Karen Nathiely Ramirez-Guzman (Eds.)

6.1 INTRODUCTION

Reduced rainfall, closed depressions are a characteristic of arid zones, and the desertification process diminishes the ecological and biological potential of this region (Alipur et al., 2016). Arid regions are important because they absorb and reflect solar radiation and keep the components of the atmosphere in balance. Arid, semiarid, and super-arid regions cover about 41% of the global terrestrial ecosystem, on which 2.5 billion people depend for their survival. 44% of the global food crop occurs in these regions (Gaur and Squires, 2017).

The vegetation of these areas, however, is not fully exploited and some of its endemic plants are categorized as succulents, because they can store water, which allows them to survive the long periods of drought (Shishoo, 2012). Some of these plants interest industries, one of them is the textile industry. Some of these plants grow in arid zones with cotton used for the production of canvas, tapestry, and clothing (Attala and Elsonbaty, 2021). Another example is the use of sisal for polishing fabrics and reinforcing glass fibers for the production of plastics used in car construction (FAO, 2021; FENG et al., 2022).

As an example, the different applications in the food industry are the nopal, which contains fruits and stalks that are rich in water, fiber, and antioxidants (Galicia-Villanueva et al., 2017). In the tequila industry a variety of cactus species (Agave) are used, such as the *Agave tequilana weber* used for tequila production, the species *A. salmiana, A. cupreata, A. duranguensis, A. fourcroydes, A. angustifolia* y *A. potatorum* used to produce mescal (Mancha Cáceres, 2017). In this chapter, we will talk about the use of some arid zone plants in the textile, food, and tequila industries.

6.2 APPLICATION PLANTS IN THE TEXTILE INDUSTRY

The textile industry is one of the oldest industries in the world, originating in prehistoric times, initially it was limited to India, China, Western Europe, and Turkey, which grew and expanded throughout the world (Antczak et al., 2019; Ranasinghe and Jayasooriya, 2021). Currently, the global textile market is an industry that has a large economic capital, this grows along with the population increase, in 2018 the net valuation of this industry was $920 billion and with an estimate for 2024 of $1230 billion (Mordor Intelligence, 2019; Rahman et al., 2022).

Textile manufacturing is composed of a complex system of sub-ministers, due to its industrial diversity that includes the production of upholstery, clothing, and industrial utensils. The complexity of textile products is due to the variety of substrates, processes, equipment, and different types of raw materials such as natural and synthetic fibers (Vhanbatte, 2022). The natural fibers used in the textile industry are divided into two: natural fibers of plant origin and those of animal origin. Natural fibers of plant origin are cotton, linen, hemp, bamboo, ramie, agave sisal, jute, coconut, and kapok (Attala and Elsonbaty, 2021).

As mentioned before, cotton is one of the most cultivated plants in the world; it is the most used species in the textile industry for the production of towels, woven fabrics, clothing, among others (Furstenberg, 2021). Cotton belongs to the Mallow family (Malvaceae) of the botanical genus Gossypium, can be grown in arid, semiarid, tropical, and subtropical zones, being a versatile plant that adapts to different climates and soil types (FENG et al., 2022; Verma et al., 2022). For its growth, a period of 140–160 days of sunshine and approximately 700–1300 mm of rain is required; however, in periods of little rainfall, its roots grow and extend deep into the soil allowing it to find water sources (Beltrão et al., 1981; Bouchaud et al., 2011).

Among the cotton-producing countries are China with a production of 26.4%, India (20.5%), USA (13.9%), Pakistan (8.5%), Brazil (6.3%), Uzbekistan (4.1%), Australia (3.8%), and Turkey (3.3%). UU (13.9%), Pakistan (8.5%), Brazil (6.3%), Uzbekistan (4.1%), Australia (3.8%), and Turkey (3.3%) (Baydar et al., 2015; Esteve-Turrillas and de la Guardia, 2017; FAOSTAT, 1947). Mexico produces approximately 1,162,603 tons of cotton and its cultivation covers 241,000 hectares, grown in six states—Tamaulipas, Baja California, Sonora, Chihuahua, Coahuila, and Durango—for the national and international textile, cosmetics, and food industries (SADER, 2020).

Another example is the genus Agave from the Agavaceae family, a characteristic plant of arid zones that includes about 136 species, which are native to North America (Mexico). These plants are divided into agaves producing fermented beverages of distilled beverages and fiber (Brown, 2002; Cervantes, 2005). The fiber-producing *Agaves* are *Agave lecheguilla, A. fourcroydes, A. sisalana, A. angustifolia.* A. sisalana fiber is commonly used in the textile industry for polishing fabrics and steel, in the reinforcement of glass fibers used for plastic production, due to its high strength and insulating potential. The use of this fiber in the textile market is growing every day because its residues do not harm the environment and are fully used in the production of different raw materials. The largest cultivations of this fiber are in Angola, Brazil, China, Cuba, Haiti, Indonesia, Kenya, Madagascar, Mozambique, Mexico, South Africa, Tanzania, and Thailand. Brazil produces approximately 120,000 tons and of these 100,000 are exported to the United States, Tanzania (30,000) and Kenya (25,000) both carry between 20,000 and 15,000 tons (Dev et al., 2010; FAO, 2021).

6.3 EDIBLE PLANTS FROM ARID ZONES: THEIR COMMERCIALIZATION

Arid and semiarid systems comprise approximately 41% (approximately 47 million hectares) of the world's ecosystems, of which 20% are considered degraded (Ashfaq et al., 2018; Bateman et al., 2018). These zones are characterized by reduced nutrients and low soil fertility, high temperatures, short rainy periods, and a desert landscape. However, it contains a rich biodiversity and among them are endemic plants of interest to the food industry because they have a notorious nutritional contribution for human and animal consumption (Finlayson et al., 2014; Hunter and Gibbs, 2021).

The consumption of edible plants from the arid system has benefits for the soil and human and animal health. These plants have a high nutritional value, are rich in fiber, water, and bioactive compounds that help prevent the development of some diseases (Nabhan et al., 2020; Whitfield, 2020). In view of the above, as an example we have the native Mexican plant Opuntia spp. (Nopal), a species of cactus known for its fruits (tunas) and stalks (cladodes), are usually consumed as food for their nutritional contribution and as a source of water. The nopal is found in Mexico, Brazil, South Africa, Sicily, Middle East, and North African countries, this plant has about 188 species and several genera (Ciriminna et al., 2019; Nations, 2017; Stintzing and Carle, 2005).

The nopal is normally intended for food, however, it has other applications, from the fruit and cladodes it is possible to produce fodder, alcohol, dyes, soap, pectins, and oils (Sáenz, 2007; Torres-Ponce et al., 2017). The species of the family *Opuntia* spp. may or may not have thorns, are of different sizes and their fruits are different. The prickly pear fruit contains a high content of vitamin C, potassium, calcium, and phosphorus and the chemical composition of Opuntia spp. is 91% water, 0. 66% protein, 0.11% lipids, 5.55 carbohydrates, 1.15% cellulose, 1.58% ash, fiber, phytochemicals and with a low sodium concentration (Becerril et al., 2006; Yahia and Mondragon-Jacobo, 2011).

In Mexico the nopal is commonly consumed, due to its low-calorie content it is starting to be consumed in the United States, Canada, Turkey, and Italy. However, its high humidity index is an issue that reduces its shelf life; however, in recent years, technologies have been developed that allow producing packaging capable of controlling humidity by having an active atmosphere. One of the strategies to increase its shelf life is to dehydrate it and turn it into octopus, thus preserving its nutritional components (Guevara-Arauza et al., 2006; Osorio-Córdoba et al., 2011). Another application of nopal in the food industry is as natural additives through its mucilage, since it is composed of thickeners, lipid substitutes, stabilizers for emulsions, and edible films that increase the shelf life of fresh to processed foods (González-gonzález, 2011; Torres-Ponce et al., 2017).

Other examples are walnuts (*Carya illinoensis and Junglans nigra*), oregano (*Lantana involucrata, Lippia graveolens*), chile piquin (*Capsicum annuum var. Aviculare*), and the mesquite tree (Prosopis glandulosa). Oregano is grown in arid and semi-arid areas; it is a plant with bushy growth reaching between 0.5 and 1.5 meters in height and producing seeds. (Mario and Espinosa, 2012). This plant is commercialized worldwide because its dried leaves are used as condiments and essential oil is extracted from its seeds. Oregano essential oil has antimicrobial and antioxidant activity, which allows it to be used in the food industry in the synthesis of stretch film used for the production of packaging, thus providing food protection (Balan et al., 2021; Ribeiro-Santos et al., 2017a, 2017b).

The mesquite plant belongs to the genus Prosopis, these are plants that grow in arid and semiarid zones, contain sweet fruits composed of 70% to 75% pericarp and 25% to 30% seeds. In former times, the Algarrobo pod was consumed as a staple food for the inhabitants of arid zones, using milling techniques it was possible to produce flour that is destined to the production of bread and cookies. This flour is composed of 40% saccharose, 25% dietary fiber, 7.8% protein, 0.8 to 2% lipids and has no allergenic molecules that are used for the production of gluten-free food with low price in the food market (Felker et al., 2012; Jaimes-Morales et al., 2022). The Algaroba tree produces a gum rich in D-galactose and L-arabinose with small concentrations of D-mannose, D-glucuronate, and D-xylose. This gum has properties that make it better than gum Arabic, because gum arabic has a lower viscosity, thus presenting a potential dietary fiber. This gum has a few applications, one of them is as a film former and increases the shelf life of fruits on the commercial shelf (Mudgil and Barak, 2020).

6.4 TEQUILA INDUSTRY

The tequila industry has achieved great success today, conquering the international market. Mexico has approximately 1380 bottled tequila brands, exported to 120 countries, with an approximate production of 202% (Neri, 2017). Mexico has about 159 species of Agave sp., and of them 119 are endemic, their flowers, fruits, stems, and pendulums are used for food by the Mexican population and from these plants beverages are produced such as tequila, honey water, mescal, and pulque (Martínez--Hernández et al., 2007).

Agave tequilana weber is rich in soluble dietary fiber, and its fructans when fermented by the intestinal microbiota, produce short-chain fatty acids, which stimulates the growth of the *Bifidobacterium* bacterial family (Koenen et al., 2016; Sáyago-Ayerdi et al., 2021). Tequila is produced from the head of the maguey tequila cactus plant (*Agave tequilana weber*). The next step is fermentation using yeast, and finally the distillation of tequila takes place (Bautista-Justo et al., 2001). Tequila production generates the residual bagasse, this residue is composed of fermentable sugars, 21.3% cellulose and 40.5% glucan, which makes it a potential biofuel (Montiel Corona and Razo-Flores, 2018; Perez-Pimienta et al., 2016).

The plants *A. salmiana, A. cupreata, A. duranguensis, A. fourcroydes, A. angustifolia*, and *A. potatorum* are species of Agave, used to produce mescal, a Mexican craft alcoholic drink. The production of mescal is done in four steps: boiling of the stalk, maceration, fermentation, and distillation (DOF—Diario Oficial de la Federación, 2020; Vera-Guzmán et al., 2018). Pulque is a drink produced from the plants *A. atrovirens, A. americana, A. salmiana*, and *A. mapisaga*, like tequila and mescal, it is a Mexican drink made by fermenting honey water (Escalante et al., 2016; Mora-López et al., 2011).

KEYWORDS

- **plant resources**
- **arid zones**
- **industrial applications**
- **biological potential**
- **commercialization**

REFERENCES

Alipur, H.; Zare, M.; Shojaei, S. Assessing the Degradation of Vegetation of Arid Zones Using FAO–UNIP Model (Case Study: Kashan Zone). *Model. Earth Syst. Environ.* **2016,** *2* (4), 1–6. https://doi.org/10.1007/S40808-016-0258-8/TABLES/7

Ashfaq, S.; Zafar, M.; Ahmad, M.; Sultana, S.; Bahadur, S.; Khan, A.; Shah, A. Microscopic Investigations of Palynological Features of Convolvulaceous Species from Arid Zone of Pakistan. *Microsc. Res. Techniq.* **2018,** *81* (2), 228–239. https://doi.org/10.1002/JEMT.22971

Attala, K.; Elsonbaty, A. Advanced Eco-Friendly UV Spectrophotometric Approach for Resolving Overlapped Spectral Signals of Antihypertensive Agents in Their Binary and Tertiary Pharmaceutical Dosage Form. *Spectrochim. Acta—Part A: Mol. Biomol. Spectrosc.* **2021,** *258*, 119855. https://doi.org/10.1016/j.saa.2021.119855

Balan, G. C.; Paulo, A. F. S.; Correa, L. G.; Alvim, I. D.; Ueno, C. T.; Coelho, A. R.; Ströher, G. R.; Yamashita, F.; Sakanaka, L. S.; Shirai, M. A. Production of Wheat Flour/PBAT Active Films Incorporated with Oregano Oil Microparticles and Its Application in Fresh Pastry Conservation. *Food Bioprocess Technol.* **2021,** *14* (8), 1587–1599. https://doi.org/10.1007/S11947-021-02659-2/FIGURES/4

Bateman, A.; Lewandrowski, W.; Stevens, J. C.; Muñoz-Rojas, M. Ecophysiological Indicators to Assess Drought Responses of Arid Zone Native Seedlings in Reconstructed Soils. *Land Degrad. Dev.* **2018,** *29* (4), 984–993. https://doi.org/10.1002/LDR.2660

Bautista-Justo, M.; García-Oropeza, L.; Salcedo-Hernández, R.; Parra-Negrete, L. A. AZÚCARES EN AGAVES (*Agave tequilana* Weber) CULTIVADOS EN EL ESTADO DE GUANAJUATO. *Acta Universitaria* **2001,** *11* (1), 33–38. http://www.redalyc.org/articulo.oa?id=41611105

Baydar, G.; Ciliz, N.; Mammadov, A. Life Cycle Assessment of Cotton Textile Products in Turkey. *Resour. Conserv. Recycl.* **2015,** *104*, 213–223. https://doi.org/10.1016/J.RESCONREC.2015.08.007

Becerril, A.; Valdivia, P.; Beatriz, C.; Mexicana, R. F.; Fitogenética, S. M. De, Revistas, R. De, América, C. De, Aguilar, G.; Beatriz, C.; Valdivia, P. Physiological Alterations Induced By Drought Stress on Prickly Pear (*Opuntia ficus-indica*). *Revista Fitotecnia Mexicana* **2006,** *29* (3), 231–237.

Beltrão, N. E. de M.; Vale, L. S. do, Marques, L. F.; Cardoso, G. D.; Silva, F. V. de F.; Araújo, W. P. de. *O Cultivo Do Algodão Orgânico No Semi-Árido the Organic Cotton Cultivation in Brazilian Semi-Arid,* 1981; pp 8–13.

Bouchaud, C.; Tengberg, M.; Prà, P. D. Cotton Cultivation and Textile Production in the Arabian Peninsula During Antiquity; The Evidence from Madâ'in Sâlih (Saudi Arabia) and Qal'at al-Bahrain (Bahrain). *Veg. Hist. Archaeobot.* **2011,** *20* (5), 405–417. https://doi.org/10.1007/s00334-011-0296-0

Brown, K. *Agave sisalana* Perrine. *Wildland Weeds* **2002,** *5*, 18–21.

Cervantes, M. Plantas de importancia económica en zonas áridas y semiáridas de Mexico. *X Encuentro de Geógrafos de América Latina* 2005; pp 3388–3407. http://www.observatoriogeograficoamericalatina.org.mx/egal10/Procesosambientales/Usoderecursos/08.pdf

Ciriminna, R.; Chavarría-Hernández, N.; Rodríguez-Hernández, A. I.; Pagliaro, M. Toward Unfolding the Bioeconomy of Nopal (*Opuntia* spp.). *Biofuels Bioprod. Biorefining* **2019,** *13* (6), 1417–1427. https://doi.org/10.1002/BBB.2018

Dev, V. R. G.; Venugopal, J. R.; Kumar, T. S.; Miranda, L. R.; Ramakrishna, S. *Agave sisalana*, a Biosorbent for the Adsorption of Reactive Red 120 from Aqueous Solution. *J. Text. Inst.* **2010,** *101* (5), 414–422. https://doi.org/10.1080/00405000802445289

DOF—Diario Oficial de la Federación. *NORMA Oficial Mexicana NOM-070-SCFI-***2016,** *Bebidas alcohólicas-Mezcal-Especificaciones*. NORMA Oficial Mexicana NOM-070-SCFI-**2016,** Bebidas Alcohólicas-Mezcal-Especificaciones, 2020. https://dof.gob.mx/nota_detalle.php?codigo=5472787&fecha=23/02/2017#gsc.tab=0

Escalante, A.; López Soto, D. R.; Velázquez Gutiérrez, J. E.; Giles-Gómez, M.; Bolívar, F.; López-Munguía, A. Pulque, a Traditional Mexican Alcoholic Fermented Beverage: Historical, Microbiological, and Technical Aspects. *Front. Microbiol.* **2016,** *7*, 1026. https://doi.org/10.3389/FMICB.2016.01026

Esteve-Turrillas, F. A.; de la Guardia, M. Environmental Impact of Recover Cotton in Textile Industry. *Resour. Conserv. Recycl.* **2017,** *116*, 107–115. https://doi.org/10.1016/J.RESCONREC.2016.09.034

FAO. *Fibers of the Future: Sisal*, 2021. https://www.fao.org/economic/futurefibres/fibres/sisal/en/

FAOSTAT. Food and Agriculture Organization of the United Nations. *Int. Org.* **1947,** *1* (2), 350–353. https://doi.org/10.1017/S0020818300006160

Felker, P.; Takeoka, G.; Dao, L. Pod Mesocarp Flour of North and South American Species of Leguminous Tree Prosopis (Mesquite): Composition and Food Applications. **2012,** *29* (1), 49–66. https://doi.org/10.1080/87559129.2012.692139

Feng, L.; Chi, B. j.; Dong, H. z. Cotton Cultivation Technology with Chinese Characteristics Has Driven the 70-Year Development of Cotton Production in China. *J. Integr. Agric.* **2022,** *21* (3), 597–609. https://doi.org/10.1016/S2095-3119 (20)63457-8

Finlayson, B. L.; Peel, M. C.; McMahon, T. A. Climate and Rivers☆. *Reference Module in Earth Systems and Environmental Sciences*, 2014. https://doi.org/10.1016/B978-0-12-409548-9.09438-0

Furstenberg, D. Von. What Is Cotton? A Complete Guide to the History, Characteristics, and Uses of Cotton. *Articles* 2021. https://books.google.com.mx/books?hl=es&lr=&id=VsBQAwAAQBAJ&oi=fnd&pg=PP1&dq=cotton&ots=T2bjH5e1ek&sig=eAK4tWvTajLycIWecNE_OvXPaV4&redir_esc=y#v=onepage&q=cotton&f=false

Galicia-Villanueva, S.; Escamilla-García, P. E.; Alvarado-Raya, H.; Aquino-González, L. V.; Serna-Álvarez, H.; Hernández-Cruz, L. M. Plantación experimental de nopal para evaluación de sistemas de fertilización y extracción de mucílago. *Revista Mexicana de Ciencias Agrícolas* **2017,** *8* (5), 1087–1099. https://doi.org/10.29312/remexca.v8i5.110

Gaur, M. K.; Squires, V. R. Geographic Extent and Characteristics of the World's Arid Zones and Their Peoples. *Clim. Variab. Impacts Land Use Livelihoods Drylands* **2017,** 3–20. https://doi.org/10.1007/978-3-319-56681-8_1/COVER/

González-gonzález, L. R. Desarrollo y evaluación de una película comestible obtenida del mucílago de nopal (Opuntia ficus-indica) utilizada para reducir la tasa de respiración de nopal verdura. *Congreso Iberoamericano de Ingeniería de Alimentos* **2011,** *3*, 131–138.

Guevara-Arauza, J. C.; Yahia, E. M.; Cedeño, L.; Tijskens, L. M. M. Modeling the Effects of Temperature and Relative Humidity on Gas Exchange of Prickly Pear Cactus (*Opuntia* spp.) Stems. *LWT—Food Sci. Technol.* **2006,** *39* (7), 796–805. https://doi.org/10.1016/J.LWT.2005.05.007

Hunter, E. A.; Gibbs, J. P. Habitats. *Galapagos Giant Tortoises* **2021,** 281–298. https://doi.org/10.1016/B978-0-12-817554-5.00001-0

Jaimes-Morales, J.; Marrugo-Ligardo, Y. A.; Acevedo-Correa, D. Analysis of Mesquite (*Prosopis juliflora*) Protein Concentrate for Possible Use as Supplementary Protein. *Int. J. Food Sci.* **2022,** *2022*. https://doi.org/10.1155/2022/7621818

Koenen, M. E.; Cruz Rubio, J. M.; Mueller, M.; Venema, K. The Effect of Agave Fructan Products on the Activity and Composition of the Microbiota Determined in a Dynamic In Vitro Model of the Human Proximal Large Intestine. *J. Funct. Foods* **2016,** *22*, 201–210. https://doi.org/10.1016/J.JFF.2016.01.018

Mancha Cáceres, O. I. Divided Spirits. Tequila, Mezcal, and The Politics of Production, de Sarah Bowen. *AIBR. Revista de Antropología Iberoamericana* **2017,** *12* (2), 246–250. https://doi.org/10.11156/aibr.v12i2.68162

Mario, L.; Espinosa, T. *Cultivo de Orégano, opción productiva para las zonas semidesérticas Ecología y distribución* 2012.

Martínez-Hernández, A.; Pastrana-Chávez, J.; Sánchez-Villarreal, A.; Lara-Reyna, J.; Herrera-Estrella, L.; Herrera-Estrella, A.; de la Vega, O.; Simpson-Williamson, J. Genómica de Agave tequilana: Identificación de genes útiles para la industria tequilera y el desarrollo de usos alternativos del agave. Jal. Sep, January, 2007.

Montiel Corona, V.; Razo-Flores, E. Continuous Hydrogen and Methane Production from *Agave tequilana* Bagasse Hydrolysate by Sequential Process to Maximize Energy Recovery Efficiency. *Bioresour. Technol.* **2018,** *249*, 334–341. https://doi.org/10.1016/J.BIORTECH.2017.10.032

Mora-López, J. L.; Reyes-Agüero, J. A.; Flores-Flores, J. L.; Peña-Valdivia, C. B.; Aguirre-Rivera, J. R. Variación morfológica y humanización de la sección Salmianae del género Agave. *Agrociencia* **2011,** *45* (4), 465–477. http://www.scielo.org.mx/scielo.php?script=sci_arttext&pid=S1405-31952011000400006

Mordor Intelligence. *Global Textile Industry—Growth, Trends, and Forecast (2020–2025)*, 2019. https://www.mordorintelligence.com/industry-reports/global-textile-industry–growth-trends-and-forecast-2019—2024

Mudgil, D.; Barak, S. Mesquite Gum (*Prosopis gum*): Structure, Properties & Applications—A Review. *Int. J. Biol. Macromol.* **2020,** *159*, 1094–1102. https://doi.org/10.1016/J.IJBIOMAC.2020.05.153

Nabhan, G. P.; Riordan, E. C.; Monti, L.; Rea, A. M.; Wilder, B. T.; Ezcurra, E.; Mabry, J. B.; Aronson, J.; Barron-Gafford, G. A.; García, J. M.; Búrquez, A.; Crews, T. E.; Mirocha, P.; Hodgson, W. C. An Aridamerican Model for Agriculture in a Hotter, Water Scarce World. *Plants People Planet* **2020,** *2* (6), 627–639. https://doi.org/10.1002/PPP3.10129

Nations, U. *From Thorny to Tasty: UN Agriculture Agency Looking at Cactus as Climate Resilient Food*, 2017. https://news.un.org/en/story/2017/11/637911

Neri, J. C. *Estrategias básicas de competitividad* 2017.

Osorio-Córdoba, J.; Pelayo-Zaldívar, C.; Verde-Calvo, J. R.; Ponce-Valadez, M.; Díaz de león-Sánchez, F.; Bosquez-Molina, E.; Rodríguez-Huezo, E. Conservación de nopal verdura "Milpa Alta" (*Opuntia ficus mdica* mill.) desespinado en envases con atmosfera modificada. *Revista Mexicana de Ingeniera Qumica* **2011,** *10* (1), 93–104. http://www.scielo.org.mx/scielo.php?script=sci_arttext&pid=S1665-27382011000100010

Perez-Pimienta, J. A.; Poggi-Varaldo, H. M.; Ponce-Noyola, T.; Ramos-Valdivia, A. C.; Chavez-Carvayar, J. A.; Stavila, V.; Simmons, B. A. Fractional Pretreatment of Raw and Calcium Oxalate-Extracted Agave Bagasse Using Ionic Liquid and Alkaline Hydrogen Peroxide. *Biomass Bioenergy* **2016,** *91*, 48–55. https://doi.org/10.1016/J.BIOMBIOE.2016.05.001

Rahman, S. S.; Siddiqua, S.; Cherian, C. Sustainable Applications of Textile Waste Fiber in the Construction and Geotechnical Industries: A Retrospect. In *Cleaner Engineering and Technology*, Vol. 6; 2016. https://doi.org/10.1016/j.clet.2022.100420

Ribeiro-Santos, R.; Andrade, M.; Sanches-Silva, A. Application of Encapsulated Essential Oils as Antimicrobial Agents in Food Packaging. *Curr. Opin. Food Sci.* **2017,** *14*, 78–84. https://doi.org/10.1016/J.COFS.2017.01.012

Ribeiro-Santos, R.; Andrade, M.; Sanches-Silva, A.; de Melo, N. R. Essential Oils for Food Application: Natural Substances with Established Biological Activities. *Food Bioprocess Technol.* **2017,** *11* (1), 43–71. https://doi.org/10.1007/S11947-017-1948-6

SADER. *El algodón de Mexico, fibra suave y cultivo generoso*, 2020. https://www.gob.mx/agricultura/articulos/el-algodon-de-mexico-fibra-suave-y-cultivo-generoso?idiom=es

Sáenz, C. *Utilización agroindustrial del nopal—Carmen Sáenz, Horst Berger—Google Libros*, 2007. https://books.google.com.br/books?hl=es&lr=&id=llaxlnmJjFoC&oi=fnd&pg=PR10&dq=Utilización+agroindustrial+del+nopal.+Boletín+de+Servicios+Agrícolas+de+la+FAO&ots=AfH1BZCm1u&sig=G8cZW3TQXQmGKWa-XvMQiF6q1sM#v=onepage&q=Utilización agroindustrial del nopal. B

Sáyago-Ayerdi, S. G.; Venema, K.; Tabernero, M.; Sarriá, B.; Bravo, L.; Mateos, R. Bioconversion of Polyphenols and Organic Acids by Gut Microbiota of Predigested *Hibiscus sabdariffa* L. Calyces and Agave (*A. tequilana* Weber) Fructans Assessed in a Dynamic In Vitro Model (TIM-2) of the Human Colon. *Food Res. Int.* **2021,** *143*, 110301. https://doi.org/10.1016/J.FOODRES.2021.110301

Shishoo, R. Introduction: Trends in the Global Textile Industry. In *The Global Textile and Clothing Industry: Technological Advances and Future Challenges*; **2012**; pp 1–7. https://doi.org/10.1533/9780857095626.1

Stintzing, F. C.; Carle, R. Cactus Stems (*Opuntia* spp.): A Review on Their Chemistry, Technology, and Uses. *Mol. Nutr. Food Res.* **2005,** *49* (2), 175–194. https://doi.org/10.1002/MNFR.200400071

Torres-Ponce, R. L.; Morales-Corral, D.; Ballinas-Casarrubias, M. de L.; Nevárez- Moorillón, G. V. El nopal: planta del semidesierto con aplicaciones en farmacia, alimentos y nutrición animal. *Revista Mexicana de Ciencias Agrícolas* **2017,** *6* (5), 1129–1142. https://doi.org/10.29312/remexca.v6i5.604

Vera-Guzmán, A. M.; Guzmán-Gerónimo, R. I.; López, M. G.; Chávez-Servia, J. L. Volatile Compound Profiles in Mezcal Spirits as Influenced by Agave Species and Production Processes. *Beverages* **2018,** *4* (1), 9. https://doi.org/10.3390/BEVERAGES4010009

Verma, P.; Hiremani, N. S.; Gawande, S. P.; Sain, S. K.; Nagrale, D. T.; Narkhedkar, N. G.; Prasad, Y. G. Modulation of Plant Growth and Antioxidative Defense System Through Endophyte Biopriming in Cotton (*Gossypium* spp.) and Non-Host Crops. *Heliyon* **2022,** *8* (5), e09487. https://doi.org/10.1016/J.HELIYON.2022.E09487

Vhanbatte, S. B. Background and Type of Textile Materials Processed in Textile Industry. *Appl. Biotechnol. Sustain. Text. Product.* **2022,** 1–22. https://doi.org/10.1016/B978-0-323-85651-5.00004-9

Whitfield, M. *Eating in Aridamerica: The Desert Food of the Future*. https://www.newphytologist.org/blog/eating-in-aridamerica-the-desert-food-of-the-future/

Yahia, E. M.; Mondragon-Jacobo, C. Nutritional Components and Anti-Oxidant Capacity of Ten Cultivars and Lines of Cactus Pear Fruit (*Opuntia* spp.). *Food Res. Int.* **2011,** *44* (7), 2311–2318. https://doi.org/10.1016/J.FOODRES.2011.02.042

CHAPTER 7

Agriculture in the Arid Zones

DIANA URESTI DURAN[1], HÉCTOR DANIEL INURRETA AGUIRRE[1], EDUARDO ALBERTO LARA REIMERS[2], DORA ALICIA GARCÍA GARCÍA[1], and NORMA PAOLA MELÉNDEZ RENTERÍA[3]

[1]National Institute of Forestry, Agricultural and Livestock Research, Saltillo, Mexico

[2]Departamento de Forestal in Universidad Autonoma Agraria Antonio Narro

[2]CICBEC, Autonomous University of Coahuila, Cuatrocienegas, Mexico

ABSTRACT

Mexico's drylands occupy about 101.5 million hectares, a little more than half of the Mexican territory. Of this surface, arid zones represent 15.7% and about 80% of the soils are derived from calcareous materials.

It is usual in arid zones to have the wind and water flow as the main mechanisms of sediment transport. As soil cohesion increases the quantity and quality of soil, organic matter improves. These conditions cause the development of low palatability species and deep roots, thus contributing more to the accumulation of minerals rather than to the rapid mineralization that is characteristic of these zones.

Agriculture in arid zones has rainfall as a limiting factor; however, the water stored underground and available to plants is the best parameter to indicate success in crop yields than rainfall itselfs. These species are a primary component of agroecosystems in arid zones and directly and indirectly, support both the local economy and livelihood. In most of the arid and semiarid zones of Mexico, crop production is based on a combination of hydraulic and soil conservation practices. In addition, integrated management

Exploration and Valorization of Natural Resources from Arid Zones. Cristobal Noé Aguilar, Norma Paola Meléndez-Renteria, & Karen Nathiely Ramirez-Guzman (Eds.)

practices (soil, water, flora, and wildlife) constitute viable projects to continue improving agriculture and thus the standard of living of producers. However, most are not domesticated and grow wild around communities. Many of these species do not represent an economic value, but their value is intangible for the community and the survival of the locals.

It is necessary to develop research projects in these areas that allow for sustainable environmental development and economic development of the inhabitants.

7.1 INTRODUCTION

7.1.1 BIODIVERSITY IN ARID AND SEMIARID ZONES

A quarter of the world's total land surface corresponds to arid and semiarid zones. In America, arid and semiarid zones occupy large extensions; it is estimated that one-third of North America's surface is arid and semiarid zones (Mazuela Aguila, 2013).

Approximately 806,633.44 km^2 equivalent to 41% of the Mexican territory is arid or semiarid, in which an estimated 6000 useful plant species can be used. These areas are mostly inhabited by rural populations and harbor a large number of endemic species (about 50% of the useful flora) and as many threatened ones (González Medrano, 2012; Villaseñor, 2012; Solano-Picazo, 2018).

The largest arid zone in Mexico is the Chihuahuan Desert, the largest in North America, with a surface area of 507,000 km^2, shared between Mexico and the United States. In Mexico, most of its territory covers the states of Chihuahua and Coahuila and partially the states of San Luis Potosí, Zacatecas, Nuevo León, Tamaulipas, Durango, Guanajuato, Hidalgo, and Querétaro. The Chihuahuan Desert is considered one of the 10 most important ecoregions in North America; it is the most outstanding in terms of diversity, and one of the 37 worldwide priority conservation areas (Hernández, 2004).

The life forms that considered characteristics of the desert of Chihuahua are shrubs and semi- shrubs. There is no evidence of winter herbaceous species, germination, or flowering during the winter, and most of the plants preserve their complete leaves, or at least partially during this period. Tree species are scarcely present and are confined to the proximity of water sources or rocky slopes; succulent stem species are abundant, but not conspicuous. Agostadero grassland and scrub are common. The Fabaceae and Asteraceae families are the ones that have more types of species.

Despite the environmental conditions, the Chihuahuan Desert has a great diversity of species such as microphyll desert scrubland, rosette desert scrubland, crasicaule desert scrubland, oak groves, grasslands, riparian vegetation, and piñon pine forests. It also has a wide variety of Cactaceae species unique in the world. The number of species is approximately 324, besides, it also contains five hybrids; all of these are distributed in 39 genera. Endemic plants constitute 43.6 of the total population of plant species, distributed in 17 genera and 229 species of cacti (69.8% of the total).

There are other arid and semiarid regions in Mexico, such as the Tamaulipas semiarid region located in northeastern Tamaulipas, northern Nuevo León, and northeastern Coahuila. This is known as the Tamaulipeca semiarid zone, here the thickets are larger and the entry of moisture through the Gulf of Mexico is greater. The Tamaulipas scrub has a higher average height than other scrubs of the country, such as the microphyll scrub of Baja California, Sonora, and Chihuahua.

In the center of Mexico, the Hidalgo Semiarid Zone is located in the states of Querétaro and Hidalgo. Finally, located in Puebla and Oaxaca is the Poblano-Oaxacan Semiarid Zone (González Medrano, 2012). These arid regions have been occupied by different ethnic groups since the arrival of the Spanish, most of the groups that inhabited these regions were mainly nomadic or semi-nomadic, who moved from the center to the north of the country at different times of the year.

In one way or another, the different inhabitants of the semi-desert satisfied their needs with the natural resources that the different regions have provided throughout history.

7.1.2 CHARACTERISTICS OF ARID AND SEMIARID ZONES

7.1.2.1 SOIL

Mexico's drylands occupy about 101.5 million hectares, a little more than half of the Mexican territory. Of this surface, arid zones represent 15.7%, semiarid zones 58%, and the remaining 26.3% correspond to dry subhumid zones, about 80% of the soils are derived from calcareous materials (Sutton, 2000). Figure 7.1 shows the arid, semiarid, and subhumid zones present in the country.

Soils in arid and semi-arid zones occur on plains or low slope surfaces, the chemical properties of these soils are strongly influenced by the properties of the parent rock from which they originated. Poorly drained areas tend to be high in salts, mainly chlorides, and sulfates, and this hinders water penetration and movement through the soil profile. Even the rain itself can hinder water penetration, due to the shock produced by the drops on the soil surface, which compacts the soil and reduces its porosity and in turn reduces water penetration through the soil. In addition, rain disperses the surface layer of the soil, causing erosion.

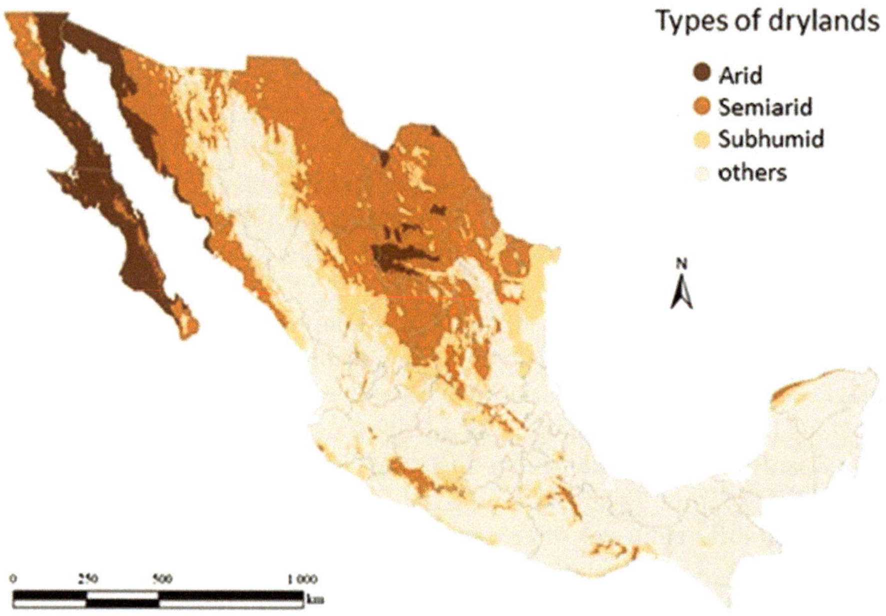

FIGURE 7.1 Distribution of drylands in Mexico.

Source: AUCh. Update of the delimitation of arid, semiarid, and subhumid zones of Mexico.

The soils of the arid and semiarid zones are those in which their formation has been conditioned by aridity, show little development, and are mostly calcareous soils with some leptosols in the mountainous zones. In general, it can be said that the soils of the arid and semiarid zones of Mexico are grouped into lithosols, yermosols, xerosols, solonchak, and planosols, latter of which can be located in plains.

Evaporation of water from the soil and groundwater, which is transported to the surface by capillarity, produces a wide variety of crusts on the rocks. The superficial layer of sandstone is formed by the hard crust originating from the deposition of carbonates. In desert soils (aridisols) the precipitation of calcium carbonate often forms a hard layer, the petrocalcic horizon (caliche). When this caliche layer is eroded, it gives rise to reliefs similar to cliffs, mesas, and hills, witnesses of the eroded sedimentary strata; crusts formed by gypsum are also common (Godínez Alvarez, 1998; Clasificación fisonómica de la vegetación de la Sierra de Catorce, 2003). The most common soils are alluvial soils, limestone rocks, and soils of igneous nature.

Usually, in arid zones, soils are closely integrated between nutrient cycling and the development of organisms. In the highlands, the soils are usually poor and exposed due to the scarce vegetation cover. While in the gentle and prolonged slopes the soil accumulates due to weathering and transport of materials, which develops vegetation across these gradients (Ugalde Ávila et al., 2009). Likewise, the concentration of soil nutrients changes along these gradients, an example being phosphorus, which is low in the higher elevations and relatively abundant in the lower elevations.

7.1.2.2 LANDSCAPE AND SOIL PROCESS

It is usual in arid zones to have the wind and water flow as the main mechanisms of sediment transport. As soil cohesion increases the quantity and quality of soil, organic matter improves. These conditions cause the development of low palatability species and deep roots, thus contributing more to the accumulation of minerals rather than to the rapid mineralization that is characteristic of these zones (van Noordwijk and Ong, 1999).

Despite the adverse conditions of the arid zones and the Chihuahuan desert, its landscapes and views are of unparalleled beauty. Both humans and animals have given in to the harsh living conditions found in all desertic regions of the world. However, we have learned to understand and adjust these same regions in order to obtain not only economical benefits, but also to be able to admire these regions.

Finally, many soils are rich in microorganisms such as bacteria, algae, and some fungi. Some soils contain Azotobacter, which fixes nitrogen in the soil.

7.1.2.3 CLIMATE

The concept of aridity that defines arid zones is deficient rainfall and atmospheric humidity and in which potential evaporation is much greater than annual rainfall. Also, rainfall in arid and semiarid zones always can be irregular, untimely, and in some cases, seasonal.

World media annual rainfall is 840 mm. In Mexico, the average annual rainfall in arid and semiarid zones ranges from 250 to 780 mm (González Medrano, 2012; Comisión Nacional de Zonas Áridas; Instituto Nacional de Ecología, 1994).

The areas that should be considered semiarid, sub-humid, or humid are those with more than 400 mm, and those that receive less than 100 mm should be considered a desert.

From the agricultural point of view, arid areas are those in which crops have low yields, specifically cereals, due to moisture deficiency; meanwhile, arid areas are those in which it is impossible to obtain crops with good yields unless they have irrigation systems.

One of the most outstanding characteristics of the arid and semiarid regions is the lack of humidity at surface level. Not only is the soil lacking in humidity, but also the air, due to the high rate of evaporation, and since evaporation is greater than precipitation, this leads to a continuous loss of humidity. In some arid areas, precipitation vaporizes before it even reaches the soil, making it practically impossible to collect water at the surface level. Another characteristic that makes the arid climate stand out is the contrast between heat and cold that can occur during the seasons, ranging from very cold winters to very hot summers. The rainy season is 65 to 80% during the summer months from July to September with an average annual rainfall ranging from 175 mm in the valleys to 300–400 mm in the high mountains. The dry season is from January to May.

One of the reasons for the aridity in northern Mexico is due to the position between 14° south latitude and 32° north latitude, and is affected by two anticyclonic cells, the Atlantic and the North Pacific, which fall in northern Mexico and southern states causing aridity in these areas. These anticyclonic conditions are formed due to the formation of high-pressure bands and descending air currents (from high altitudes to the earth's surface); the air is heated by compression (1°C per 100 m) so that when it reaches the earth's surface it is hot and dry and incapable of producing precipitation.

Mexico's position on the planet and the particular orography of northern Mexico is another factor that determines the type of level of aridity. The mountain massifs perpendicular to the direction of the winds blowing from the sea are like a barrier that prevents the passage of humid winds, these winds discharge their moisture as they rise and cool down the mountains, or the winds can also be diverted elsewhere taking the moisture with them (González Medrano, 2012).

7.1.2.4 WATER

Typically, in many arid and semiarid zones, the prospects of increasing crop production depend on irrigation systems and therefore on the availability of water, as it happens with almost any human activity.

In arid and semiarid zones, the average annual rainfall is between 350 and 600 mm, the average annual rainfall is less than the maximum annual potential evaporation, evidencing a water deficit. In addition, the distribution of rainfall is highly erratic, and few torrential events occur. This limits the development of agricultural activities.

The inhabitants of arid and semiarid zones have water as their most limiting resource, for both crop production and consumption. Currently, there are specialized techniques that allow them to collect rainwater from roofs. This type of system, together with storage and purification, allows them to have drinking water for consumption throughout the year (Tarango Arámbula, 2005).

It should be mentioned that several authors such as López Martínez (López Martínez and Guerrero Alvarado) and Goos et al. (1984) mention that rainfall is not always the best parameter to be used in the characterization of areas in terms of water availability. They consider that the water stored in the soil available to the plants is a better parameter to be used as an element to predict the success or failure of the crop.

Currently, at least 50% of the drinking water used by the world population is from a groundwater, source with just a little less percentage used for agricultural activities, 43% to be more accurate (Goos et al., 1984). This source of water is fundamental in the distribution of water, and it plays a key role in mantaining the ecology of both arid and semiarid zones. One of the reasons for this is that, due to the presence of both low precipitation and also high temperatures, the availability of water at the surface level is low at best, and nonexistent at worst.

In most of the places in the world, the amount of freshwater that is available for human consumption has diminished, making it more difficult to continue to maintain the daily activities of the people. Due to this reason, the amount of groundwater extraction has risen in recent years, particularly in this type of regions like the arid and semiarid regions, with the logic conclusion that the storage of this source of water has diminished considerably in the last decades.

The level of groundwater can be naturally recovered if enough time is provided for the groundwater to recharge, but with the continued increase in demand for water, not only the freshwater at surface level will decrease, but it is practically impossible for the groundwater to recover enough to maintain the levels that used to have.

7.1.2.5 VEGETATION

Most deserts have some characteristics in common, such as the spaced and open covers. The greatest proportion of useful plants in Mexico corresponds to herbs, trees, and shrubs (Granados-Sánchez et al., 2011). These plants can be used for a wide variety of uses, from the creation of tools to the preparation of traditional medicines, as well as their use in the preparation of traditional dishes.

In arid and semiarid zones, only specialized plants (xerophytic plants) that are adapted to the adverse conditions of arid and semiarid zones can develop. The degree to which a plant can tolerate rain deficit is given by several mechanisms, by which plants can withstand drought; these mechanisms are morphological and physiological.

Among the morphological characteristics, we find that desert plants usually have a highly developed root system, which is wide and deep; they have water-storing tissues; the size of the plants is relatively small and they are compact in the area (microphyllous vegetation); they have protective coverings with thick cuticles, hairs, or waxes; photosynthesizing stems and stomata located in depressions, crevices, furrows, etc.; storage of water reserves; reduction of cell size and tolerance of the tissues to a reduced amount of water.

Some physiological characteristics are high osmotic pressure, which allows them to extract moisture from the soil; growth adapted to the seasons when water is available; reduction of the vegetative period to very short periods; loss of foliage or all aerial organs during the dry season; stomatal closure that allows them to regulate transpiration and water loss; capacity of the protoplasm to subsist in the anhydrobiosis state, with the ability to recover when moisture is available.

The determining mechanism for the survival of a plant in arid zones is drought resistance, since it would be impossible for a plant that requires a high amount of water to survive in a region deprived of such a liquid. It is possible to classify the plants that have adapted to survive in such environments into two groups:

a) Plants that escape drought. These plants are ephemeral, their life cycle is short, they usually develop and finish their life cycle before they present severe water deficiencies, they do not present morphological or physiological mechanisms resistant to drought.
b) Drought-resistant plants. They maintain high water potentials, even when exposed to external water deficiencies. And they can be both water spenders and water savers. Water spenders can lose up to five times their weight per day, while water savers have morphological and physiological characteristics that allow them to withstand extreme droughts (González Medrano, 2012).

7.1.2.6 ROOT, LEAF, AND SEED ADAPTATION

In deserts, most of the biomass is not possible to observe at the surface, since they usually live below ground. Roots: as there is not enough water available, plants develop their root system as aridity increases. These are usually very extensive and superficial. Three strata can be distinguished in the soil profile in which the availability and retention of water are differentiated. In the superficial layer from 0 to 20 cm, being on top, it is quickly saturated but also evaporates quickly as it is more exposed to the environment, plants that can absorb water quickly, such as annuals or succulents, develop. The intermediate layer from 20 cm to 2 m contains less water, but at the same time there is less evaporation; this water is available for a longer time, shrubs and perennial grasses develop. The deep layer is greater than 2 m, in this layer the water is located at a greater depth, as phreatic mantles, the water is usually abundant and remains. Shrubs and trees develop that remain green and with deep root systems. This type of vegetation is specific and is called phreatophytes.

The leaves of xerophytic plants have very characteristic features, they are usually very small and even nonexistent by reducing their size to the maximum while they are protected with a strong cuticle; this causes a smaller transpiring surface and little risk of suffering excessive heating. In many plants the leaves are usually replaced by thorns; they are also usually fleshy which allows them to keep the humidity inside.

The abundance and distribution of seeds are very important in arid and semiarid zones since they are the structure of plant communities and are also a source of food for wildlife. The seeds are generally found within 5 cm deep in the soil, where they are usually preyed upon by ants or rodents such as rabbits, birds, or lizards. Seeds have developed mechanisms to subsist such as the presence of cyanoglucosides that function as a defense against predators. Besides, seeds, as an adaptive strategy, delay their germination. If they germinate immediately in response to favorable conditions, they are subject to greater competition with other individuals that respond immediately to the availability of water (González Medrano, 2012).

The arid zones of Mexico have a wide environmental and biological variation ranging from xerophytic vegetation to grasslands, oak woodlands, shrublands, and low deciduous forests (Rzedowski, 1978). This vegetation is susceptible to microclimatic variations, soil type, land use, the incidence of fire, etc. In the extensive arid and semiarid plains, they are intermingled with non-desert areas and mix xerophytic vegetation with more mesic vegetation, in which pine, juniper, and oak forests can be found. Topography and the presence of streams determine the variation and distribution of plant species in the arid zones. On the slopes of the hillsides the xerophytic scrub predominates, the dominant woody species vary from one site to another.

The Chihuahuan Desert in the north of Mexico is considered one of the most species-rich dry regions in the world (Sutton, 2000; Wake and Morafka, 1978; Hoyt, 2002). In the arid and semiarid zones, it is possible to find several endemic species from the *Cactaceae* family (Hernández et al., 1983).

The predominant vegetation in the Chihuahuan Desert is classified as microphilous desert scrub, desert rosetophytic material, and desert crassicaule material. The microphilous desert scrub is distinguished by small-leaf shrubs such as *Larrea tridentata* and is typical of flat terrain and the lower part of the mountains. In the rosetophilous desert scrub, we find shrub or subshrub species with elongated and narrow leaves such as *Agave lechuguilla, Dasylirion spp*., or what is known as calcareous desert scrub, which is made up of mezquital, izotal, and magueyal. In addition, the crassicaule desert scrub includes grassland, chaparral, nopales, dune vegetation, and halophyte vegetation (Granados-Sánchez et al., 2011).

The Rosette scrub is located on hills and hillsides derived from sedimentary material, mainly limestone and, to a lesser extent, on shale. Such is the frequency of its distribution on limestone that in some works it has been referred to as "calcareous desert scrub" (Rzedowski, 1965). It is widely distributed in the center and south of the Chihuahuan Desert; the hillsides on which it is established are steep slopes with shallow and stony soils, and there are often outcroppings of bedrock (limestone), which can outcrop up to 30 or 35%.

7.2 AGRICULTURE AND ECONOMY OF THE ARID AND SEMIARID ZONES OF MEXICO

Contrary to popular beliefs, the arid and semiarid zones of Mexico have significant agricultural production, including corn, wheat under irrigation, and beans under rainfed conditions.

The arid and semiarid zones are more than deserts; they are an inexhaustible source of sustainable forest resources that are highly exploitable and contribute to improving the economy of families in arid and semiarid zones.

These ecosystems cover 56.92 million of the country's 138 million hectares of forest. Using the information from the 2010 census in Mexico, the country's arid zones are home to nearly 33.6 million people (30% of the total population), 18% of whom live in rural areas, while the remaining 82% live in urban areas.

The population of rural communities, especially the indigenous, has always had a very vast knowledge of the use of the natural resources that surround them, particularly the plants that exist, since it is a cultural part of their traditions that they continue to practice and protect until today.

However, the arid regions of northern Mexico have been affected in different ways, from a strong alteration in the distribution and density of their plant genetic resources, to the loss of knowledge in the use of their plants; this is due to a process of transculturation due to factors such as migration, climate change, overgrazing, globalization, the establishment of new more technical crops and the improvement of agricultural plants, the latter established worldwide to ensure food sovereignty.

Many rural communities that decided to stablish themselves both in arid and semiarid zones have steadily changed their economy, health, nutrition, economic activity, and culture. Many of their needs are met from outside their environment and the requirement for inputs from outside is increasing (e.g., fertilizers, pesticides, agrochemicals, fuels, substrates, seeds, irrigation technology, etc). This has left the local economy unprotected and forced a greater external dependency.

However, they still depend largely on spontaneous plant resources for their economy, either directly or indirectly, through the exploitation of wild fauna or that under domestication, emphasizing that as resources are limited and survival is more difficult, the idea of sustainable management is excluded in most of the Mexican territories. To this we combine the historical idea that the north of the country is directed toward livestock; this is due to the quality of the forages, which do have a higher concentration of nutrients than in the tropical zones, but the density of these pastures is lower and the semi-desert soils are severely affected due to grazing and their recovery time is longer than in a normal ecosystem. Most of the inhabitants of the semi-desert produce cattle, and we can see how in the most degraded places, the ecosystem can no longer support bovine and equine animals, drastically changing to smaller species such as sheep and goats. Overgrazing since colonial years has destroyed the grasslands and degraded the land to a point where the semi-arid ecosystems became so poor that their vegetation changed to herbaceous species or thorny shrubs of very low size, low biomass, little palatable and little quantity of nutrients, with goats being the only ones capable of browsing and continuing to take advantage of these species.

7.2.1 AGRICULTURAL PRODUCTION

Some of the most demanding environments are arid and semiarid zones, where local people have developed strategies that involve the diversified and self-sufficient use of the environment (Irigoyen-Rascón, 2016; Luque et al., 2020).

The arid and semiarid zones of Mexico have provided resources and environmental services to the country, including the production of fodder for livestock, food and medicinal products, freshwater, energy, and tourism. Many rural communities in these areas depend heavily on desert resources, either directly or indirectly through the exploitation of wild flora and fauna. Their needs range from fodder for livestock, firewood for fuel, wood for construction, plants for food and drink, and even for industrial uses such as tanning, coloring, medicinal, nectar-producing, and even soil-fixing and soil-forming plants. The substances produced by these plants, which range from gums, resins, latex, starch, grains and proteins, fatty acids, to secondary metabolic compounds, some with pharmacopeia possibilities, are of economic interest. These substances are found in fruits, seeds, stems, roots, rhizomes, or bulbs of plants (González Medrano, 2012).

The development of the arid zones was due to the technologies derived from the industrial revolution, governmental and economic needs, which made possible the development of extensive areas for the production of crops through irrigation systems, with them the agricultural frontier in the north of Mexico was able to expand as never before. The purpose of this agricultural expansion was threefold: (1) to increase cultivated areas, to assure harvests, (2) creation of small property (small owners) by fractioning irrigated land, (3) to economic liberation of the peasant class. Then there was an acceleration in the yield of Mexico's agricultural production.

Also, three fundamental facts influenced the construction of large dams, and with them, the most extensive irrigation systems, in the north of the country: (1) geographic-ecological, due to the presence of extensive desert territory, close to 60% of the Mexican territory, whose main characteristic is aridity; (2) the scarce population due to the characteristics of the desert, and (3) the location that has almost direct contact with the United States market.

Agriculture in arid zones has rainfall as a limiting factor; however, the water stored underground and available to plants is the best parameter to indicate success in crop yields than rainfall itself (López Martínez and Guerrero Alvarado). In the state of Chihuahua, agriculture is successfully carried out (Manzanares Rivera, 2016) due to the extraction of water from underground aquifers; this strategy enables the operation of high-yield agricultural developments, and these developments occupy an area of 221,000 ha and specialize in crops of high economic value, such as cotton.

7.2.2 MAIN CROPS

The authors have agreed that the unpopulated and distant desert began to receive resources dedicated to irrigated agriculture mainly in the states of Sinaloa, Tamaulipas, Sonora, Baja California, and Chihuahua. Table 7.1 (González Medrano, 2012) shows the irrigation districts and the main historical crops that prevailed in those territories.

TABLE 7.1 Historical Crops in Arid and Semiarid Zones (González Medrano, 2012).

Irrigated area	Hectares under irrigation	Main crops
Yaqui Valley, Sonora	220,000	Wheat, rice, cotton
Mayo Valley, Sonora	100,000	Chickpea, wheat
Fuerte Valley, Sinaloa	230,000	Sugarcane, vegetables
Culiacan Valley, Sinaloa	95,000	Vegetables, cotton
Valley of Mexicali, Baja California	180,000–200,000	Cotton
Comarca Lagunera, Coahuila and Durango	100,000	Cotton, grapevine, alfalfa
Bajo Rio Bravo, Tamaulipas	200,000–350,000	Cotton, sorghum

During the green revolution, rice and wheat were the predominant crops in the north, along with agro-industrial development. However, in the long term, wheat would be defined as the historical crop of the Yaqui Valley plains, and later cotton would abruptly displace rice production. It is worth noting the agricultural and agro-industrial impact that wheat had in these areas and how it managed to position itself as the predominant crop in the entire region.

The success of cotton made it a strategic raw material, as it was a conditioning factor for subsidies in government development plans. It is one of the strategic crops developed in the north. It had and still has presence in "La Laguna," "Valle de Juarez," "Mexicali," "Matamoros," "Delicias," "Culiacan," "Valle Fuerte," and the coast of "Hermosillo."

In the state of Coahuila, the most important crops are corn (*Zea mays*), apple (*Malus domestica*), pumpkin (*Cucurbita pepo*), peach (*Prunus persica*), beans (*Phaseolus vulgaris*), and oats (*Avena sativa*), most of which are grown in plots and others in backyards. They are mainly used for human feed.

The introduction of new species should be tested in as many sites as possible and the value of the product should have a high economic value. On the other hand, use endemic plant resources that have a high agro-industrial potential and that can give good yields even in the soil and climatic conditions of arid zones, thus optimizing agricultural productivity in arid zones.

The endemic plant resources that have been exploited in a traditional way and the sustained development has been restricted to a small group of species among the many that have potential are candelilla (*Euphorbia antisyphiloica*), to obtain wax; guayule (*Parthenium argentata*) to obtain rubber; lechuguilla (*Agave lechuguilla*) to obtain fiber for cordage, pita or palm; yucca (*Yucca filifera*) to use it as a fence and to obtain edible flowers and fruits; samandoque (*Hesperaloe funifera*) for fiber; mesquite (*Prosopis laevigata* and *Prosopis juliflora*) to use pods and foliage as fodder and wood for construction and fuel, among others with less exploitation such as: cow's rib (*Atriplex canescens*), jojoba (*Simmondsia chinensis*); piñon pine trees (*Pinus cembroide*, P. *nelsonii*, P. *Pinaceae*); garambullo (*Myrtillocactus geometrizans*); cardón (*Pachycereus pringlei*); pitaya (*Stenocereus griseus*); mezcal (*Agave spp*); bacanora (*Agave pacifica*); sotol (*Dasylirion leiophyllum*); cañagria (*Rumex hymenosepalus*); oregano (*Lippia palmeri* and *L. graveolens*) (González Medrano, 2012).

On the other hand, the plants that develop in saline habitats and can grow in arid zones also have a high degree of industrialization and due to these qualities they allow for optimizing productivity in the field (Murillo-Amador et al., 2006; Rueda-Puente et al., 2009). Halophyte plants are around 3000 species (Glenn et al., 1998) ranging from grasses, shrubs, and bushes. These plants can be cultivated with traditional agriculture and develop adequately in arid zones.

7.2.3 PLANTS OF ECONOMIC IMPORTANCE IN ARID AND SEMIARID ZONES

The species that stand out for their food use in arid areas that grow wild are: *Yucca carnerosana, Flurencia cernua, Tagetes lucida, Agave salmiana, Opuntia streptacantha, Arctostaphylus pungens, Hedeoma drummondi, Mentha spicata, Poliomintha longiflora, Pinus cembroides,* and *Ruta graveolens.* For medicinal purposes: *Artemisia ludoviciana, Flourencia cernua, Tegetes lucida, Trixis californica, Opuntia streptacantha, Arctostaphylos pungens,*

Hedeoma drummondii, Mejorana hortensis, Mentha spicata and *Poliomintha longiflora.* For fodder purposes: *Agave salmiana, Yucca carnerosana, Yucca filifera, Opuntia streptacantha* and *Chenopodium graveolens.*

These species are a primary component of agroecosystems in arid zones and directly and indirectly, support both the local economy and livelihood. However, most are not domesticated and grow wild around communities. Many of these species do not represent an economic value, but their value is intangible for the community and the survival of the locals. Species such as the ones mentioned in Table 7.2 already have an industrial demand, but the purchase prices are very low for the locals, many of these products such as wax cerote have an international demand and are even priced in dollars. Table 7.2 shows the economically important species and their uses (Comisión Nacional Forestal, 2009).

TABLE 7.2 Economically Important Species (Comisión Nacional Forestal, 2009).

Species	Uses
Agave lechuguilla	Industrial: manufacturing of brushes, mats, and car filters.
	Craft: bags, sandals, baskets, cosmetics, shampoo, and soap.
Dasylirion cedrosanum	Fodder: for cattle.
	Medicinal: bad urine and fever. Industry: soaps, glue for plywood, and cardboard compressed.
Lippia berlandieri Schauer, Lippia graveolens, Poliomintha longiflora	Food: consumption of dried leaves as seasoning.
	Industrial: oil extraction.
	Medicinal: antioxidant, anti-inflammatory, antiseptic, and antiparasitic.
	Agroindustrial: fungicide and insecticide
Jatropha dioica	Medicine: treatment for hair loss, fungal removal, bleeding gums, and cauterize wounds.
	Cosmetic: as it tempts to give a dark color to the hair

7.3 PROBLEMS OF AGRICULTURE IN ARID ZONES

Worldwide, crop production in arid zones is a function of the expression of the factors that occur naturally in the environment. To take advantage of resources and reduce production risks, it is necessary to develop methodologies that integrate soil, climate, and plant factors (Hernández et al., 1983). One of the most outstanding features of arid zones is the low availability of water, high evaporation due to incident solar radiation, low humidity, vapor pressure, high daytime, nighttime temperature differences, and wind speed, all of which contribute to making agriculture in arid zones a challenge.

7.3.1 PLANT SPECIES OF TRADITIONAL AGRICULTURAL USE

The Ministry of Agriculture, Livestock, Rural Development, Fisheries and Food (SAGARPA) reported only for the state of Coahuila sowings of more than 276 thousand ha with more than 44 crops, of which 33 are annuals and 11 perennials; rainfed or rainfed crops represent 42.2%, and of these, 27.5% correspond to crops of corn, beans, sorghum and apple; the rest to the planting of forage crops, various pastures, and forage nopal (Secretaría de Agricultura et al., 2013).

Martínez-Burciaga and Lara-Guajardo (2003) mentioned that in the state of Coahuila there are approximately 131,085 hectares open to seasonal agriculture. The district with the largest seasonal agricultural area is Saltillo, which is located in the south and southeast of the state, considered the most important for the area planted with basic crops, mainly corn and beans. It can be understood that this agriculture is subsistence and market prices are not enough to cover one-way costs, since, in northern Mexico, life is more expensive, including supplies and processes. The abandonment of the land is increasingly evident and the migration to the cities, in this case to Saltillo and Ramos Arizpe, has increased and the peasants leave their agricultural activities because there are better wages to work in the industry; the lack of protection for the countryside is seen as a growing problem toward increasing dependence on other regions of the country.

Another aspect that deserves special attention is that related to agricultural mechanization. Its implementation has been the direct result of rural capitalist development and has been stimulated by the "green revolution" and by the transnational companies that have been operating in Mexico for years and monopolize the market and the products that are extracted from the countryside and/or are applied to it.

Agricultural mechanization encounters serious technical and economic obstacles, but mainly economic ones, with the small farmers. But it grows more dynamically than the agricultural product, shows its structural nature in this system, and finds all possibility of increasing its yields through mechanization closed, but the variable costs in agrochemicals and external technology affect the price in a direct and hit the small producer.

The northwest is the most mechanized region of the country, almost 98% of the cultivated area is totally or partially mechanized; on the other hand, the less productive regions of the center and south are the least mechanized. The dominant types of cultivation have a considerable influence on the degree of mechanization since forages and cereals allow complete mechanization. On the other hand, cotton and sugar cane still do not have efficient machinery that can advantageously substitute labor.

From the social point of view, mechanization aggravates the problem of joblessness in the rural area. In the northwest region, technology has come to displace labor; ejidatario (rural workers) only works, on average, 120 days a year on his plot, and as agriculture becomes more mechanized, this problem becomes notably more acute. There is government support that subsidizes farm products; however, this does not solve the real problems of the peasants.

7.3.2 CONSERVATION OF ARID AND SEMIARID ZONES

In general, arid and semiarid zones are fragile ecosystems, very susceptible to the influence of climatic and human factors, with limited humidity due to low precipitation and high evaporation rates, coupled with thin soils and low primary productivity, together with other factors that make these areas vulnerable to degradation (Reid et al., 2005).

However, due to the agricultural exploitation of these resources, there is concern about the accelerated loss of resource biodiversity. The conservation of ecosystems and their biodiversity has received much interest in recent years and is currently a priority issue worldwide.

Subway water resources have also been overexploited, and it is common to extract water at greater depths. In addition, the salinity of agricultural land is a recurrent phenomenon in these areas, which leads to low crop production capacity under these conditions.

Migration of people to other regions, even outside the country, to reach opportunities, is another of the problems that plague arid and semiarid zones. Migration has been the cause of disintegration of domestic units and community and political structures (Alarcón-Cháires et al., 2013). It is due to male migration that it is recurrent to see in these regions the feminization of minor agricultural activities. In rural areas, communities tend to be very small and scattered, which leads to deficient public services, such as drinking water, electricity, poor medical services and infrastructure, and inadequate communication.

Climate conditions in the world are subject to what has been called climate change, which according to the World Bank, is a consequence of global warming, which could reach up to 4°C by 2060, if the release of greenhouse gases is not stopped (World Bank, 2013). The situation in Latin America and the Caribbean is different from that of the developed countries; while the latter are the ones that mainly generate the global externality resulting from emissions and suffer from it, the Latin American and Caribbean region contributes little to generating it, but suffers disproportionately from it (Samaniego, 2008). The existing relationship between the resources of arid zones and the inhabitants in the face of socioecological and climatological phenomena on a local to the global scale must be considered and analyzed (Reid et al., 2005; Casas et al., 2017).

7.4 IMPORTANCE OF GRASSLANDS

Grasslands serve as productive areas for fauna, microfauna, recreational use, and cattle grazing, as well as watersheds (Gurrola Reyes, 2005). Viewed strictly from the perspective of economics, grasslands are extremely important, since they constitute the most suitable natural environment for livestock exploitation with species such as cattle, goats, horses, and sheep, although the nutritional preferences of goats and sheep tend to be concentrated in other types of vegetation (Rzedowski, 1978).

The grasslands of northern Mexico were a source of great wealth for the haciendas that were dedicated to livestock at the end of the 19th century and the beginning of the 20th century, and after the agrarian distribution, the best grassland areas were established as agricultural areas for the cultivation of rain-fed corn (Shreve, 1942), initiating agricultural activity in the study area at the end of the 19th century (Favret-Tondato, 2013).

7.4.1 OVERGRAZING IN THE GRASSLANDS OF MEXICO

The use of natural grasslands in Mexico in many cases is not optimal and the lack of good grassland management causes overgrazing that does not allow maximum yield and they are increasingly vulnerable due to unplanned grazing, opening of agricultural areas, invasion of exotic species, as well as the drought that is causing its decline (Rzedowski, 1978). Overgrazing is common in many plant communities, increasing the risks of erosion, degradation, and plant invasion (Encina-Domínguez et al., 2014).

According to Rzedowski (1978), grasslands are generally of medium height (20 to 70 cm), although, due to intense grazing, they remain at lower heights. Overgrazing and excessive trampling prevent the growth and development of plants that are desired by livestock, favoring the establishment of plants that animals do not eat, and that are often poisonous and reduce soil cover, exposing it to erosion.

7.5 DEVELOPMENT ALTERNATIVES

7.5.1 THE IMPORTANCE OF WEED, NEGLECTED AND UNDERUTILIZED SPECIES FOR SUSTAINABLE AGRICULTURE

Currently, in all cultivated lands around the world, there are several species that have become associated with certain crops. Many of these species are local and some others are introduced, brought by man over the years and expanding their distribution directly or indirectly.

These plants, in context with industrial agriculture, are plant species that affect different interests of farmers by significantly lowering the production and quality of different agricultural crops. In some cases, their presence can alter the local biodiversity, changing the structure and chemistry of soil components, the flow of nutrients, and water conditions, some have allopathic effects that inhibit the growth of crops, for example, the giant tabacote (*Nicotiana glauca*), This plant has spread throughout Latin America and even Europe and is considered an invasive exotic species. Every part of the plant is toxic, except the mature seeds, because they contain malate and citrate of the alkaloid nicotine, analogous to *Nicotiana tabacum* L., as well as isinicotine, nicotoine, pyrrolidine, oxalic acid, etc., this species is easily found in all areas with anthropogenic disturbances and establishes very easily, even in places with very desertic conditions and poor soils such as northern Mexico.

Another common example is clover (*Trifolium repens*) which is found in perennial forage crops such as alfalfa; it can spread without any complication due to the planting conditions given to alfalfa. This causes problems as it covers the soil without problem and competes for available water, thus increasing production costs and decreasing the quality of the crop as it goes for animal fodder, usually cattle. Another native species that is also considered an invasive opportunistic species in Mexico and that presents a production problem for farmers in arid areas of northern Mexico, such as northern Zacatecas, southern Coahuila, and neighboring areas of Nuevo Leon, is the picoso grass (*Amelichloa clandestina*), an opportunistic species that due to its aggressiveness to colonize impacted terrain forms dense zacatales, scrub, zacatales and pinyon pine forests between 800 and 2100 m altitude, grows isolated through the grassland, especially in overgrazed terrain and its control is very difficult (Kettenring and Adams, 2011).

One of the greatest sources of knowledge of weed damage comes from agricultural crop loss assessments. Weeds are known to cause a direct loss of approximately 10% of annual agricultural production. In cereals, this loss has been estimated to affect more than 150 million tons globally. However, such losses are not the same in different countries, regions of the world and affected crops.

In the 1980s, it began to be recorded and assessed the effects of these weeds in different crops, and it was estimated that the alarming losses of agricultural production caused by weeds, which decreased crop production in different countries such as Europe, where it decreased by 7%, relatively low compared to Africa with 16%. The main large-scale crops affected were sugar cane (15.1%), rice (10.6%), and cotton (5.8%) (Fernández, 1982). These diseases are caused by weeds that compete with agricultural crops for space, water, light and nutrients, thus increasing the cost of pesticide control and damaging the ecosystem in the use of pesticides.

However, in small-scale agriculture in many communities around the world, and especially in the semiarid zones of Mexico, there are some species that are considered weeds in mid-level and industrial agriculture but play an inverse role in the local or regional economy, having little or low commercial value, but playing a very important role in the semidesert communities. There are species listed as weeds, but the use of these species has been undervalued and, in some cases, their uses have been forgotten. The use of these species can be gastronomic, cultural, and economic, with low or almost null scale, causing them to go unnoticed for lacking commercial interest.

Often, these species are forgotten even by researchers, policy makers, and traders, even though their different uses are known to the locals in different ways, such as construction material, fibers, utensils, firewood, etc. They may have a food character and their existence means a valuable local resource, especially for those rural communities that are in arid and marginalized conditions. These forgotten, underutilized, or unknown species are now considered by some as strategic for food security.

Since these species have great nutritional value and/or can complement the nutrition of the population, as well as can be used as medicinal or even have some use as a pesticide, in some cases, these species can be minor crops, which makes them visible locally, within society, or culturally speaking within some sporadic regional dishes or holidays, so they do not get to be considered or economically visible because of their lack of market due to lack of knowledge of the use and sometimes due to lack of information on their management, They are being forgotten and excluded due to the deculturization faced by the communities when adopting improved species with a wider or even universal use in the last few decades.

It should be clarified that most of these species are not domesticated. But their quick adaptation to drastic climate changes due to global warming, as well as their easy establishment in disturbed soils, even in very poor soils such as arid soils, they manage to complete their phenological cycle. These local species have a low incidence of diseases or pests in these plants, which could reduce the use of pesticides during their production. It should be noted that these species can be very nutritious. Unfortunately, the domestication of these species is generally difficult, complicating their orderly production.

Currently, these weeds are classified as follows:

Annual herbs

These herbs are those that do not live more than 1 year. They are fast-growing and short-lived, reaching their phenological cycle quickly. Some of them grow so fast that they flower even in autumn. An example of this would be purslane, quelite, yellow oxalis, chickweed, stinkweed (Datura stramonium), etc.

Biannual herbs

The developmental cycle of these species usually takes longer, between one and 2 years. Their seeds germinate in spring or autumn, but they do not flower or bear fruit until the following year. This group is not very numerous, the wild carrot and certain thistles, etc. stand out.

Perennial herbs

These plants flower and produce fruit for several consecutive years. In addition to seeds, they generally have other means of dispersal. There are examples such as huizache, palo dulce (blue stick), ochote in melliferous or honey matters, retama (tecoma stans), pingüica, mimbre (Chilopsis linearis), maguey (Agave spp), rosa de castilla, plantain, sorrel, higuerilla (ricinus comunis), curly, ground ivy, and dandelion.

In the north of Mexico, it is quite common to find these ignored species, many of them are easily found and used in regional dishes such as: yucca flower, purslane, epazote, squash flower, wild tomatillo, cabuches, and quelites, which are consumed very locally in regional dishes, most of them do not reach a commercial value. Purslane itself does not have a high commercial value and a high nutritional value; however, this species is seen as a weed.

Some people in the semi-desert grind and use the seeds of the higuerilla, applying it to their crops; this is due to its pesticide properties, which are used to fumigate and control fungi, bacteria, and aphids due to its astringent power. Although in Mexico it is considered a weed, in the industry, oil is extracted from its seed, called castor oil and has numerous industrial applications for the production of plastic and synthetic fibers; also this oil is utilized in the production of inks, enamels, and lubricants. It is considered highly toxic when ingested by people although intoxication is not serious if the seeds are swallowed as a whole.

Species such as mesquite have been used due to their potential as forage food, as well as the use of its pods, which have a high nutritional content, although this has not prevented its disuse in gastronomy; mesquite flour used to be used as a sweetener for cakes or coffee.

Another of its uses in gastronomy was the preparation of "mezquitamal," a version of the tamale, a typical Mexican dish, produced in the arid region of the country. This version was smaller in size, and due to its high number of sugars it is considered a delicacy, especially if it is prepared in an artisanal way in the milling, where the process of pulverizing the mesquite with a wooden stick in the middle of a trunk is carried out for several minutes, leaving the product of this process to rest, which was repeated until obtaining a flour derived from the mesquite. The use of mezquitamal is limited to festivities, especially in the south of Zacatecas.

Another undervalued species is the huizache, which is considered a pest for the bean-growing region of Zacatecas and Durango. It has a great adaptation value to arid zones, rapidly establishing itself everywhere, and its high environmental value as a species that helps prevent soil erosion. This species is key for the elaboration of honey, since in general the flowers provide better properties during the production of specific mesquite honey, beekeepers begin to understand that this species supports a better quality of honey when the bees have diversification of species. The honey quality is modified according to the relation on mesquite and huizache that the bees visit during the production; and has even begun to make honey from the hedionda plant. The huizache has even been used for some home remedies, having ethnobotanical uses, in the community of the black mascogos, Seminoles. They use the most tender parts of the huizache for the elaboration of tea and energizing infusions; this community uses it in substitution of the cinnamon or the coffee.

Medicinal plants such as arnica, stafiate, mullein, laurel, and epazote remain far below a real management in relationship to their potential use, since their abundance and low price in the market become unattractive to the locals and are not taken advantage of. But they are used from time to time in ailments such as sore throats, colds, diarrhea, colic, stomach infections and as relaxants.

Factors such as the threat of climate change, the high costs of production in modern agriculture, the increase of pests and the use of agrochemicals for their control, the degradation of agricultural soils and globalization, have changed paradigms in the research of new sources of food to satisfy food sovereignty.

Possible uses of these species considered as weeds, and that until recently have been considered even strategic for the survival of many communities, due to the nutritional contributions to supply part of the diets of farmers living in communities with low living resources, a nutritional contribution to supply part of the diets of farmers living in communities with low living resources, a nutritional contribution to the survival of many communities, Recently, they have even been considered strategic for the survival of many communities, because they have been found nutritional contributions to supply part of the diets of farmers living in communities with low living resources, a food use, medicinal, forage, and even as a herbicide, thus helping the economy and nutrition of the local population.

7.6 ARID PLANTS USED TO PRODUCE BIOFUELS

The rise in the usage of biofuels in the arid and semiarid regions of Mexico is a foreseeable and necessary goal, given the great geological geographic diversity of these areas of the country. It is necessary that the ways of obtaining bioenergy are adequate and sustainable. There are alternative and supplementary sources of bioenergy and biofuels from vegetable oils and fats; obtaining these bioenergies is highly viable in the arid and semiarid areas of northeastern Mexico and can even be transferable to the rest of the national territory, and this allows advantages of renewability, environmental and economic sustainability, which is of high strategic priority.

In most of the arid and semiarid zones of Mexico, crop production is based on a combination of hydraulic and soil conservation practices. In addition, integrated management practices (soil, water, flora, and wildlife) constitute viable projects to continue improving agriculture and thus the standard of living of producers.

TABLE 7.3 Arid Zone Species with Bioenergy Potential (Castellanos et al., 2012).

Groups of species with bioenergy potential	
Biodiesel	**Solid biofuel**
Cnidosculos angustidens	*Atriplex canescens*
Cucurbita foetidissima	*Prosopis spp*
Cucurbita digitata	*Salsola kali*
Cucurbita argyrosperma	*Yuca filifera*
Cuphea wrigthii	**Liquid biofuel**
Encelia farinosa	*Agave angustifolia*
Jatropha cinérea	**Biocrude oil**
Jatropha cordata	*Asclepias albicans*
Jatropha cardiophylla	*Asclepias erosa*
Lesquerella fendleri	*Asclepias subulata*
Proboscidea parviflora	*Chrysothamus paniculatus*
Prosopis velutina	*Pedilanthus macrocarpus*
Simmondsia chinensis	

There is lack of knowledge of the inhabitants of certain endemic species in the arid and semiarid zones, which are extremely rich and are even exploited by people outside the zone. This situation has not allowed the inhabitants to use these resources as an economic alternative, some examples:

Mezquite (*Prosopis spp.*), Ocotillo (*Fouquieria splendens*), and different types of yuccas (*Yucca spp.*), among others; plants with forage potential such as Grasslands, nopal (*Opuntia spp.*), Costilla de Vaca (*Atriplex spp.*), Guayacán (*Guaiacum sanctum*), Vara Dulce (*Eysenhardtia polystachya*), Palo Verde (*Parkinsonia spp.*), Rosa de Castilla (*Rosa centifolia*), Jojoba (*Simmondsia chinensis*), Guajillo (*Acacia berlandieri*), and Ramoncillo (*Trophis mexicana*), among others.

Plants with industrial potential: Lechuguilla (*Agave lecheguilla*), Candelilla (Euphorbia antisyphilitica), Guayule (*Parthenium argentatum*), Palma Samandoca (*Yucca carnerosana*) and Palma China (*Yucca decipiens*), among others. Plants with medicinal potential: Albarda (*Fouquieria splendens*), Peyote (*Mirtillocactus geometrizans*), Chaparro Prieto (*Acacia rigidula*), Cenizo (*Atriplex canescens*), Candelilla (*Euphorbia antisyphilitica*), Damiana, (*Turnera difusa*), toloache (*Datura stramonium*), anacahuita (*Cordia boissieri*), hojasén (*Flourensia cernua*), guayacán (*Guaiacum sanctus*), gobernadora (*Larrea tridentata*), and oregano (*Lippia spp.*), among others. Plants with ornamental potential include all desert species, especially *Cactaceae* family.

Finally, it is worth mentioning that despite the great potential that these species have for use by local people, their utilization has been scarce and their exploitation has not been the most adequate. It is necessary to develop research projects in these areas that allow for sustainable environmental development and economic development of the inhabitants. Focused on taking advantage of, protecting and conserving desert resources, the resource management strategies that are carried out for agricultural production reveal the way of perceiving the territory by those who inhabit them, this helps to understand how socioecological factors affect the management responses (Blancas et al., 2016). The management of these resources is the expression of the interactive and harmonious relationships between biological diversity and cultural diversity (Bogaert et al., 2014).

Therefore, the perspective of sustainability and balance with the environment is important, so understanding these changes requires having parameters on the degree of pressure exerted on water and plant resources.

Recently in Mexico, regulatory initiatives have been implemented in the use of non-surface water, which allows weighing environmental aspects before the anthropogenic impacts associated with economic activities. This is due to the relative problems caused by the decrease of water in the watercourses, as a consequence of the competition between uses and regulation focused on sustainability that did not exist until a few years ago.

KEYWORDS

- **agricultural and arid zones**
- **landscapes, grasslands**
- **ethnobotanical plants**
- **arid shrubs**
- **irrigation systems**

REFERENCES

Alarcón-Cháires, P.; Chávez, T.; Chávez, C. *WIRIKUTA Defensa del territorio ancestral de un pueblo originario*; Frente en Defensa de Wirikuta Tamatsima-Wahaa: Morelia, 2013.

Blancas, J.; Casas, A.; Moreno-Calles, A. I.; Caballero, J. Cultural Motives of Plant Management and Domestication. In *Ethnobotany of Mexico*; Springer: New York, 2016; pp 233–255. DOI: 10.1007/978-1-4614-6669-7_10

Bogaert, J.; Hong, S.-K.; Min, Q. *Biocultural Landscapes: Diversity, Functions and Values*; Springer, 2014.

Casas, A.; Torres, I.; Delgado-Lemus, A.; Rangel-Landa, S.; Ilsley, C.; Torres-Guevara, J.; Cruz, A.; Parra, F.; Moreno-Calles, A. I.; Camou, A. et al. Ciencia para la sustentabilidad: investigación, educación y procesos participativos. *Revista Mexicana de Biodiversidad* **2017,** *88*, 113–128. DOI: 10.1016/j.rmb.2017.10.003

Castellanos, A. E.; Sotelo, J. L.; Esqueda, M.; Téllez, O. *Uso de la biodiversidad para bioenergía y biocombustibles en las zonas áridas de Mexico*; Departamento de Desarrollo y Producción Editorial, Universidad de Sonora: Hermosillo, Sonora, 2012.

Clasificación fisonómica de la vegetación de la Sierra de Catorce. *Sociedad Mexicana de la Ciencia del Suelo, A.C* **2003,** *21* (3), 321–332. https://www.redalyc.org/pdf/573/57321303.pdf

Comisión Nacional de Zonas Áridas; Instituto Nacional de Ecología. *Mezquite (Prosopis spp.). Cultivo Alternativo de las Zonas Aridas y Semiáridas de Mexico*; Comisión Nacional de Zonas Áridas: Mexico, 1994.

Comisión Nacional Forestal. *Catálogo de recursos forestales maderables y no maderables*; Mexico, 2009.

Encina-Domínguez, J. A.; Valdés-Reyna, J.; Villarreal-Quintanilla, J. A. Estructura de un zacatal de toboso (*Hilaria mutica*: Poaceae) asociado a sustrato ígneo en el Noreste de Coahuila, Mexico. *J. Bot. Res. Inst. Texas* **2014,** *8* (2), 583–594. https://www.jstor.org/stable/26549410

Favret-Tondato, R. Apropiación de los recursos naturales y producción en el territorio de la Sierra de Zapaliname. En *Guía para conocer y valorar el área protegida de la Sierra de Zapalinamé*; Arizpe-Narro, A. J., Ed.; Elementocero: Saltillo, 2013; pp 89–102.

Fernández, O. A. *Manejo integrado de malezas*; Centro de Recursos Naturales Renovables de la Zona Semiárida y Departamento de Ciencias Agrarias: Campinas, 1982.

Glenn, E. P.; Jed Brown, J.; O'Leary, J. W. Irrigating Crops with Seawater. *Sci. Am.* **1998,** *279* (2), 76–81. DOI: 10.1038/scientificamerican0898-76

Godínez Alvarez, H. Los desiertos mexicanos. Sus características e importancia. *Ciencia y desarrollo* **1998,** *24* (143), 17–22.

González Medrano, F. *Las zonas áridas y semiáridas de Mexico y su vegetación*; Instituto Nacional de Ecología: Mexico, 2012.

Goos, R. J.; Johnson, B. E.; Sobolik, F. J.; Schneider, R. P. Stored Available Soil Water and the Fallow/Recrop Decision-Critical Level Approach. *Soil Science Society of America Journal* **1984,** *48* (5), 1134–1137. DOI: 10.2136/sssaj1984.03615995004800050036x

Granados-Sánchez, D.; Sánchez-González, A.; Granados-Victorino, R. L.; Borja-de la Rosa, A. Ecología de la vegetación del desierto chihuahuense. *Revista Chapingo Serie Ciencias Forestales y del Ambiente* **2011,** *XVII* (Especial), 111–130. DOI: 10.5154/r.rchscfa.2010.10.102

Gurrola Reyes, J. N. *Consumo de biomasa y preferencia alimenticia de Brachystola magna (Girard) Boopedon nubilum (Say) y Melanoplus lakinus (Scudder). (Orthoptera: Acrididae), en pastizales de Durango*; Universidad Autonoma de Zacatecas: Durango, 2005.

Hernández, H. M.; Gómez-Hinostrosa, C.; Goettsch, B. Checklist of Chihuahuan Desert Cactaceae. *Harvard Papers in Botany* **2004,** *9* (1), 51–68. http://www.ibiologia.unam.mx/pdf/directorio/h/hernandez_hec/Checklist.pdf

Hernández, Y. C.; Villa, C. M.; Jiménez, L. T. Crop Production Regional Risk Analysi. In *Conferencia Internacional para el Desarrollo de Zonas Árida*; Mexico, Mexico, May 10, 1983.

Hoyt, C. A. The Chihuahuan Desert: Diversity at Risk. *Endangered Species Bulletin*, April 3, **2002,** 16–17.

Irigoyen-Rascón, F. *Tarahumara Medicine: Ethnobotany and Healing among the Rarámuri of Mexico*; University of Oklahoma Press, 2016.

Kettenring, K. M.; Adams, C. R. Lessons Learned from Invasive Plant Control Experiments: A Systematic Review and Meta-Analysis. *J. Appl. Ecol.* **2011,** *48* (4), 970–979. DOI: 10.1111/j.1365-2664.2011.01979.x

López Martínez, J. D.; Guerrero Alvarado, J. Contenido de humedad del suelo para decidir la siembra en agricultura de zonas áridas. *Terra Latinoamericana 17* (1), 51–57. https://www.redalyc.org/pdf/573/57317107.pdf

Luque, D.; Martinez-Yrizar, A.; Burquez, A.; López-Cruz, G.; Murphy, A. D. Indigenous People of Sonora: Biocultural complexes, Climate Change and Political Ecology. En *Complejos bioculturales de Mexico. Bienestar comunitario en escenarios de cambio climático*; Benemérita Universidad Autónoma de Puebla: Puebla, 2020; pp 120–144.

Manzanares Rivera, J. L. Hacer florecer al desierto: Análisis sobre la intensidad de uso de los recursos hídricos subterráneos y superficiales en Chihuahua, Mexico. *Cuadernos de Desarrollo Rural* **2016,** *13* (77), 35. DOI: 10.11144/javeriana.cdr13-77.hfda

Martínez Burciaga, O. U.; Lara Guajardo, G. J. *Potencial productivo de áreas de temporal en el estado de Coahuila. Una propuesta de conversión productiva*; INIFAP-CIRNE: Saltillo, 2003.

Mazuela Aguila, P. C. Agriculture in Arid and Semiarid Zones. *Revista de Agricultura en Zonas Áridas* **2013,** *31* (2), 3–4. DOI: 10.4067/S0718-34292013000200001

Murillo-Amador, B.; Jones, H. G.; Kaya, C.; Aguilar, R. L.; García-Hernández, J. L.; Troyo-Diéguez, E.; Ávila-Serrano, N. Y.; Rueda-Puente, E. Effects of Foliar Application of Calcium Nitrate on Growth and Physiological Attributes of Cowpea (*Vigna unguiculata* L. Walp.) Grown Under Salt Stress. *Environ. Exp. Bot.* **2006,** *58* (1–3), 188–196. DOI: 10.1016/j.envexpbot.2005.08.003

Reid, W.; Mooney, H.; Cropper, A.; Capistrano, D.; Carpenter, S.; Chopra, K.; Dasgupta, P.; Dietz, T.; Kumar, A.; Hassan, R. et al. *Millennium Ecosystem Assessment*; UN Report; UN: Washington, 2005.

Rueda-Puente, E. O.; Villegas-Espinoza, J. A.; Gerlach-Barrera, L. E.; Tarazón-Herrera, M. A.; Murillo-Amador, B.; García-Hernández, J. L.; Troyo-Diéguez, E.; Preciado-Rangel, P. Effect of Inoculation of Plant Growth Promoting Bacteria on Germination of Salicornia bigelovii. *Terra Latinoamericana* **2009,** *27* (4), 345–354. http://www.scielo.org.mx/pdf/tl/v27n4/v27n4a9.pdf

Rzedowski, J. *Vegetación de Mexico*; Limusa: Mexico, 1978.

Rzedowski, J. *Vegetación del Estado de San Luis Potosí*; Universidad Autónoma de San Luis Potosí: San Luis Potosí, 1965.

Samaniego, J. L. *Cambio climático y desarrollo en América Latina y el Caribe*; United Nations Organization: Santiago, 2008.

Secretaría de Agricultura, Ganadería, Desarrollo Rural, Pesca y Alimentación; Servicio de Información y Estadística Agroalimentaria y Pesquera. *Anuario estadístico de la producción agrícola de los Estados Unidos Mexicanos. Estadísticas por estado*; Secretaría de Agricultura, Ganadería, Desarrollo Rural, Pesca y Alimentación: Mexico, 2013.

Shreve, F. The Desert Vegetation of North America. *Bot. Rev.* **1942,** *8* (4), 195–246. DOI: 10.1007/bf02882228

Solano-Picazo, C. Etnobotánica de wirikuta: uso de recursos vegetales silvestres en el desierto de San Luis Potosí, Mexico. *Etnobiología* **2018,** *16* (3), 54–77. https://revistaetnobiologia.mx/index.php/etno/article/view/121/120

Sutton, A. *El Desierto Chihuahuense, nuestro desierto*, 1 April 2000. http://www.pronatura.org.

Tarango Arámbula, L. A. Problemática y alternativas de desarrollo de las zonas áridas y semiáridas de Mexico. *Revista Chapingo Serie Zonas Áridas* **2005,** *IV* (2), 17–21. https://www.redalyc.org/pdf/4555/455545052003.pdf

Ugalde Ávila, J.; Granados-Sánchez, D.; Sánchez-González, A. Succession in the microphyll desert scrub of Larrea tridentata (DC) Cov. in the Sierra de Catorce, San Luis Potosi, Mexico. *Terra Latinoamericana* **2008,** *26* (2), 153–160. https://www.terralatinoamericana.org.mx/index.php/terra/article/view/1361/1564

van Noordwijk, M.; Ong, C. K. *Agroforestry Systems* **1999,** *45* (1/3), 131–158. DOI: 10.1023/a:1006245605705

Villaseñor, J. L. Checklist of the Native Vascular Plants of Mexico. *Revista Mexicana de Biodiversidad* **2016,** *87* (3), 559–902. DOI: 10.1016/j.rmb.2016.06.017

Wake, D. B.; Morafka, D. J. A Biogeographical Analysis of the Chihuahuan Desert through Its Herpetofauna. *Copeia* **1978,** *1978* (2), 369. DOI: 10.2307/1443590

World Bank. *Turn Down the Heat*, 2013. DOI: 10.1596/978-1-4648-0055-9

CHAPTER 8

Potential for the Recovery of Agro-Industrial Wastes from Arid Zones: Oregano Wastes as Prebiotics Source

NORMA PAOLA MELENDEZ-RENTERÍA[1],
MIGUEL ÁNGEL DE LEÓN-ZAPATA[1], ELAN IÑAKY LAREDO-ALCALÁ[1],
RAUL RODRIGUEZ HERRERA[2],
GUADALUPE VIRGINIA NEVAREZ MOORILLON[3], and
CRISTÓBAL NOÉ AGUILAR[2]

[1]*CICBEC, Autonomous University of Coahuila, Cuatrocienegas, Mexico*

[2]*Faculty of Chemistry, Autonomous University of Coahuila, Saltillo, Mexico*

[3]*Faculty of Chemistry, Autonomous University of Chihuahua, Chihuahua, Mexico*

ABSTRACT

Oregano is a culinary spice, traditional in Mexican food, but used all over the world. Oil extraction is the main industrial activity in which this plant is used; the residues from this activity accumulate in the production plants, causing contamination. Thanks to its agroindustrial nature, oregano residues (bagasse or oregano wastes) can be used in biotransformation processes.

Functional foods are modified products or food ingredients, that confer more benefits that the traditional foods. Prebiotics, probiotics and symbiotics are considering functional foods. Probiotics are viable microorganisms who are added to the human diet, in adequate quantities, gives benefits to the host; while prebiotics are non-digestible food ingredients, that increase the cell growth or the activity from the intestinal microbiota, and this new bacterial

Exploration and Valorization of Natural Resources from Arid Zones. Cristobal Noé Aguilar, Norma Paola Meléndez-Renteria, & Karen Nathiely Ramirez-Guzman (Eds.)

population confers benefits to the host. The conjunction between probiotics and prebiotics gives as result a symbiotic product.

The chemical characterization of the aqueous extracts, from fermented Mexican oregano wastes, reported principally carbohydrates and polyphenols. The polyphenols presence suggests that these molecules can contribute to the lactic acid strains growth and/or probiotics; which use those quantities as growth inductor instead that carbon source. After the growth evaluation, from the different strain of the Lactobacillus genera, using as culture media MRS and MRSm (without glucose as carbon source), we concluded that the aqueous extracts stimulate the bacterial development, specifically Lactobacillus casei ATCC 393, plus the 120 h fermentation time extract and supplementing with 1 % to the MRS media. We demonstrate that the arid zones' plant residues, such as the oregano, offer several applications to assure the zero residues accomplishment in order to promote biotechnological applications in a sustainable and ecological environment that enhance the human health and environmental preservation.

8.1 INTRODUCTION

Oregano is a culinary spice, traditional in Mexican food, but used all over the world. More than 60 species and subspecies belonging to the families *Lamiaceae* and *Verbenaceae* are known under this name, but the most important are the European or Mediterranean oregano (Origanum vulgare subsp. *hirtum* and *Origanum vulgare* subsp. *gracite*) and Mexican oregano (*Lippia graveolens*, *L. palmeri* y *Poliominta longiflora*) (Ocampo-Velázquez et al., 2009).

Mexican oregano plants are endemic shrubs of the semi-desert region of Mexico because they tolerate drought stress (less than 300 mm of rain per year) (Mata-González and Meléndez-González, 2005).

Among the reported uses of Mexican oregano, in addition to its culinary use, it is used in traditional medicine as an analgesic, anti-inflammatory, antipyretic, antispasmodic, remedy for diarrhea, dysentery, menstrual disorders, respiratory diseases, diabetes, or even as an abortifacient (Pascual et al., 2001). Although the most reported use in the literature is the antimicrobial activity of the plant and of the essential oil extracted from the latter, it has been reported that it has antimicrobial activity (Burt and Reinders, 2003; Nostro et al., 2007; Paredes-Aguilar et al., 2007, 2008).

Because oil extraction is the main industrial activity in which this plant is used, the residues from this activity accumulate in the production plants, causing contamination. Thanks to its agroindustrial nature, oregano residues (bagasse or oregano wastes) can be used in biotransformation processes, such as solid-state fermentation (Mussatto et al., 2012), from which value-added compounds such as phenolic antioxidants can be produced or released (Correia et al., 2004; Martins et al., 2010), cellular protein (Lateff et al., 2008), enzymes (Bogar et al., 2003; Hajji et al., 2008; Selwal et al., 2011), among others.

It has already been reported that aqueous extracts from the solid-state fermentation of oregano residues induce the growth of lactic acid bacteria (Meléndez-Rentería et al., 2014), so it is necessary to continue research to corroborate the mechanism by which this phenomenon occurs. Functional food is a modified food product or food ingredient that can provide greater benefits than those offered by traditional foods. The effect can be both in the maintenance of health status and in the reduction of the risk of suffering from disease (De las Cagigas-Reig and Blanco-Anesto, 2002). It should be noted that a functional food may be beneficial for the general population or only for a certain race of people or a certain group of consumers, but the concept of having health benefits is maintained.

Functional foods include prebiotics, probiotics, and symbiotic foods. Probiotics are defined as viable microorganisms that when added to the diet, in adequate amounts, provide benefits to the host (Borchers et al., 2009), whereas prebiotics are non-digestible food ingredients that increase the growth or activity of certain members of a gut microbial community and this new community provides benefits to the host (Burr et al., 2010). With a combination of prebiotics and probiotics, it is possible to produce symbiotic products (Castro and Rodríguez, 2005). The effect they show in the host is a synergism, where the prebiotic favors the development of the probiotic, and the latter generates certain benefits in the host. For symbiotic food to work as such, it is necessary to find a good combination of its main components.

In previous research, Meléndez-Rentería et al. (2010) found that aqueous extracts obtained from the solid-state fermentation of oregano bagasse favor the growth of some lactic acid bacteria. Therefore, the present project is proposed to evaluate the prebiotic capacity of the substances that could be present in this extract, evaluating different probiotic strains.

According to the definition given by FAO, prebiotics is non-living components of food that confer a health benefit to the host, associated with the modulation of the microbiota (Piñeiro et al., 2007). But to be considered within this concept (Piñeiro et al. 2007), the following precepts must also be met—it must not be hydrolyzed or absorbed in the upper gastrointestinal tract; it must be selectively fermented in the colon, by a limited number of bacteria with beneficial potential (mainly, *Lactobacillus* and *Bifidobacterium*); and finally, it should be able to alter the microflora of the colon toward healthier conditions (Niba et al., 2009).

Compounds that are considered prebiotics include inulin, fructooligosaccharides (FOS), galactooligosaccharides (GOS), soyoligosaccharides, xylooligosaccharides, pyrodextrins, isomaltooligosaccharides, and lactulose. But there is also another group known as new prebiotic compounds, including pecticoligosaccharides, lactosucrose, sugar alcohols, glucooligosaccharides, levanos or fructans, resistant starch, xylosaccharides, and soy-oligosaccharides (Piñeiro et al., 2007).

Although it has already been mentioned that the main prebiotics is carbohydrate-type compounds, it has been found that they are not the only molecules with such activity. Phenolic compounds such as epicatechin, catechin, 3-O-methyl-gallic acid, gallic acid, and caffeic acid have been shown to favor the development of intestinal bacteria (Cheng-Lee et al., 2006).

Other compounds, to which prebiotic activity is also attributed, are molecules known as sugar alcohols, including xylitol and lactitol (Kalui et al., 2010), as well as polysaccharides obtained from marine algae, such as fucoidan, laminarin, alginate, and their derivatives (O'Sulivan et al., 2010).

As mentioned by Geier et al. (2007), the potential benefits of symbiotic therapies are obvious, but the big challenge, as with probiotics or prebiotics alone, is to determine the best combination for each disease and each person.

This document details the potential prebiotic effects of aqueous extracts, obtained by solid-state fermentation, of Mexican oregano wastes, which can be included in various formulations where the objective is to promote the growth of lactic acid bacteria, especially *Lactobacillus casei*, as well as the additional benefit that prebiotics provides to the consumer.

8.2 EXPERIMENTAL METHODS AND MATERIALS

8.2.1 OREGANO WASTES, FERMENTATION, AND EXTRACTS OBTENTION

The bagasse was obtained from the "Oreganeros del centro-sur de Chihuahua" industry in Chihuahua, Mexico. The bagasse was composed of a mixture of leaves and stems from which the essential oil had been extracted by steam dragging.

The bagasse was placed in black plastic bags and transported to the lab, where it was dried at 60 °C for 24 h, then pulverized with the help of a blender and stored in black plastic bags until use.

To obtain the aqueous extract (EAB), 20 g of sample and 500 mL of distilled water were placed in a precipitated beaker, protected from light, and magnetically shaken for 4 h. After the time elapsed, they were filtered through blanket cloth and thick pore filter paper sequentially, storing the filtrate in containers protected from light at −20 °C until its characterization.

The bioreactors used (250 mL Erlen-Meyer flask) were prepared by adding 1×10^7 spores/g of sterile dry material as inoculum, in 10 g of the mixture (3 g of bagasse and 7 mL of Pontecorvo culture medium). To obtain the inoculum, *A. niger* PSH was propagated on PDA and after four days of incubation at 30 °C, the spores were harvested with 10 mL of sterile 0.1% Tween 80 and a magnetic stirrer, counting was performed in the central grid of a Neubauer chamber, and the spore concentration was calculated using the following equation

$$\frac{Spores}{mL} = (count\ number)(25)(1 \times 10^4)(dilution\ factor) \tag{8.1}$$

The already inoculated bioreactors were incubated at 30 °C for 120 h, taking samples at the initial and final times. The assay was carried out in triplicate.

To obtain the aqueous extracts at 0 h of fermentation (EAF 0 h) and at 120 h of fermentation (EAF 120 h), 44 mL of deionized distilled water were added to each flask and subjected to magnetic agitation for 20 min.

After this time, they were filtered sequentially by muslin fabric and fine pore filter paper. They were stored in vials protected from light at −20 °C until chemical characterization.

8.2.2 FERMENTATION EXTRACTS CHARACTERIZATION

The characterization was performed for several compounds of different nature and the methodologies performed were as follows:

8.2.2.1 TOTAL SUGARS

The content of total sugars in the samples was determined by the phenol-sulfuric method, as reported by Dubois et al. (1956).

In a test tube, 250 μL of the sample were placed, then 250 μL of 5% phenol were added and placed in an ice bath for 5 min; after that time 1000 μL of concentrated H_2SO_4 (previously cooled) were slowly added; after that, it was gently shaken and placed in a boiling water bath for 5 min, and then allowed to cool; finally the absorbance was read in a spectrophotometer, at a wavelength of 488 nm. The obtained results were analyzed using the regression equation, obtained from the calibration curve for total sugars made with glucose as standard.

8.2.2.2 REDUCING SUGARS

The used method was reported by Somogyi (1926) and Nelson (1944); 250 μL of sample and 250 μL of Somogyi's reagent were placed in a tube, then placed in a water bath for 5 min and allowed to cool, subsequently, 250 μL of Nelson's reagent was added and the solution was shaken. The reaction was diluted with 2 mL of distilled water and the absorbance was read at 660 nm. The results obtained were analyzed using the regression equation obtained from the calibration curve for reducing sugars, carried out with glucose as standard.

8.2.2.3 TOTAL POLYPHENOLS

The methodology used was that reported by Waterman and Mole (1994) with the Folin-Ciocalteu reagent (phosphomolybdic-phosphotungstic acid) and is based on an oxidation–reduction reaction; 400 μL of the sample and 400 μL of Folin-Ciocalteu's reagent were placed in a test tube, shaken, and allowed to stand for 5 min; then 400 μL of 0.01 M sodium carbonate was added, shaken, and allowed to stand for an additional 5 min.

Then 2.5 mL of distilled water was added, shaken again, and the absorbance was read at 725 nm in a spectrophotometer, taking as blank a tube that carried the same treatment, but using styled water instead of a sample. The concentration of the samples was calculated using the linear regression equation generated from a standard curve of gallic acid, carried out with the same treatment given to the samples.

8.2.2.4 QUANTIFICATION OF CONDENSED TANNINS AND CATECHINS

The methodology is the HCl-butanol methodology reported by Swain and Hillis (1959). The technique is based on the oxidation of the condensed tannins, or proanthocyanidins, forming then the anthocyanidins, which are the ones that give the violet color of the solutions.

To carry out the quantification reaction, 500 μL of sample and 100 μL of ferric reagent were placed in a test tube, then diluted with 3 mL of HCl-Butanol (1:9). The tubes were closed with screw caps and heated for 1 h in boiling water, then allowed to cool and the absorbance was read at 460 nm. Samples' concentration was calculated using the linear regression equation generated from a catechin standard curve, carried out with the same treatment given to the samples.

For the specific quantification of catechin, the extracts were filtered through 0.22 μm Millipore membranes, and 2 mL were placed in special HPLC vials. The equipment used was a Varian Pro Star high-performance liquid chromatograph (Palo Alto, CA, USA), with a diode array detector (PDA Pro Star 330) at 280 nm. The column used for the separation was Varian C18 ODS of 25 cm in length. The working temperature was 30 °C and 3% acetic acid (prepared in H_2O MiliQ with a conductivity of 18 mΩ, filtered by 0.45 μm membranes) was used as the mobile phase with a flow rate of 1 mL/min. Total run time per sample was 25 min, equilibration time was 15 min, with limiting pressures of 6–400 atm, while the injection volume was 10 μL. Samples concentration was calculated by the lineal regression equation obtained by the catechin standard curve.

To complete the characterization of the extracts, in terms of polyphenolic compounds, HPLC separation was performed under the following conditions. A run time of 65 min eluting in gradients, using acidified water as solvent A and methanol as solvent B, the ratio of both is shown in Table 8.1. The flow rate used was 0.4 mL/min in a zorbax SB-C18 sovent saber 3.0 × 250 mm 5-μ Agilent (USA) column. The detection was carried out at 280 nm to obtain the chromatographic profile of total phenols, at 320 nm for the profile of hydroxycinnamic phenolic acids, and at 360 nm for the profile of flavanols.

8.2.2.5 STATISTICAL ANALYSIS

Samples were analyzed with three replicates for spectrophotometric determinations and four replicates for antioxidant activity determinations.

The results of all determinations were analyzed by analysis of variance and the means obtained were compared with Tukey's test, all this with SAS statistical software version 9.

TABLE 8.1 Solvent Ratios for the HPLC Separation of Phenolic Compounds in Aqueous Extracts of Oregano Wastes.

Time (min)	Acidic water (%)	Methanol (%)
0.00	90.0	10.0
7.00	85.0	15.0
23.00	70.0	30.0
30.00	60.0	40.0
45.00	40.0	60.0
50.00	20.0	80.0
55.00	00.0	100.0
60.00	00.0	100.0
65.00	90.0	10.0

8.2.2.6 SELECTION, REACTIVATION, AND PROPAGATION OF STRAINS

Based on the growth response obtained in the results of Meléndez-Rentería et al. (2014), the strains *Lactobacillus casei* ATCC 393, *Lactobacillus paracasei* ATCC 27092, *Lactobacillus acidophilus* (isolated from a pharmaceutical product, Sinuberase) and *Lactobacilllus casei* 01 71.7 (from UAM-I collection) were selected. The first two are reference strains, the third is a commercial probiotic and the last one is a strain capable of growing on polyphenol-rich substrates.

The strains were in cryopreservation, for this reason, they were reactivated by inoculation in MRS broth; whose composition is (g/L) 18.5 of glucose, 10 of gelatin peptone, 8 of meat extract, 4 of yeast extract, 2 of sodium acetate, 2 of K_2HPO_4, 2 of ammonium citrate, 1 of Tween 80, 0.2 of $MgSO_4$-$7H_2O$ and 0.05 of $MnSO_4$-$4H_2O$; then they were incubated under carboxylic conditions for 72 h at 30°C. Subsequently, tubes were inoculated with MRS in flute beak and incubated in carboxyphilic conditions for 48 h at 30 °C; after incubation, Gram staining was performed and upon confirming that they corresponded to the strains of interest, they were kept at room temperature for 1 month.

For propagation before each experiment, a colony from the MRS tubes was inoculated in MRS broth and incubated at 30 °C for 48 h in carboxyphilic conditions.

8.2.2.7 MICROPLATE CULTURE SUPPLEMENTING THE MEDIUM WITH AQUEOUS EXTRACTS

Bacterial cultures were adjusted to an approximate cell number of 1.5×10^8 bacteria/mL (Mc Farland's standard 0.5) in phosphate solution.

At the same time, solutions with 1 and 10% (v/v) of each extract or fraction were prepared in MRS broth and in MRSm broth (without glucose), to evaluate the possibility of the bacteria using the components of the aqueous extract of fermented bagasse or the fractions derived from the latter as a carbon source.

For the culture, 250 μL of the corresponding supplemented culture broth and 25 μL of the strain to be inoculated were placed in one well of the microplate. As a negative control, 275 μL of each of the supplemented culture broths were placed, as a positive control 250 μL of culture broth (MRS and MRSm) were placed with 25 μL of each of the strains, and 275 μL of each of the culture broths were placed as a blank. Once the above was completed, they were incubated for 24 h at 30 °C. Once the incubation time had elapsed, the optical density of each culture was determined using a microplate reader (Bio-Rad, model 550, at 595 nm).

8.2.2.8 PREBIOTIC EFFECT MEASURE

Of the 17 treatments where a higher number of cells was obtained than the respective control, after 24 h of incubation, the measure of the corresponding prebiotic effect was analyzed for each treatment, according to the equation proposed by Vulevic et al. (2004), at 24 and 48 h of incubation. The main factors in the equation are substrate consumption, prebiotic index, and the production of short-chain fatty acids. To quantify substrate consumption based on the quantification of reducing sugars, the DNS reagent was used (Miller, 1959). The prebiotic index was calculated with the difference of μ_{max} of the probiotics (*L. casei* ATCC 393, L. casei 01 71.7, *L. paracasei* ATCC 27092, and *L. acidophilus*) and *E. coli* ATCC 25922. The production of short-chain fatty acids was carried out using the methodology proposed in Agilent technical notes 228–398.

Each of the treatments was performed in screw-capped glass tubes. A working volume of 4 mL was used by placing 3960 μL of culture medium, corresponding to each treatment, 40 μL of the necessary extract in case the final concentration was 1%; for the final concentration to be 10%, 3600 μL of culture medium and 400 μL of the necessary extract were placed. Each treatment was inoculated with 200 μL of a bacteria solution (according to each treatment) adjusted to a concentration of approximately 1.5×10^8 cells/mL. Bacterial growth was monitored by O.D. at 600 nm on an Epoch microplate reader (Biotek), bacterial concentration was determined by a calibration curve for each strain in MRS or MRSm medium, as appropriate. Sampling times were at 24 and 48 h of incubation.

For each sampling time, a 1.5 mL sample of each treatment was taken in a sterile Eppendorf tube, centrifuged at 10,000 rpm for 8 min, and the supernatant was placed in 0.5 mL aliquots and frozen until analysis. Subsequently, quantification of the main short-chain fatty acids (acetic acid, lactic acid, propionic acid, and butyric acid) produced by the probiotic strains during prebiotic fermentation was performed. For this purpose, 20 μL of HCl 3 N were added to a 0.5 mL aliquot to ionize all the acids and have them protonated in the solution. For the quantification a Varian CP-3800 mass coupled gas equipment was used, an HP-INNOWax 30 m × 0.32 mm × 0.25 μm column, the carrier gas was helium, the injection volume of 1 μL, a split of 20:1, the injector and detector temperatures were 260 °C, while the oven temperature was by a gradient starting with 80 °C (1 min) and then rising to 120 °C (20 °C/min), after this it had another increased to 205 °C (10 °C/min) and held at this temperature for 2 min.

For the external calibration curve, Sigma-Aldrich standards were used, which were as follows: propionic acid (>99.5% purity), butyric acid (>99% purity), lactic acid (D and L isomers in equal concentration), and glacial acetic acid (99.7% purity). All four were diluted in an aqueous mixture in a 1:4 ratio. Prebiotic effect measurement values were performed in triplicate and analyzed using t-tests and categorical data comparisons. All analyses were performed with SAS statistical software version 9.0.

8.3 RESULTS AND DISCUSSION

8.3.1 EXTRACTS CHARACTERIZATION

The quantification of total sugars had the highest concentration in the EAF 0 h sample with 66.52 mg/g material (Table 8.2), while the EAB sample had 53.63 mg/g material and the EAF 120 h sample had 19.72 mg/g material. Only the EAF 0 h sample is significantly different from the rest, but this is because the culture medium, added for the fermentation to take place, is supplemented with glucose, so the concentration of total sugars was considerably elevated.

TABLE 8.2 Chemical Characterization of Oregano Waste Extracts—Non-Fermented and Fermented.

	EAB (mg/g)	EAF 0 h (mg/g)	EAF 120 h (mg/g)
Total sugars	53.63 ± 6.57	66.52 ± 4.99	19.72 ± 7.12
Reducing sugars	19.47 ± 0.76	97.16 ± 9.11	5.40 ± 0.23
Phenolic compounds	13.47 ± 0.07	4.57 ± 0.11	5.57 ± 0.24
Catechin	2.34 ± 1.09	2.61 ± 0.34	0.00 ± 0.00
Catechin equivalents	34.86 ± 2.20	17.34 ± 0.34	8.52 ± 0.27

Regarding reducing sugars, the results were similar (Table 8.2), as expected. The EAF 0 h sample had a significant difference (97.16 mg/g material) from the EAB (19.47 mg/g material) and EAF 120 h (5.40 mg/g material) samples because the carbohydrate supplement added to the culture medium was made with glucose.

Phenolic compounds were quantified to confirm their presence as possible molecules responsible for the growth of probiotic bacteria. The EAB extract contained 13.47 mg/g of gallic acid equivalents, being significantly higher than the 4.57 and 5.57 mg/g of gallic acid equivalents of the EAF 0 h and EAF 120 h extracts, respectively (Table 8.2). It has already been reported that lactic acid bacteria are able to grow, tolerate, and even degrade polyphenolic compounds, which is why they are usually added as inoculants in silage processes or as probiotics for ruminants (Guzman-Lopez et al., 2009).

Some authors claim that high concentrations of polyphenolic compounds can impair the development of lactic acid bacteria and probiotics (Khalil, 2010), so the amount found in aqueous extracts of fermented oregano bagasse may not inhibit growth, but on the contrary function as a supplement to the culture medium.

The results also agree with those reported by Kovatcheva-Apostolova et al. (2008) who found a low amount of polyphenolic compounds (7.9 mg/g of dry extract) in *Rosa damascena* extracts; however, Pandey and Rizvi (2009), mention that some fruits or fresh vegetables can have up to 10 mg/kg, so the amounts present in the aqueous extracts of fermented bagasse are higher than those mentioned in the reference.

Catechins and their derivatives belong to the family of condensed tannins or complex tannins. An effective reaction used for the determination of complex tannins or also called proanthocyanidins is the reaction in acid medium with butanol. When the above-mentioned reaction was carried out with the aqueous extracts of fermented bagasse, it was found that the highest amount of proanthocyanidins, expressed as catechin equivalents, was in the EAB extract (34.86 mg/g) showed in Table 8.2.

EAB extract (34.86 mg/g), followed by the EAF 0 h extract (17.34 mg/g) and the EAF 120 h extract (8.52 mg/g) (Table 8.2); statistically, all samples are different. Thus, it was proved that the research samples possess compounds belonging to the tannin family described above. To complement the afore-mentioned results, the amount of catechin in each extract was determined by HPLC, and it was found that only in the EAB and EAF extracts 0.5% catechin was present.

EAB and EAF 0 h extracts showed signals corresponding to the molecule in question (2.34 and 2.61 mg/g respectively, Figure 8.1). These results demonstrate that the polyphenolic type molecules found in the extracts are from different families, so it can be assumed that the mixture of compounds is complex, or the compounds are large molecules.

One of the main biological activities attributed to polyphenolic compounds is their antioxidant capacity. In addition, the capture of free radicals would favor the carboxylation necessary for the better development of lactic acid bacteria.

To have a clearer idea of the polyphenolic composition of the oregano bagasse extracts, it was decided to carry out the separation in HPLC with a second methodology, with which it was possible to identify the group of polyphenols to which it corresponds, according to the wavelength at which the signal was detected (Figures 8.1 and 8.2).

In the polyphenol fraction of each of the extracts, the intensity of the signals increased, because with the fractionation a concentration of the polyphenolic compounds was achieved. A higher intensity (which could be translated into concentration) of compounds was found in the EAF 0 h, however, in the EAF 120 h, more signals were observed, which could be attributed to the degradation reactions of the compounds or because of the metabolism of the fungus.

This is related to the degradation of the compounds since previous results showed that the main components of the extracts were hydrolyzable tannins and a small amount of flavonols, while phenolic acids can be obtained by hydrolysis of the hydrolyzable tannins.

The comparison with standards to identify the names of each one of the signals was not performed because the samples with which we worked were

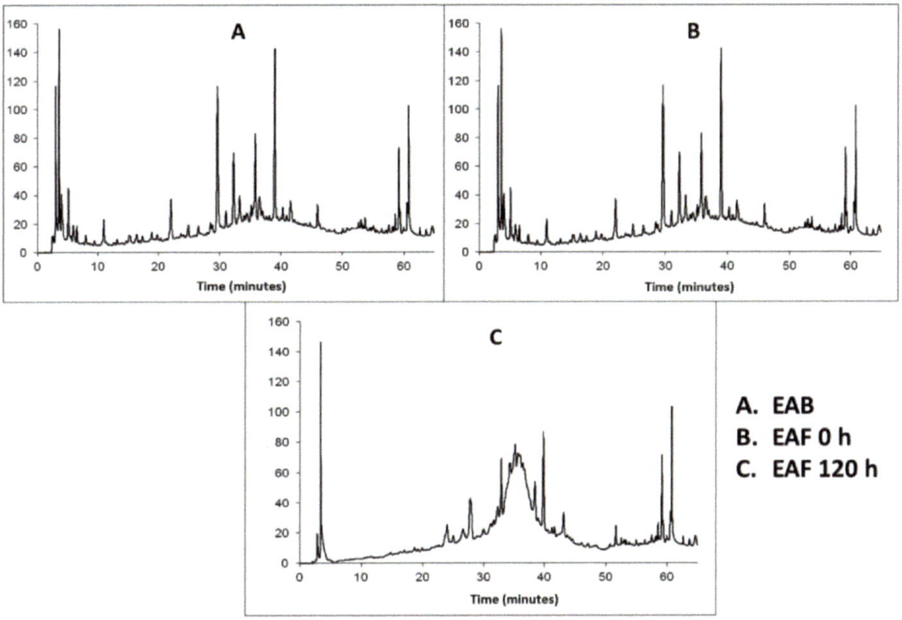

FIGURE 8.1 HPLC profile from oregano waste extracts. (A) Extract without fermentation, (B) Extract at 0 h fermentation, (**C**) Extract at 120 h fermentation.

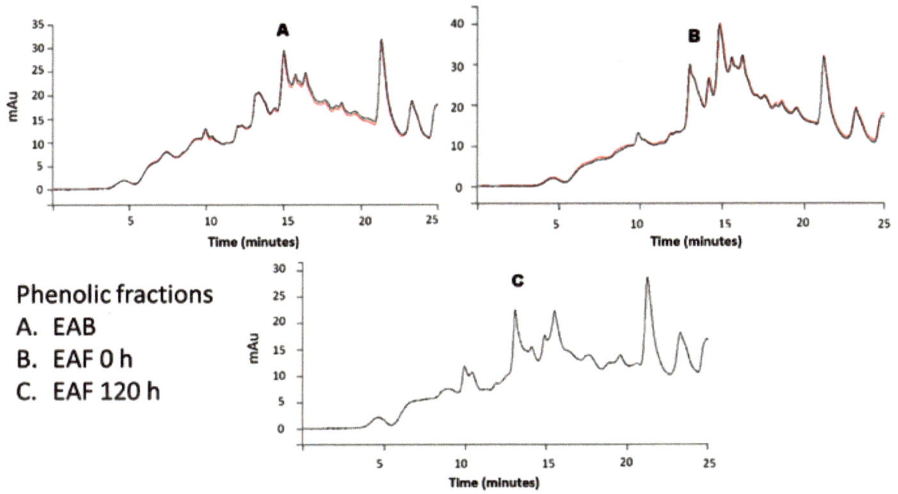

FIGURE 8.2 HPLC profile from the phenolic fraction of oregano fermented wastes extracts. (A) Phenolic extract without fermentation, (B) Phenolic extract at 0 h fermentation, (C) Phenolic extract at 120 h fermentation.

the same ones with which the not well-defined chromatograms were obtained; that is to say, fresh extracts of polyphenols were not obtained and since these are quite reactive and sensitive molecules, it is very possible that oxidation and/or degradation processes of the molecules present in the extracts and their fractions have occurred, but it was proved that this methodology is adequate for the analysis of the polyphenolic components.

The reactivity of polyphenols is mainly attributed to the position of the hydroxyl groups present in the aromatic nucleus; the ortho-hydroxyl groups (present in the B-ring of flavonoids) promote oxidation while the meta-hydroxyl groups (present in the A-ring) induce electrophilic aromatic substitution (Sreejalekshmi and Krishnan, 2011).

Within the characterization of low molecular weight tannin molecules, Aguilera-Carbo et al. (2011) mention that the carbon-carbon bonds that integrate them are not susceptible to be formed again after their hydrolysis; the reason for which the compounds detected with the second HPLC methodology are small molecular weight molecules derived from possible degradation processes.

In procyanidins (oligomers and polymers of flavan-3-ol units), the degradation process is not well characterized due to the lack of purified substrates (Contreras-Domínguez et al. 2011), but it is known that in flavonoids nucleophilic addition can be experienced in the central C ring when they have a net positive charge. It has also been shown that flavanols are readily oxidized to their corresponding ortho-quinone (o-quinone) (Sreejalekshmi and Krishnan, 2011), which are highly reactive molecules as they act in turn as hydrogen acceptors or donors; while the reaction of formation of o-quinones is catalyzed by polyphenol oxidase enzymes (Indu and Jayachandran, 2011).

In studies focused on elucidating the possible mechanism of procyanidin biodegradation, Ramirez-Coronel and Augur (2011) mention that with epicatechin-(4β→8)-epicatechin molecules used as a carbon source from ascomycetes the mechanism is based on the opening of the catechol ring of the terminal unit of the dimer by intra-diol cleavage of an oxygenase enzyme; the degradation of (+)-catechin is instead mediated by a monooxygenase enzyme, which introduces an oxygen atom to the molecule, favoring the opening of the heterocycle allowing subsequent degradation; but in a further experiment, where catechin was the main substrate, strains of *Aspergillus* sp. and *Penicillium* sp. the degradation products identified were mostly phloroglucinol carboxylic acid and protocatechuic acid. These results agree with the findings of the present investigation where the primary polyphenolic characterization suggests the presence of flavanol molecules in the extracts and the presence of phenolic acids detected by HPLC confirms the hypothesis of degradation of complex molecules, as well as the presence of hydrolyzable tannins, but the mechanism by which this degradation occurs is not clear as well as the mechanism by which such degradation occurs is still unknown.

To corroborate the composition of the extracts, as far as polyphenolic compounds are concerned, it is proposed to obtain new extracts and from these make a new HPLC run to fully characterize each of the polyphenolic molecules present in the samples and corroborate that the phenolic acids and flavanols detected belong to polyphenol degradation products.

8.3.2 BACTERIAL GROWTH RESPONSE IN SUPPLEMENTED CULTURES

Bacterial growth when supplementing the culture medium with the different extracts and their fractions did not show any pattern according to the concentrations or the strains used.

Comparing culture media (complete and modified), bacterial growth was higher with the complete medium. These results confirm that lactic acid bacteria are microorganisms with strict nutritional requirements, and supplementing the culture medium with the aqueous extracts and their fractions was an extra for the composition of the culture medium. But it is worth mentioning the growth of bacterial strains in the culture medium without glucose (modified), where it is proved that the extracts have some compounds that function as a carbon source and/or inducers for their development and reproduction.

It is worth mentioning that of the four strains used in the study only *L. acidophilus* and *L. paracasei* ATCC 27092 were the only ones that were able to grow more than their respective control in the culture medium without glucose; but supplementing the modified culture medium with the aqueous extracts or their fractions did not show additional benefits in growth; on the contrary, only the EAB F0 fraction, at a concentration of 10%, favored growth in *L. paracasei* ATCC 27092 (1.45 ± 0.037 OD) with respect to the control (0.272 ± 0.023 OD) (Table 8.3).

The bacterial growth observed in the modified culture medium suggests that the strains have the capacity to develop under these conditions, although at 24 h when the number of viable cells was counted, these did not exceed the respective control, perhaps if the culture is left longer, the result could be better and exceed the count of the control. This is due to the fact that in this culture medium, the strains do not have carbon sources available so that it is necessary that before growing exponentially, they must synthesize a series of enzymes or other metabolites that allow them to assimilate the components added in the oregano bagasse extracts and thus complete their growth phase.

Regarding the extracts, the fermented extracts were better than the EAB and depending on the strain, the fermentation time was more adequate for the growth to exceed the control. Examples of the afore mentioned are the cases of *L. casei* 71.7, where the best result was with EAF 0 h (1.531 ± 0.040 vs. 1.497 ± 0.106 OD of the control); for *L. casei* ATCC 393, a good response was obtained with EAF 0 h (1.6 ± 0.037 OD) and EAF 120 h (1.613 ± 0.002 OD), with 1.541 ± 0.032 OD being the control value (Table 8.3). The best result was obtained in *L. paracasei* ATCC 27092, whose control obtained an amount of 1.489 ± 0.104 OD when the culture medium was supplemented with EAF 0 h (1.564 ± 0.048 OD), and finally, in *L. acidophilus* the growth in the supplemented media did not improve in any of the cases to the control (1.517 ± 0.009 OD). Molan et al. (2010) found that supplementing culture media with aqueous extracts of blueberry favored the growth of *Lactobacillus strains*, results similar to those found in the present investigation.

In fractions, the best results were those of EAF 0 h, both F0 and F4, with three strains exceeding the growth of their respective controls, *L. casei* ATCC 393, *L. paracasei* ATCC 27092 and *L. casei* 71.7. *L. casei* ATCC 393 exceeded the 1.541 ± 0.032 OD of the control in MRS 1% medium with both fractions (1.628 ± 0.016 OD for F0 and 1.574 ± 0.070 OD for F4), while in MRS 10% only the F0 fraction had the result of interest with 1.551 ± 0.030 OD. *L. paracasei* ATCC 27092, in complete culture medium, responded with the F0 fraction at the two concentrations tested, the respective counts for 1 and 10% being 1.623 ± 0.007 OD and 1.514 ± 0.051 OD. In comparison with the strains described above, *L. casei* 71.7 showed higher growth than the control only in the modified culture media and with the F0 fraction, corresponding to 0.410 ± 0.018 OD for MRSm 1% and 0.457 ± 0.084 OD for MRSm 10% (Table 8.3).

The growth of bacterial cells in the fractions is an expected result. In the F0 fraction, the concentration of carbohydrates present in the extract suggests that these molecules are used as a carbon source for the development of the lactic acid strains since the growth of most lactic acid strains is based mainly on the metabolism of mono- and disaccharides (De Vos, 1996). But recent studies claim that such strains can also metabolize compounds of polyphenolic nature such as ρ-coumaric, caffeic, ferulic (Curiel et al., 2010), gallic (Guzman-Lopez et al., 2009; Curiel et al., 2010), protocatechuic (Curiel et al., 2010), m-coumaric (Rodríguez et al. 2008), and tannic

TABLE 8.3 Best Incubation Treatments of Probiotic Strains in MRS and MRS Medium Without Carbon Source (MRSm), with and Without Supplementation of Aqueous Extract of Fermented Oregano Bagasse or Its Fractions.

Extract	Strain	Culture media	Treatment number	Cellular growth (OD ± SD)	Control growth (OD ± SD)
EAB	*L. casei* 71.7	MRSm 1%	1	0.332 ± 0.023	0.263 ± 0.024
EAB F0	*L. paracasei* ATCC 27092	MRSm 10%	2	1.495 ± 0.037	0.272 ± 0.023
EAF 0 h	*L. casei* ATCC 393	MRS 1%	3	1.600 ± 0.037	1.541 ± 0.032
	L. paracasei ATCC 27092	MRS 1%	4	1.564 ± 0.048	1.489 ± 0.104
	L. paracasei ATCC 27092	MRS 10%	5	1.548 ± 0.030	1.489 ± 0.104
	L. casei 71.7	MRS 1%	6	1.531 ± 0.040	1.497 ± 0.106
EAF 0 h F0	*L. casei* ATCC 393	MRS 1%	7	1.628 ± 0.016	1.541 ± 0.032
	L. paracasei ATCC 27092	MRS 1%	8	1.623 ± 0.007	1.489 ± 0.104
	L. paracasei ATCC 27092	MRS 10%	9	1.514 ± 0.051	1.489 ± 0.104
	L. casei 71.7	MRSm 1%	10	0.410 ± 0.018	0.263 ± 0.024
	L. casei 71.7	MRSm 10%	11	0.457 ± 0.084	0.263 ± 0.024
EAF 0 h F4	*L. casei* ATCC 393	MRS 1%	12	1.574 ± 0.070	1.541 ± 0.032
EAF 120 h	*L. casei* ATCC 393	MRS 1%	13	1.613 ± 0.002	1.541 ± 0.032
	L. casei ATCC 393	MRS 10%	14	1.613 ± 0.007	1.541 ± 0.032
	L. casei ATCC 393	MRSm 1%	15	0.356 ± 0.043	0.277 ± 0.009
	L. casei 71.7	MRSm 1%	16	0.357 ± 0.027	0.263 ± 0.024
EAF 120 h F4	*L. casei* ATCC 393	MRSm 1%	17	0.290 ± 0.024	0.277 ± 0.009

acids (Rodríguez et al. 2009); this metabolism is not well characterized yet, but the literature reports that it can be attributed to different enzymes such as tannase, phenolic acid decarboxylase (PAD) or alcohol benzyl dehydrogenase (Rodríguez et al. 2009).

There were strains where the complete extract showed higher growth than the control, but in fractions, this did not happen; that is, the synergy of the components favored the bacteria. But there was also the opposite behavior, where the complete extract did not favor growth, but once fractionated, it exceeded the growth of the control.

As a summary of the results described above, the lactic acid strain with the best response to all extracts was *L. casei* ATCC 393 (Table 8.3); *L. casei* 71.7 outperformed the control with complete extracts and only with the EAF 0 h F0 fraction in MRSm; *L. acidophilus* alone outgrew the control in MRS and with the fractions, EAF 0 h F4 and EAF 120 h F0 and finally *L. paracasei* ATCC 27092 outperformed the control in EAF 0 h and EAF 0 h F0. These metabolic profiles demonstrate that lactic acid bacteria have a more flexible metabolism than proposed by De Vos (1996). It is with these combinations that the analysis of the measurement of the prebiotic effect will be continued.

In the results obtained for the measurement of the prebiotic effect (Table 8.4), only treatments 1, 2, 3, 4, 6, 13, and 14 were significantly better than their respective controls at 24 and 48 h of incubation.

Differences can be observed between the values obtained using the complete culture medium (MRS) and the medium without glucose (MRSm). The values obtained for MPE using MRSm medium are similar to those reported by Cardarelli et al. (2007), whose results range from 0.25 to 0.84, but are measured to a metabolized symbiotic product in an in vitro fermentation model; they are also similar to the values reported by Vulevic et al. (2004), who obtained MPE from −0.6 to 1.4, after simulated gastrointestinal fermentation of oligosaccharide compounds. All these results mentioned above are different from the values obtained in the MRS medium, this is directly attributed to the concentration of glucose contained in this culture medium, which is significantly higher than in the MRSm medium; together with the addition of the aqueous extract of oregano, which contains additional sugars, so that the assimilation of the substrate is different from the conditions used under the conditions of the other authors.

The best results were found at 48 h of culture, this is mainly attributed to the concentration of short-chain fatty acids quantified, which were higher at 48 h of culture. The amount of SCFA produced by probiotics is important for the selection of a prebiotic compound, this is because according to what is

TABLE 8.4 Measurement of the Prebiotic Effect (MPE) in Each Treatment at 24 and 48 h of Incubation.

Strain	Culture media	Treatment	24 h				48 h			
			x	y	z	MPE	x	y	z	MPE
L. casei ATCC 393	MRS	3	15061.10	0.197	0.0	2967.536*[a]	15643.35	0.197	1.009	16082.673*[b]
		7	15845.03	0.155	0.0	2456.480*	15737.77	0.155	0.719	11575.908
		12	15138.14	0.174	0.0	2634.536*	−2323.18	0.174	0.973	2296.815*
		13	17564.87	0.133	0.0	2336.628*	17595.52	0.133	0.734	13125.919*
		14	16203.55	0.157	0.0	2544.457*	15582.14	0.157	0.772	12276.147
		Control	14989.25	0.093	0.0	1394.500	14987.54	0.093	0.822	12398.858
	MRSm	15	3.00	0.002	0.0	0.506*[a]	−6.03	0.002	0.000	0.512*[a]
		Control	−63.35	0.006	0.0	0.880	−71.81	0.006	0.000	0.931
L. casei 71.1	MRS	6	17386.76	0.129	0.0	2243.392*[a]	17444.65	0.129	0.762	13482.459[a]
		Control	15562.14	0.105	0.0	1634.525	15805.60	0.105	0.771	12299.104
	MRSm	1	−27.67	0.004	0.0	0.617[a]	−48.57	0.004	0.000	0.704*[a]
		10	10.34	−0.005	0.0	0.552	−3.98	−0.005	0.000	0.520
		11	−44.34	0.003	0.0	0.633*	1.87	0.003	0.000	0.506
		16	1.67	0.000	0.0	0.500	−7.47	0.000	0.000	0.500
		Control	−4.33	0.001	0.0	0.504	−4.66	0.001	0.000	0.505
L. paracasei ATCC 27092	MRS	4	17193.03	0.122	0.0	2098.050*[a]	16852.53	0.122	0.935	15891.188*[b]
		5	13032.63	0.083	0.0	1082.208*	12923.67	0.083	0.890	11552.472
		8	16841.89	0.141	0.0	2375.207*	16838.49	0.141	0.741	12701.697
		9	3350.74	0.069	0.0	231.701*	13692.40	0.069	0.809	11117.867
		Control	14992.72	0.109	0.0	1634.707	15113.60	0.109	0.751	11469.741
	MRSm	2	−42.34	0.004	0.0	0.669*[a]	−40.76	0.004	0.000	0.663*[a]
		Control	−15.34	0.006	0.0	0.592	−2.64	0.006	0.000	0.516

**Significant differences between treatment and control; columns with different letters at 24 and 48 h indicate significant differences between incubation times. All results are the average of three replicates* reported in the literature SCFA gives a beneficial systemic effect (Pavlovi et al., 2012). In 2010, Al-Lahham (2010) reported that low concentrations of propionic acid in the liver and plasma reduce food consumption, exert immunosuppressive actions and probably also improve tissue sensitivity to insulin; likewise Thirabunyanon and Hongwittayakorn (2013), found that using a 16 ppm dose of propionic and butyric acids, it is possible to induce colon cancer cell death with rates of 1.6–13.8% and 6.2–23.3% for each acid, respectively. On the other hand, Lisitskaya et al. (2012) mentions that organic acids produced by certain strains of lactobacteria have inhibitory effects. In order for aqueous extracts of fermented bagasse of oregano to be proposed for the formulation of a symbiotic food, it is necessary that, in addition to the prebiotic effect already demonstrated, the number of probiotic cells at the end of the culture is considerable. Comparing treatments 3, 4, 6, 13, and 14, performed in MRS, cultures 13 (9.90 and 10.28 Log cell/mL at 24 and 48 h, respectively) and 14 (10.01 and 10.17 Log cell/mL at 24 and 48 h, respectively) yielded the highest cell counts, but the latter treatment was not significantly different from the control at 48 h of culture.

Taking into consideration all the aforementioned data, treatment 13, *L. casei* ATCC 393 in MRS with 1% of the aqueous extract of 120 h of solid-state fermentation of oregano bagasse, was selected as the best pre-probiotic combination.

8.4 CONCLUSION

With the results described above, it is possible to conclude that the aqueous extracts of fermented bagasse of oregano possess sugars and polyphenolic compounds as main components.

It was demonstrated that the aqueous extracts of fermented oregano bagasse possess prebiotic activity. The *L. casei* ATCC 393 strain obtained the best response in all the extracts; while the extract of 120 h of fermentation, supplemented at 1%, was the best pre-probiotic combination. With the results obtained in the cultures in a modified medium, it is confirmed that the lactic acid strains possess metabolic machinery to assimilate the components present in the oregano bagasse extracts, but it is necessary to carry out more specific tests to decipher the mechanism by which this phenomenon is carried out.

Also, we demonstrate that the arid zones' plant residues, such as the oregano, offer several applications to assure the zero residues accomplishment in order to promote biotechnological applications in a sustainable and ecological environment that enhance the human health and environmental preservation.

KEYWORDS

- **prebiotics**
- **polyphenols**
- **oregano wastes**
- **solid-state fermentation**

REFERENCES

Aguilera-Carbo, A. F.; Chavez-Gonzalez, M.; Rodríguez-Duran, L. V.; Prado-Barragán, L. A.; Aguilar, C. B. N.; Biodegradation of Hydrolisable Tannins: The Case of Ellagitannins. In: *Chemistry and Biotechnology of Polyphenols*; Sabu, A.; Roussos, S.; Aguilar, C. N., Eds.; India, 2011.

Al-Lahham, S. H.; Peppelenbosch, M. P.; Roelofsen, H.; Vonk, R. J.; Venema, K. Biological Effect of Propionic Acid in Humans; Metabolism, Potential Applications and Underlying Mechanisms. *Biochim. Biophys. Acta.* **2010,** *1801*, 1175–1183. DOI: 10.1016/j.bbalip.2010.07.007

Bogar, B.; Szakacs, G.; Pandey, A.; Abdulhameed, S.; Linden, J. C.; Tengerdy, R. P. Production of Phytase by *Mucor racemosus* in Solid-State Fermentation. *Biotechnol. Prog.* **2003,** *19*, 312–319. DOI: 10.1021/bp020126v

Borchers A. T.; Selmi C.; Meyers F. J.; Keen C. L.; Gershwin M. E. Probiotics and Immunity. *J Gastroenterol.* **2009**. DOI: 10.1007/s00535-008-2296-0

Burr, G.; Hume, M.; Ricke, S.; Nisbet, D.; Gatlin III, D. In Vitro and In Vivo Evaluation of the Prebiotics GroBiotic®-A, Inulin, Mannanoligosaccharide, and Galactooligosaccharide on the Digestive Microbiota and Performance of Hybrid Striped Bass (*Morone chrysops* x *Morone saxatilis*). *Microbiol. Ecol.* **2010,** *59*, 187–198. DOI: 10.1007/s00248-009-9597-6.

Burt, S. A.; Reinders, R. D. Antibacterial Activity of Selected Plant Essential Oils Against *Escherichia coli* O157:H7. *Lett. Appl. Microbiol.* **2003,** *36*, 162–167. DOI: 10.1046/j.1472-765x.2003.01285.x

Cardarelli, H. R.; Saad, S. M. I.; Gibson, G. R.; Vulevic, J. Functional Petit-Suisse Cheese: Measure of the Prebiotic Effect. *Anaerobe* **2007,** *13*, 200–207. DOI: 10.1016/j.anaerobe.2007.05.003

Castro M.; Rodríguez F. Levaduras: probióticos y prebióticos que mejoran la producción animal. *Revista corpoica* **2005,** *6*, 1. DOI: 10.21930/rcta.vol6_num1_art:33

Cheng-Lee H.; Jenner A. M.; Seng-Low C.; Kun-Lee Y. Effect of Tea Phenolics and Their Aromatic Fecal Bacterial Metabolites on Intestinal Microbiota. *Res. Microbiol.* **2006,** *157*, 876–884. DOI: 10.1016/j.resmic.2006.07.004

Contreras-Domínguez, M.; Guyot, S.; Roussos, S.; Augur, C. Enzymatic Degradation of Procyanidin B1 by *Aspergillus fumigatus*. In: *Chemistry and Biotechnology of Polyphenols*; Sabu, A., Roussos, S.; Aguilar, C.N., Eds.; India, 2011.

Correia, R. T. P.; McCue, P.; Magalhães, M. M. A.; Macêdo, G. R.; Shetty, K. Production of Phenolic Antioxidants by the Solid-State Bioconversion of Pineapple Waste Mixed with Soy Flour Using *Rhizopus oligosporus*. *Proc. Biochem.* **2004,** *39*, 2167–2172. DOI: 10.1016/j.procbio.2003.11.034

Curiel, J. A.; Rodriguez, H.; Landete, J. M.; de las Rivas, B.; Muñoz, R. Ability of *Lactobacillus brevis* Strains to Degrade Food Phenolic Acids. *Food Chem.* **2010,** *120*, 225–229. DOI: 10.1016/j.foodchem.2009.10.012

De las Cagigas-Reig A. L.; Blanco-Anesto J. Prebióticos y probióticos, una relación beneficiosa. *Revista cubana alimentación y nutricion* **2002,** *16* (1), 63–68.

De Vos, W. M. Metabolic Engineering of Sugar Catabolism in Lactic Acid Bacteria. *Antonie van Leeuwenhoek* **1996,** *70*, 223–242. DOI: 10.1007/BF00395934

Dubois, M.; Gilles, K. A.; Hamilton, J. K.; Rebers, P. A.; Smith, F.; Colorimetric Method for Determination of Sugars and Related Substances. *Anal. Chem.* **1956,** *28* (3), 350–356. https://doi.org/10.1021/ac60111a017

Geier, M. S.; Butler, R. N.; Howarth, G. S. Inflammatory Bowel Disease: Current Insights Into Pathogenesis and New Therapeutic Options; Probiotics, Prebiotics and Symbiotic. *Int. J. Food Microbiol.* **2007,** *115*, 1–11. DOI: 10.1016/j.ijfoodmicro.2006.10.006

Guzman-Lopez, O.; Loera, O.; Parada, J. L.; Castillo-Morales, A.; Martinez-Ramirez, C.; Augur, C.; Gaime-Perraud, I.; Saucedo-Castañeda, G. Microcultures of Lactic Acid Bacteria: Characterization and Selection of Strains, Optimization of Nutrients and Gallic Acid Concentration. *J. Ind. Microbiol. Biotechnol.* **2009,** *36*, 11–20. DOI: 10.1007/s10295-008-0465-z

Hajji, M.; Rebai, A.; Gharsallah, N.; Nasri, M. Optimization of Alkaline Protease Production by *Aspergillus clavatus* ES1 in *Mirabilis jalapa* Tuber Powder Using Statistical Experiment Design. *Appl. Microbiol. Biotechnol.* **2008,** *79*, 915–923. DOI: 10.1007/s00253-008-1508-0

Hernández, T.; Canales, M.; García, A. M.; Duran, A.; Meráz, S.; Dávila, P.; Ávila, J. G. Antifungal Activity of the Essential Oils os Two Verbenaceae: *Lantana achyranthifolia* and *Lippia graveolens* of Zapotitlán de las Salinas, Puebla (Mexico). *Boletín Latinoamericano y del Caribe de Plantas Medicinales y Aromáticas* **2008,** *7* (4), 203–207. https://www.redalyc.org/articulo.oa?id=85670404

Indu, C. N.; Jayachandran, K. Polyphenol Oxidase and Its Applications. In: *Chemistry and Biotechnology of Polyphenols*; Sabu, A., Roussos, S., Aguilar, C. N., Eds.; India, 2011.

Kalui C. M.; Mathara J. M.; Kutima P. M. Probiotic Potential of Spontaneously Fermented Cereal Based Food—A Review. *Afr. J. Biotechnol.* **2010,** *9* (17), 2490–2498.

Khalil, R. K. S. Influence of Gallic Acid and Catechin Polyphenols on Probiotic Properties of *Streptococcus thermophilus* CHCC 3534 Strain. *World J. Microbial. Biotech.* **2010,** *26*, 2069–2079. DOI: 10.1007/s11274-010-0393-8

Kovatcheva-Apostolova, E. G.; Georgiev, M. I.; Ilieva, M. P.; Skibsted, L. H.; Rødtjer, A.; Larsen-Andersen, M. Extracts of Plant Cell Cultures of *Lavandula vera* and *Rosa damascene* as Sources of Phenolic Antioxidants for Use in Foods. *Eur. Food Res. Technol.* **2008,** *227*, 1234–1249. https://doi.org/10.1007/s00217-008-0842-x

Lateff, A.; Oloke, J. K.; Gueguim, K.; Oyeniyi, S. O.; Onifade, O. R.; Oyeleye, A. O.; Oladosu, O. C.; Oyelami, A. O. Improving the Quality of Agro-Wastes by Solid-State Fermentation: Enhanced Antioxidant Activities and Nutritional Qualities. *World J. Microbiol. Biotechnol.* **2008,** *24*, 2369–2374. DOI: 10.1007/s11274-008-9749-8

Lisitskaya, K. V.; Nikolaev, I. V.; Torkova, A. A.; Popov, V. O.; Koroleva, O. V. Analysis of Functional Properties of Biologically Active Substances Using Eukaryotic Cell Models (Review). *Appl. Biochem. Microbiol.* **2012,** *48* (6), 525–540. DOI: 10.1134/S000368381 2060087

Martins, S.; Mercado, D.; Mata-Gómez, M.; Rodriguez, L.; Aguilera-Carbo, A.; Rodriguez, R.; Aguilar, C. N. Microbial Production of Potent Phenolic-Antioxidants Through Solid State Fermentation. In: *Sustainable Biotechnology*; Singh, O.V., Harvey, S. P., Eds., Springer Science+Business, 2010. DOI: 10.1007/978-90-481-3295-9_12

Mata-González, R.; Meléndez-González, R. Growth Characteristics of *Mexican oregano* (*Lippia berlandieri* Schauer) Under Salt Stress. *Sourtwestern Naturalist* **2005,** *50* (1), 1–6. DOI: 10.1894/0038-4909(2005)050<0001:gcomol>2.0.co;2

Meléndez-Rentería N. P.; Silva-Vázquez R.; Nevárez-Moorillón G. V.; Aguilar-González C. N. Rodríguez-Herrera R. Diversidad Genética y Actividades Biológicas de *Lippia berlandieri* Schauer. Tesis de maestría. Departamento de investigación en alimentos, Facultad de Ciencias Químicas, Universidad Autónoma de Coahuila, 2010.

Meléndez-Rentería, P.; Rodríguez-Herrera, R.; Aguilar, C. N.; Nevárez-Moorillón, G. V. Microbiological Effect of Fermented *Mexican oregano* (*Lippia berlandieri* Schauer) Waste. *Waste Biomass Valorization,* **2014,** 5, 57-63. DOI: 10.1007/s12649-013-9222-2

Miller, G. L.; Use of Dinitrosalicylic Acid Reagent for Determination of Reducing Sugar. *Anal. Chem.* **1959,** *31* (3), 426–428. https://doi.org/10.1021/ac60147a030.

Molan, A. L.; Liu, Z.; Kruger, M. The Ability of Blackcurrant Extracts to Positively Modulate Key Markers of Gastrointestinal Function in Rats. *World J. Microbiol. Biotechnol.* **2010,** *25,* 1243–1249. DOI: 10.1007/s11274-010-0352-4

Mussatto, S. I.; Ballesteros, L. F.; Martins, S.; Teixeira, J. A. Use of Agro-Industrial Wastes in Solid-State Fermentation Processes. In: *Industrial Waste*; Show, K.-Y., Ed.; 2012. ISBN: 978-953-51-0253-3. Doi: 10.5772/36310. http://www.intechopen.com/books/industrial-waste/use-of-agro-industrial-wastes-in-solid-state-fermentation-processes.

Nelson, N. *A Photometric Adaptation of the Somogyi Method for the Determinations of Biological Chemistry*, 1944. https://doi.org/10.1016/S0021-9258(18)71980-7

Niba, A. T.; Beal, J. D.; Kudi, A. C.; Brooks, P. H. Bacterial Fermentation in the Gastrointestinal Tract of Non-Ruminants: Influence of Fermented Feeds and Fermentable Carbohydrates. *Trop. Anim. Health Prod.* **2009,** *41*, 1393–1407. DOI: 10.1007/s11250-009-9327-6

Nostro, A.; Saudano-Roccaro, A.; Bisignano, G.; Marino, A.; Cannatelli, M. A.; Pizzimenti, F. C.; Cioni, P. L.; Procopio, F.; Blanco, A. R.; Effects of Oregano, Carvacrol and Thymol on *Staphylococcus aureus* and *Staphylococcus epidermidis* Biofilms. *J. Med. Microbiol.* **2007,** *56,* 519–523. DOI: 10.1099/jmm.0.46804-06.

O'Sulivan L.; Murphy B.; McLoughlin P.; Duggan P.; Lawlor P. G.; Hughes H.; Gardiner G. E.; Prebiotics from Marine Macroalgae for Human and Animal Health Applications. *Marine Drugs* **2010,** *8*, 2038–2064. DOI: 10.3390/md8072038

Ocampo-Velázquez, R. V.; Malda-Barrera, G. X.; Suárez-Ramos, G. Reproductive Biology of *Mexican oregano* (*Lippia graveolens* Kunth) in Three Exploitation Conditions. *Agrociencia* **2009,** *43*, 475–482.

Pandey, K. B.; Rizvi, S. I. Plant Polyphenols as Dietary Antioxidants in Human Health and Disease. *Oxid. Med. Cell Longevity* **2009,** *2* (5), 270–278. DOI: 10.4161/oxim.2.5.9498.

Paredes-Aguilar, M. C.; Gastélum-Franco, M. G.; Silva-Vázquez, R. Efecto antimicrobiano del orégano mexicano (*Lippia berlandieri* Schauer) y de su aceite esencial sobre cinco especies del género *Vibrio*. *Revista Fitotecnia Mexicana* **2007,** *30* (3), 261–267. https://www.redalyc.org/articulo.oa?id=61003008

Pascual, M. E.; Slowing, K.; Carretero, E.; Sánchez-Mata, D.; Villar, A. Lippia: Traditional Uses, Chemistry and Pharmacology: A Review. *J. Ethnopharmacol.* **2001,** *76*, 201–214. DOI: 10.1016/s0378-8741(01)00234-3

Pavlovi, N.; Stankov, K.; Mikov, M.; Probiotics-Interaction with Bile Acids and Impact on Cholesterol Metabolism. *Appl. Biochem. Biotech.* **2012,** *168*, 1880–1895. DOI: 10.1007/s12010-012-9904-4

Piñeiro, M.; Asp, N. G.; Brunser, O.; Macfarlane, S.; Morelli, L.; Reid, G.; Tuohy, K. FAO Technical Meeting on Prebiotics. 15–16 Sept. 2007. http://www.fao.org/ag/agn/agns/files/Prebiotics_Tech_Meeting_Report.pdf

Ramirez-Coronel, M. A.; Augur, C. Biodegradation of Condensed Tannins. In: Chemistry and Biotechnology of Polyphenols; Sabu, A., Roussos, S., Aguilar, C. N., Eds.; India, 2011.

Rodriguez, H.; Curiel, J. A.; Landete, J. M.; de las Rivas, B.; Lopez de Felipe, F.; Gomez-Cordoves, C.; Mancheño, J. M.; Muñoz, R. Food Phenolics and Lactic Acid Bacteria. *Int. J. Food Microbiol.* **2009,** *132*, 79–90. DOI: 10.1016/j.ijfoodmicro.2009.03.025

Rodríguez, H.; Landete, J. M.; de las Rivas, B.; Muñoz, R. Metabolism of Food Phenolic Acids by *Lactobacillus plantarum* CECT 748T. *Food Chem.* **2008**. DOI: 10.1016/j.foodchem.2007.09.067

Selwal, M. K.; Yadav, A.; Selwal, K. K.; Aggarwal N. K., Gupta, R.; Gautam, S. K.; Tannase Production by *Penicillum atramentosum* Under SSF and Its Applications in Wine Clarification and Tea Cream Solubilization. *Braz. J. Microbiol.* **2011,** *42*, 374–387. DOI: 10.1590/s1517-83822011000100047

Somogyi, M. Notes on Sugar Determination. *J. Biol. Chem.* **1926**. DOI: 10.1016/S0021-9258(19)50870-5

Sreejalekshmi, K. G.; Krishnan, K. A. Chemical Aspects of Polyphenols—A Short Survey from Molecular Structure to Reactivity. In: *Chemistry and Biotechnology of Polyphenols*; Sabu, A., Roussos, S., Aguilar, C. N., Eds.; India, 2011.

Swain, T.; Hillis, E. The Phenolic Constituents of *Prunus domestica*: The Quantitative Analysis of Phenolic Constituents. *J. Sci. Food Agric.* **1959,** *10*, 63–68. http://dx.doi.org/10.1002/jsfa.2740100110

Thirabunyanon, M.; Hongwittayakorn, P. Potential Probiotic Lactic Acid Bacteria of Human Origin Induce Antiproliferation of Colon Cancer Cells via Synergic Actions in Adhesion to Cancer Cells and Short-Chain Fatty Acid Bioproduction. *Appl. Biochem. Biotechnol.* **2013,** *169*, 511–525. DOI: 10.1007/s12010-012-9995-y

Vulevic, J.; Rastall, R. A.; Gibson, G. R. Developing a Quantitative Approach for Determining the In Vitro Prebiotic Potential of Dietary Oligosaccharides. *FEMS Microbiol. Lett.* **2004,** *236*, 153–159. DOI: 10.1016/j.femsle.2004.05.036

Waterman P. G.; Mole, S. *Analysis of Phenolic Plant Metabolites*; Blackwell Scientific Publication: USA, 1994.

CHAPTER 9

Venomous Animals and Their Toxins in the Arid Zones of Mexico

DAVID RAMIRO AGUILLÓN-GUTIÉRREZ

Laboratorio de Bioindicadores, Centro de Investigación y Jardín Etnobiológico, Universidad Autónoma de Coahuila, Viesca, Coahuila, Mexico

ABSTRACT

A toxin is a damaging substance made within living cells or organisms (microorganisms, fungus, plants, or animals). The word "biotoxin" is used to confirm the biological origin. Biotoxins have two functions: predation, such as in the spider, snake, centipede, and scorpion, and defense as in the bee, ant, and wasp. "Toxin" refers to a singular molecular compound with a particular action or binding site; venom must be injected into an organism by an apparatus like a sting, and poison is transferred by ingestion or absorption. Not all animals produce their own poison, some of them obtain their venoms or poisons from substances that are smeared, ingested, or accumulated in their tissues.

In the Mexican desert we can find several species of venomous animals and insects, some of them are rattlesnakes (genera *Crotalus* and *Sistrurus*, subfamily *Crotalinae*, family *Viperidae*); lizards as *Heloderma*; toads; spiders (genus *Latrodectus and Loxosceles);* scorpions; centipedes; bees and wasps and ants.

Poisons and toxins are diverse in nature and many of them have or could have biotechnological applications, mainly in the field of pharmacology. The high value for humans of these toxins and their possible applications should be reflected in the conservation of these species and in arid zones in general, since as species are lost, the ecosystem is degraded, directly affecting human beings.

Exploration and Valorization of Natural Resources from Arid Zones. Cristobal Noé Aguilar, Norma Paola Meléndez-Renteria, & Karen Nathiely Ramirez-Guzman (Eds.)

9.1 VENOMS, POISONS, AND TOXINS

A toxin is a damaging substance made within living cells or organisms (microorganisms, fungus, plants, or animals) (Repetto, 1987). The word "biotoxin" is used to confirm the biological origin. Biotoxins have two functions: predation, such as in the spider, snake, centipede, and scorpion, and defense as in the bee, ant, and wasp. "Toxin" refers to a singular molecular compound with a particular action or binding site (Brodie, 2009).

Venom must be injected into an organism by an apparatus like a sting (Casewell et al., 2013; Arbuckle, 2015; Harris and Arbuckle, 2016), and poison is transferred by ingestion or absorption. It is remarkable that venoms are more flexible compared to poisons, as they can be used for all the mentioned functions (predation, defense, and competition), whereas poisons can only be used for defense to deter predators (Arbuckle, 2015; Harris and Arbuckle, 2016).

Not all animals produce their own poison, some of them obtain their venoms or poisons from substances that are smeared, ingested, or accumulated in their tissues. For example, some species of European or African porcupines have the ability to capture toads, and before eating them, they chew them until they obtain toxic secretions from their skin, and later these are mixed with their saliva that they will spread through their spines when licking them. Others, like insects, extract and accumulate toxins from the plants they eat, and poison dart frogs obtain their toxins from the ants and mites they ingest (Cifuentes and Cupul, 2010).

9.2 DESERTIC AND SEMIDESERTIC ZONES

In general, an area with low rainfall is considered a desert (<90–500 mm of precipitation per year), with low soil moisture content, low atmospheric humidity, a bare, salty and dry soil, with air movements greater than in humid climates, small cyclonic disturbances, and high daytime temperatures. During the day, the desert absorbs 90% of the solar radiation that warms the soil and the lower layer of the air, because dust particles and clouds only deviate 10%, while, at night, heat accumulated during the day is lost, as 90% of that heat escapes into the upper layers of the atmosphere.

There are different criteria to classify deserts, for example, the geomorphological criterion, according to the type of soil, the criterion based on precipitation and temperature, geographic criteria, biological, and agricultural criteria, among others. In general, arid zones are classified as extremely arid or hyper arid zones, arid, and semi-arid zones. The biological potential of deserts is severely limited by water scarcity, for example, xerophytic vegetation is adapted to this condition. Deserts are currently threatened ecosystems, since anthropogenic activities such as livestock (overgrazing), the agriculture, the introduction of exotic species, pollution, climate change, and the overexploitation of natural resources, puts these ecosystems at serious, promoting desertification, and forming areas that do not have the natural elements of deserts (Cloudsley-Thompson, 1977; Leopold, 1979; Hernández, 2006; Granados-Sánchez et al., 2012; León-de la Luz et al., 2013; Saygin, 2017).

9.2.1 THE ARID AND SEMIARID ZONES OF MEXICO

If it is considered that more than 60% of the Mexican Republic is between 20° and 40° north latitude, then a large part of its surface is subject to arid conditions, which vary regionally due to a set of physical and biological factors (Granados-Sánchez et al., 2012). According to Miranda (1955), there is a close relationship between the climate and the vegetation that determines the conditions in a place. The arid and semi-arid regions of the country have both a flora and fauna richness that is favored by the variety of environments, from those influenced by the global strip of arid zones to those influenced by the tropical strips (Granados-Sánchez et al., 2012). This allows the identification of nine arid regions in Mexico: Sonorense, Chihuahuense (Figure 9.1), Hidalguense, Poblana, Tamaulipeca, Guerrerense, Tehuantepeca, Veracruzana, and Yucateca (Miranda, 1955).

9.3 ANIMALS AND THEIR TOXINS IN THE ARID ZONES OF MEXICO

9.3.1 VERTEBRATES

Phylum: Chordata

Class: Reptilia or Sauropsida

Rattlesnakes

Rattlesnakes are venomous snakes (genera *Crotalus* and *Sistrurus,* subfamily Crotalinae) with distribution in the American continent (Figure 9.2). Rattlesnakes have a poison apparatus, which consist in the venom gland, the venom

FIGURE 9.1 Some typical landscapes of the Chihuahuense Desert of Mexico.

FIGURE 9.2 Some rattlesnakes from the arid zones of Mexico.

duct and the fangs. These structures allow rattlesnakes to efficiently attack their prey and ensure their food, this means that the venom is produced in sufficient quantities in their body, that it leaves the glands through the ducts and fangs and that it is successfully administered on the prey, leading to paralysis or death (Klauber, 1972).

The main function of snake venom is to immobilize or kill its prey and facilitate its digestion. The venom they produce is a mixture of dissimilar substances such as enzymes, proteins, polypeptides, anticoagulants, and others, which have a high biological specialization. The composition of the venom will depend on various factors such as the species, distribution, and age, among others (Russell, 1983; Fernández-Badillo et al., 2011; Ávila-Villegas, 2017).

Due to the high complexity of snake venoms, it is difficult to classify them. They are commonly divided according to the mode of action of their toxins, for example, neurotoxic, in which components that act on the nervous system predominate, cytotoxic, which affect various cells, myotoxic that affect the muscles or hemotoxic whose components attack blood cells (Fernández-Badillo et al., 2011; Ávila-Villegas, 2017).

Just as snake venoms cause damage to organs and tissues of their prey, such as drops in blood pressure, paralysis, internal bleeding, among others, they are invaluable resources to generate treatments against various diseases and conditions such as blood pressure problems, failure in coagulation, cardiomyopathies, Alzheimer's, and cancer. This is because, for example, poisons cause bleeding by causing coagulation failure, so they could serve as medicines against problems related to the formation of thrombi and clots in the appropriate doses (Ávila-Villegas, 2017).

Heloderma lizards

Heloderma is a genus of venomous lizards with distribution in United States of America, Mexico and Guatemala (Lemos-Espinal et al., 2015) (Figure 9.3).

The venom glands of *Heloderma* are located behind the eyes, at the end of the lower jaws. *Heloderma* venom is used in defense (Cogger and Zweifel, 1992).

Some common components of the *Heloderma* venoms are hyaluronidase, gilatoxin, phospholipase A, helothermine, exendines, gilatide, and serotonine (Mackessy, 2010). Exendin-4 is an incretin-mimetic peptide. Exendin-4 enhances the functions of β-cells and the up-regulation of GLP-1 receptors. In addition, exendin-4 can ameliorate ailments as neuropathy, nephropathy, and ventricular remodeling (Yap and Misuan, 2019).

FIGURE 9.3 Heloderma lizards from Mexico.

Class: Amphibia

Toads

Toxins are produced by toads; although, potency change with species and even between geographic areas for the same species. Toad toxin is a defensive mechanism and is secreted by glands find dorsal and caudal to the eyes and by glands disperse throughout the skin (Gwaltney-Brant, 2022) (Figure 9.4).

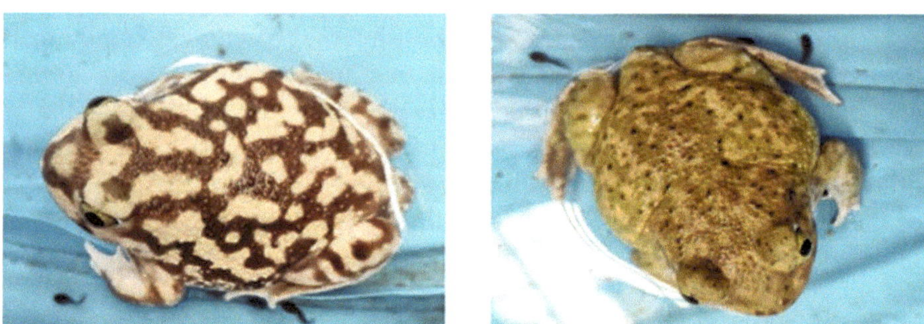

FIGURE 9.4 Toads from the arid zones of Mexico.

A special case is that of the Sonoran Desert toad (*Incilius alvarius*). The skin secretions of this toad are particularly toxic and have been known to cause paralysis and death in dogs that mouth them (Rorabaugh and Lemos-Espinal, 2016). The skin contains a number of indolealkylamines and their metabolites belonging to the series of 5-hydroxy-indolealkylamines and 5-methoxyindolealkylamines. The most abundant representative of 5-hydroxyindolealkylamines is bufotenine (Erspamer et al., 1967).

Toxins from toads contain rich chemicals with pharmaceutical potential, for example, peptides, steroids, indole alkaloids, bufogargarizanines, and organic acids. Modern studies revealed that these components can be converted into medicines for inflammatory diseases and cancer (Qi et al., 2018).

9.3.2 INVERTEBRATES

Phylum: Arthropoda

Class: Arachnida

Spiders

Genus Latrodectus:

Latrodectus is a genus of spiders known as black widows (Figure 9.5). These spiders have a potent venom containing the neurotoxin latrotoxin. These latrotoxins have been described: five insecticidal toxins, (α, β, γ, δ and ε-latroinsectotoxins), one vertebrate-specific neurotoxin, alpha-latrotoxin, and one toxin affecting crustaceans, α-latrocrustatoxin (Grishin, 1998).

Latrodectus spiders' venom can be used for Black widow spider envenomation (BWSE) and BWSE-associated priapism (Hoover and Fortenberry, 2004).

FIGURE 9.5 Black widow spider.

Genus Loxosceles:

The recluse spiders also known as violin spiders have a tissue-destroying venoms containing the dermonecrotic agent sphingomyelinase D and sulfates nucleosides (Binford et al., 2005).

Antimicrobial peptides discovered in spider venoms may be useful resources for the design of structurally new anti-infective agents effective against drug-resistant microorganisms, which works against Gram-negative bacteria (Segura-Ramírez and Silva-Júnior, 2018).

Scorpions

Scorpions are predatory arachnids of the order Scorpions that have a venomous sting used for offense and defense (Figure 9.6).

The composition of scorpion venom is complex and heterogeneous, for example, neurotoxins; hemolysins; peptides that are classified into three superfamilies: peptides containing cysteine-stabilized (CS) α/β motifs, calcins, and non-disulfide bridged peptides (NDBPs); enzymes, inorganic salts; amino acids; nucleotides; amines; and lipids (Hoffmann, 2003; Ahmadi et al., 2020).

Scorpions are a valuable source of biologically active molecules, from novel antibiotics to potential anticancer therapeutics (Ortiz et al., 2015).

FIGURE 9.6 Scorpions from the arid zone of Mexico.

Class: Chilopoda

Centipedes

Centipedes are venomous invertebrate animals (subphylum Myriapoda) (Figure 9.7). There is a species called giant desert centipede (*Scolopendra heros*) that inhabits the arid zones of Southwestern United States and

Northern Mexico (Barnes, 2002). Centipede venoms are mixtures containing 5-hydroxytryptamine, histamine, lipids, polysaccharides, and enzymes. The lipid components are squalene, fatty acids, cholesterol, triglycerides, cholesterol esters, and phospholipids. The venom of the North American giant desert centipede holds a cytolysin (Yildiz et al., 2006).

Centipede venoms are complex chemical arsenals that are rich in disulfide-constrained peptides that have novel pharmacology centipede venom proteins that might be useful from a therapeutic perspective (Undheim et al., 2016).

FIGURE 9.7 Centipedes from the arid zone of Mexico.

Class: Insecta

Order Hymenoptera

Bees and wasps

Bees and wasps are social insects and some of them are also important pollinating insects (Figure 9.8). Bee and wasp venom contains substances like histamine, acid phosphatase, mast-cell degranulation protein, hyaluronidase norepinephrine, dopamine, lysophospholipase, and serotonin. Melittin make possible ion diffusion across cell membranes and works with phospholipase to hemolyze red blood cells. Wasp venoms contain amines, peptides, and proteins, including hydrolases.

FIGURE 9.8 Bee and wasp from the arid zone of Mexico.

Honey bee venom (apitoxine) is a cytotoxic and hemotoxic liquid containing proteins, which may produce local inflammation. Diverse neurotoxins like bradykinin-related peptides, pompilidotoxins, philanthotoxins, and microbracotoxin have been isolated from wasp venom (Morgan and Wilson, 1999; Postma, 2009; Aufschnaiter et al., 2020).

Bee venom can be used to treat diseases such as multiple sclerosis, lupus, sciatica, low back pain, arthritis, rheumatoid arthritis, and tennis elbow, among others (Mahmoud, 2012). Meanwhile, wasp venom present antiarthritic activity (Gao et al., 2020).

Ants

Ant venoms are rich in linear, dimeric, and disulfide-bonded peptides and bioactive proteins, and other volatile and non-volatile compounds such as alkaloids and hydrocarbon. Ant venom peptides can be classified into cytolytic and neurotoxic peptides. These peptides have been classified by their structure into three groups: linear, dimeric, and inhibitor cystine knot (ICK)-like peptides (Touchard et al., 2016).

Ant venom may act as an anti-inflammatory agent in diseases like rheumatoid arthritis because of its activity on the proteins of the complement system (Costa-Neto, 2005).

9.4 CONCLUSION

Arid zones have a very particular biodiversity, since evolutionary mechanisms need to be developed to allow adaptation to these extreme environments. One of these mechanisms in some animals is the production of venom, poisons, and toxins that serve as defense and attack against other animals. The fauna of the Mexican desert that produces poison is made up of reptiles, amphibians, arachnids, myriapods, and insects. Poisons and toxins are diverse in nature and many of them have or could have biotechnological applications, mainly in the field of pharmacology. The high value for humans of these toxins and their possible applications should be reflected in the conservation of these species and in arid zones in general, since as species are lost, the ecosystem is degraded, directly affecting human beings.

ACKNOWLEDGMENT

I would like to express my special thanks to Dr. Jorge Alejandro Aguirre Joya, Dr. Cristian Torres León, Ing. Rodolfo Hernández Vélez, Ing. Salvador Hernández Vélez, Biol. Alexis Pardo Ramírez, Biol. Marisol Moreno Chávez, Ing. Alma Leticia Espinoza, Ing. Juan Carlos Chavarría, Ing. Isaac Gallardo, Ing. Fernando Camacho, MVZ. Cristina Villagrana Vázquez, to my students from FCB-UJED and to all the CIJE work team for providing some of the photographs.

KEYWORDS

- **venomous animals**
- **toxines**
- **arid zones**
- **Mexico**

REFERENCES

Ahmadi, S.; Knerr, J. M.; Argemi, L.; Bordon, K. C.; Pucca, M. B.; Cerni, F. A.; Arantes, E. C.; Caliskan, F.; Laustsen, A. H. Scorpion Venom: Detriments and Benefits. *Biomedicines* **2020,** *8*,118.

Arbuckle, K. *Evolutionary Context of Venom in Animals: Evolution of Venomous Animals and Their Toxins*; Springer: Dordrecht, 2015.

Aufschnaiter, A.; Kohler, V.; Khalifa, S.; El-Wahed, A. A.; Du, M.; El-Seedi, H.; Buttner, S. Apitoxin and Its Components Against Cancer, Neurodegeneration and Rheumatoid Arthritis: Limitations and Possibilities. *Toxins* **2020,** *12*, 66.

Ávila-Villegas, H. *Serpiente de cascabel: entre el peligro y la conservación*; CONABIO: Mexico, 2017.

Barnes, J. K. Giant Redheaded Centipede. *Arthr Mus Notes 13*. University of Arkansas, 2002.

Binford, G. J.; Cordes, M. H. J.; Wells, M. A. Sphingomyelinase D from Venoms of *Loxosceles* Spiders: Evolutionary Insights from cDNA Sequences and Gene Structure.*Toxicon* 2005, 45 (5), 547–560.

Brodie, E. D.; Toxins and Venoms. *Curr. Biol.* **2009,** *19* (20), 931–935.

Casewell, N. R.; Wuster, W.; Vonk, F. J.; Harrison, R. A.; Fry, B. G. Complex Cocktails: The Evolutionary Novelty of Venoms. *Trends Ecol. Evol.* **2013,** *28* (4), 219–229.

Cifuentes, J. L.; Cupul, F. G. *Venenos: armas químicas de la naturaleza*; FCE: Mexico, 2010.

Cloudsley-Thompson, J. L. *El hombre y la biología de las zonas áridas*; Orbis: Barcelona, España, 1977.

Cogger, H. G.; Zweifel, R. G. *Reptiles y anfibios*; Lunwerg: España, 1992.

Costa-Neto, E. M. Entomotherapy, or the Medicinal Use of Insects. *J. Ethnobiol.* **2005,** *25* (1), 93–114.

Erspamer, V.; Vitali, T.; Roseghini, M.; Cei, J. M. 5-Methoxy and 5-Hydroxyindoles in the Skin of *Bufo alvarius*. *Biochem. Pharmacol.* **1967,** *16*, 1149–1164.

Fernández-Badillo, L.; Morales-Capellán, N.; Goyenechea-Mayer Goyenechea, I. *Serpientes venenosas del estado de Hidalgo*; UAEH: Mexico, 2011.

Gao, Y.; Yu, W. X.; Duan, X. M.; Ni, L. L.; Liu, H.; Zhao, H. R.; Xiao, H.; Zhang, C. G.; Yang, Z. B. Wasp Venom Possesses Potential Therapeutic Effect in Experimental Models of Rheumatoid Arthritis. *Evid.-Based Complement. Altern. Med.* **2020,** 6394625.

Granados-Sánchez, D.; Hernández-García M. A.; López-Ríos G. F. *Ecología de los desiertos del mundo*; Universidad Autónoma Chapingo Texcoco: Estado de Mexico, Mexico, 2012.

Grishin, E. V. Black Widow Spider Toxins: The Present and the Future. *Toxicon* 1998, *36* (11), 1693–1701.

Gwaltney-Brant, S. M. Toad Poisoning in Dogs and Cats. MSD Veterinary Manual, 2022.

Harris, R. J.; Arbuckle, K. Tempo and Mode of the Evolution of Venom and Poison in Tetrapods. *Toxins* **2016,** *8*, 193.

Hernández, M. H. La vida en los desiertos mexicanos. Ciudad de Mexico, Mexico: Colección La Ciencia para Todos, 213, Fondo de Cultura Económica, 2006.

Hoffmann, A. *El maravilloso mundo de los arácnidos*; Mexico: FCE, 2003.

Hoover, N. G.; Fortenberry, J. D. Use of Antivenin to Treat Priapism After a Black Widow Spider Bite. *Pediatrics* **2004,** *114* (1), e128–e129.

Klauber, L. M. *Rattlesnakes*; University of California Press, 1972.

Lemos-Espinal, J. A., Smith, H. M.; Dixon, J. R.; Cruz, A. *Anfibios y reptiles de Sonora, Chihuahua y Coahuila, Mexico*; Mexico: CONABIO, 2015.

León-de la Luz, J. L.; Blázquez-Moreno, M. C.; Ortega-Rubio, A. *¿Qué se mueve en el desierto? Historias del matorral sarcocaule*' CIB: La Paz, Baja California Sur, Mexico, 2013.
Leopold, A. S. *El desierto*; Time-Life: Virginia, 1979.
Mackessy, S. P. *Handbook of Venoms and Toxins of Reptiles*; CRC: New York, 2010.
Mahmoud, A. A. M. A. Studies on Bee Venom and Its Medical Uses. *Int. J. Adv. Res. Tech.* **2012,** *1* (2).
Miranda, F. Formas de vida vegetales y el problema de la delimitación en las zonas áridas de Mexico. In*: Mesas redondas sobre problemas de las zonas áridas de Mexico*; IMRNR: Mexico, DF, 1955.
Morgan, E. D.; Wilson, I. D. Insect Hormones and Insect Chemical Ecology. *Comprehen. Nat. Prod. Chem.* 1999, *8,* 263–375.
Ortiz, E.; Gurrola, G. B.; Schwartz. E. F.; Possani, L. D. Scorpion Venom Components as Potential Candidates for Drug Development. *Toxicon* 2015, *93*, 125–135.
Postma, T. L. *Clinical Neurotoxicology: Neurotoxic Animal Poisons and Venoms—Dobbs.* Saunders: Philadelphia, 2009.
Qi, J.; Hasanat, A.; Zulfiker, M.; Li, C.; Good, D.; Wei, M. The Development of Toad Toxins as Potential Therapeutic Agents. *Toxins* **2018**, *10*, 336. DOI: 10.3390/toxins10080336
Repetto, M. *Toxicología fundamental*, 3ª edición; Diaz de Santos: Madrid, España, 1997.
Rorabaugh, J. C.; Lemos-Espinal, J. A. *A Field Guide to the Amphibians and Reptiles of Sonora*; Mexico: ECO, 2016.
Russell, F. E. *Snake Venom Poisoning*; Scholium: New York, 1983.
Saygin, S. D. Strategies to Enhance Sustainability of Land Resources in Arid Regions. In: *Arid Environments and Sustainability*; Arman, H., Ed.; InTech, 2017.
Segura-Ramírez, P. J.; Silva-Júnior, P. I. *Loxosceles gaucho* Spider Venom: An Untapped Source of Antimicrobial Agents. *Toxins* **2018,** *10* (12), 522. https://doi.org/10.3390/toxins10120522.
Touchard, A.; Aili, S. R.; Fox, E. G. P.; Escoubas, P.; Orivel, J.; Nicholson, G. M.; Dejean, A. The Biochemical Toxin Arsenal from Ant Venoms. *Toxins* **2016,** *8*, 30. DOI: 10.3390/toxins8010030.
Undheim, E. A. B.; Jenner, R. A.; King, G. F. Centipede Venoms as a Source of Drug Leads. *Natural Hist. Museum* **2016,** 12.
Yap, M. K. K.; Misuan, N. Exendin-4 from Heloderma Suspectum Venom: From Discovery Toits Latest Application as Type II Diabetes Combatant. *Basic Clin. Pharmacol. Toxicol.* **2019,** *124*, 513–527.
Yildiz, A.; Biceroglu, S; Yakut, N.; Bilir, C.; Akdemir, R.; Akilli, A. Acute Myocardial Infarction in a Young Man Caused by Centipede Sting. *Emerg. Med. J.* 2006, *23,* e30.

CHAPTER 10

Mexican Chihuahuan Desert Plants: Bioactive Compounds, Biological Properties, and Potential Applications in Organic Agriculture

ELAN IÑAKY LAREDO-ALCALÁ[1], MIGUEL ÁNGEL DE LEÓN-ZAPATA[1], CYNTHIA LIZETH BARRERA MARTINEZ[1], NORMA PAOLA MELÉNDEZ-RENTERÍA[1], and CRISTÓBAL NOÉ AGUILAR[2]

[1]*Research Center for the Conservation of Biodiversity and Ecology of Coahuila (CICBEC), Autonomous University of Coahuila, Saltillo, Coahuila, Mexico*

[2]*Department of Research (DIA-UAdeC), School of Chemistry, University Autonomous of Coahuila, Saltillo, Coahuila, Mexico*

ABSTRACT

Arid zones represent great potential, where the flora and fauna are the product of thousands of years of physiological adaptation for their survival. In Mexico, semi-arid ecosystems cover 70.79 million hectares, which are characterized by their high vulnerability to desertification and the presence of a rural population that takes advantage of non-timber forest resources of various species, such as *Flourensia cernua, Fouquieria splendens*, and *Agave Lechuguilla.*

The extracts of these endemic plants of the Mexican semi-desert contain bioactive compounds with interesting biological properties. The potential use of the plant extracts for phytopathogens control, in organic agriculture, represents a new natural and sustainable alternative to the use of synthetic pesticides.

Exploration and Valorization of Natural Resources from Arid Zones. Cristobal Noé Aguilar, Norma Paola Meléndez-Renteria, & Karen Nathiely Ramirez-Guzman (Eds.)

In Mexico, the entities that have the greatest influence on the regulation of pesticides are COFEPRIS and SEMARNAT. At the international level, the Food and Agriculture Organization of the United Nations (FAO), this being an agency that has led efforts to issue guidelines for the proper management of pesticides, in addition to the World Health Organization.

In this chapter, the bioactive compounds, and biological properties of plants of the Mexican semi-desert as *Flourensia cernua, Fouquieria splendens*, and *Agave lechuguilla* are described. In addition to the potential of extracts obtained from plants of the Mexican semi-desert as biopesticides, their application in organic agriculture, as well as the required regulations to be put into practice.

10.1 INTRODUCTION

Arid zones represent great potential, where the flora and fauna are the product of thousands of years of physiological adaptation for their survival. Mexico has relevant and extensive plant biodiversity; it is the fourth richest country in the world in this regard (Jasso-De Rodríguez et al., 2011). In parts of northern Mexico where the climate is semi-arid, there are many wild plants that grow in this extreme condition (Jasso-Rodríguez et al., 2012).

Non-wood forest resources have been used in traditional medicine, as a source of food, and as raw materials for handicrafts, which is the reason because they represent an important income source in rural areas (Meinhold and Darr, 2019). In Mexico, semi-arid ecosystems cover 70.79 million hectares, which are characterized by their high vulnerability to desertification and the presence of a rural population that takes advantage of non-timber forest resources of various species, such as *Flourensia cernua, Fouquieria splendens,* and *Agave lechuguilla* (Valdez-Villavicencio et al., 2021).

Flourensia cernua is an endemic species that grows in semi-arid zones and is found in the deserts of Chihuahua and Sonora covering different Mexican states, such as Coahuila, Chihuahua, Durango, Nuevo León, San Luis Potosí, Sonora, and Zacatecas (Jasso de Rodríguez et al., 2006). Hojasen is commonly known in different ways as it is found in both the United States of America and Mexico. The names given in the United States of America are tarbush, Hojasen, American-Tarbush, Black-brush, Barnnish-brush, and Hojasen. In Mexico, it is known as Hojasen, Escobilla Negra, and Tar Bush due to its peculiar aroma (Innes, 2022). This plant has been commonly used as an infusion for the treatment of several gastrointestinal diseases, such as stomach pain, diarrhea, and dysentery, and is used as a purgative (Jasso de Rodríguez et al., 2019), expectorant and antirheumatic.

Fouquieria splendens is commonly known as ocotillo or albarda, it is found in the arid zones of Mexico, from the northern zone of the Chihuahuan Desert to the semi-arid region of Oaxaca (Abd El-Ghani et al., 2017), but also in parts of the Sonora desert (Warnhoff and Halls, 1965; Peinado et al., 2011). Specifically, it has also been found in states, such as Baja California Norte, Baja California Sur, Chihuahua, Coahuila, Guanajuato, Nuevo León, Querétaro, San Luis Potosí, Sonora, Tamaulipas, and Zacatecas (Sharma et al., 2017). The Apache Indians used the liquid from the boiled roots to reduce tiredness and fatigue, they also used the powder from the roots for painful swellings, and the wax from the stems for waist dressings (Krochmal et al., 1954). The plant has purgative properties, respiratory, and gastrointestinal disorders, diseases, problems, infections, or even dysentery (Sharma et al., 2017).

The lechuguilla (*Agave lechuguilla* Torr., Asparagaceae) is a plant native to the arid and semi-arid zones of the southern United States of America and northeastern Mexico. Its national distribution area covers approximately 20 million hectares, including the states of Coahuila, Chihuahua, Nuevo León, Durango, San Luis Potosí, Tamaulipas, and Zacatecas (Castillo et al., 2012; Narcia et al., 2012). The lechuguilla is a non-timber forest resource, and its use has been crucial support among the economic activities of the ixtlera region residents in Mexico. Lechuguilla fibers are the most used resource, which are obtained from the plant heart composed of the most tender leaves grouped in the center of the plant (Narcia et al., 2012; Pando et al., 2004).

The extracts of these endemic plants of the Mexican semi-desert contain bioactive compounds with interesting biological properties (Jasso de Rodríguez et al., 2015; Chelinho et al., 2017).

Carry on in the research of natural products for application in agricultural crops with pesticide activity but friendly with the environment. The potential use of plant extracts of *Flourensia cernua*, *Fouquieria splendens,* and *Agave lechuguilla*, for phytopathogens control in organic agriculture represents a new natural and sustainable alternative to the use of synthetic pesticides. The importance of botanical biopesticides is attributed to their efficacy, biodegradability, diverse action modes, low toxicity, as well as the availability of raw materials (Neeraj et al., 2017).

In Mexico, the entities that have the greatest influence on the regulation of pesticides are COFEPRIS and SEMARNAT. At the international level, the Food and Agriculture Organization of the United Nations (FAO), this being an agency that has led efforts to issue guidelines for the proper management of pesticides, in addition to the World Health Organization.

In this chapter, the bioactive compounds, and biological properties of plants of the Mexican semi-desert as *Flourensia cernua, Fouquieria splendens,* and *Agave lechuguilla* are described. In addition to the potential of extracts obtained from plants of the Mexican semi-desert as biopesticides, their application in organic agriculture, as well as the required regulations to be put into practice.

10.2 FLOURENSIA CERNUA

10.2.1 MORPHOLOGICAL DESCRIPTION

Flourensia cernua is a perennial shrub that grows from a network of roots that can extend 4 m horizontally. Most are shallow, but some extend up to 5 m deep into the ground. It usually grows to a maximum height of about 1 m but can be as tall as 2 m.

It can grow erect or spread out. It has many branches, branching out from the base of the stem. The branches are covered with alternately arranged thick oval leaves up to 2.5 cm in length, sometimes reaching 4 cm. The edges of the leaves are smooth or wavy. The pendulous flower heads contain several yellow disk florets and no ray florets (Innes, 2022). The fruit is a hairy achene up to 1 cm long. Most parts of the plant are highly resinous and have a tarry or hoppy odor as well as a bitter taste (Mata et al., 2003).

Leaf production of *F. cernua* is affected by moisture levels and the plant has been observed to produce a first set of small, scaly leaves during a dry spring, and the second set of larger leaves later in the season as humidity increases. Growth occurs early in the year when rain is abundant. Flowering occurs in the fall. The plant generally produces few flowers in dry years. The root network is shallow and vast with some very deep roots, helping it to collect water from a wide area of soil, another adaptation to its dry habitat (Innes, 2022).

10.2.2 BIOACTIVE COMPOUNDS

The main bioactive compounds of *F. cernua* leaves reported in the literature are methyl orsellinate, ermanine, flourensadiol, dehydrofluorensic acid, long-chain hydrocarbons from tetracosane 4-olide to triacontane-4-olide and lactones (Jasso-De Rodríguez et al., 2007), in addition to saponins (Méndez et al., 2021), terpenes (Estell et al., 2013), condensed tannins equivalent to catechins (Méndez et al., 2012; Belmares, et al., 2009; Castillo et al., 2010; De León et al., 2013), and flavonoids such as luteolin 7-O-rutinoside and 6-C-glucosyl-8-C arabinosyl apigenin (De León-Zapata et al., 2016; Álvarez-Pérez et al., 2020).

The distribution of volatile bioactive compounds in immature, intermediate, and mature leaves of *F. cernua* has also been studied. Volatile compounds were extracted with ethanol and analyzed by gas chromatography/mass spectrometry. Obtained results showed that the age of the leaves of *F. cernua* affects the terpenes (mainly sesquiterpenes) concentration and the sampling variability, where 63 compounds differed between the age categories of the leaves. Immature leaves contained higher concentrations of 46 chemicals than intermediate or mature age categories, but intermediate and mature leaves only differed in seven compounds (Estell et al., 2013).

10.2.3 BIOLOGIC PROPERTIES

10.2.3.1 ANTIOXIDANT PROPERTIES

It has been shown that at a higher concentration of bioactive compounds in *F. cernua* leaves, the ability to inhibit free radicals is greater; therefore, there is larger antioxidant activity (De León-Zapata et al., 2016). This is consistent with the fact that the antioxidant activity of phenolic compounds is mainly due to the number of hydroxyl groups (De León-Zapata et al., 2016), as well as their redox properties (De León et al., 2013). The yield of bioactive compounds from *F. cernua* leaves is affected by the harvest season, the climatic conditions, the phenological step of the plant, the plant tissue (leaves, stems, roots, etc.) (Hyder et al., 2005), and the polarity of the solvent (Weng-Kong et al., 2012).

Tannins are highly soluble, due to the interaction of water with hydroxyl groups and carboxylic acids (Méndez et al., 2012). In the first instance, high phenolic compound content could suggest a good antioxidant capacity (Cheung et al., 2003), due to the greater amount of available hydroxyl groups. Flavonoids, terpenes (Martínez-Flórez et al., 2002), and hydrolyzable tannins contain in their chemical structure a variable number of hydroxyl groups (Martínez-Flórez et al., 2002), which are involved in the neutralization of free radicals by donating electrons and therefore influence antioxidant activity (De León et al., 2013).

The bioactive compounds of tarbush have shown greater antioxidant power for the scavenging of ABTS radicals compared with the DPPH radical

(De León-Zapata et al., 2016; Álvarez-Pérez et al., 2020). This is due to the sensitivity of the ABTS radical, as it is a structure that reacts easily with hydrophilic and lipophilic compounds (Pérez et al., 2003) and reductor sugars present in *F. cernua* leaves (Belmares et al., 2009). Instead, the DPPH radical reacts with hydrophilic compounds, such as gallic acid (Álvarez et al., 2008). The antioxidant activity of bioactive compounds from *F. cernua* leaves is higher than that reported in several varieties of raspberries rich in ellagitannins (De Ancos et al., 2000).

The bioactive compounds of *F. cernua* leaves have also been the subject of other types of studies in the medical area, such as a microencapsulation study of ethanolic extracts from three species of *Flourensia* spp. (*Flourensia cernua, Flourensia microphylla,* and *Flourensia retinophylla*) and evaluation of the controlled release of microencapsulated extracts in an in vitro gastro-intestinal system. Encapsulation was performed using the alginate gelation technique. The results showed that microencapsulation was able to preserve the antioxidant power of the extracts and protect them until they reached the gastrointestinal system in vitro. It is worth mentioning that this work was a pioneer in laying the fundamentals for the application of extracts of *Flourensia* spp. in microsystems to take advantage of its benefits for human health (Jasso de Rodríguez et al., 2019).

Currently, there are few studies to determine the toxicity of the bioactive compounds *of F. cernua* leaves. For the first time, the cytotoxic activity of pure benzopyrans and benzofurans from *F. cernua* has been reported using red blood cells and measuring the hemoglobin released in cell destruction. The results showed that benzopyrans were more active than benzofurans (Towers et al., 1980). Bioassay-directed fractionation of a CH_2Cl_2-MeOH (1:1) extract from the aerial parts of *F. cernua* has also been reported, from which three phytotoxic compounds, such as dehydrofluorensic acid, flou-rensadiol, and methyl orselinate were isolated. The results indicated that the phytotoxic compounds caused a significant inhibition in the growth of the radicle of *Amaranthus hypochondriacus* and *Echinochloa crus-galli*; in addition to interacting with bovine brain calmodulin and inhibiting the activation of the enzyme cAMP phosphodiesterase-dependent on calmodulin (Mata et al., 2003).

Furthermore, crude extracts of *F. cernua* leaves and their fractions are cytotoxic against five human breast cancer cell lines (Molina-Salinas et al., 2006). The *F. cernua* shrub has rapidly increased in prevalence within the grasslands of the Chihuahuan Desert and is comparable in nutrient content, such as protein and fiber, with the alfalfa (lucerne). The increase of *F. cernua* leaves in the diet of cattle can improve the quality of the diet, but reduces its prevalence, in addition to producing toxicosis, due to its high consumption due to the content of bioactive compounds, so it is recommended to remove the use as cattle feed (Holst and Oredsson, 2005). The acute toxicity of an ethanolic extract of *F. cernua* leaves in rats has also been evaluated using the alternative method of fixed doses, where it was found that the LD50 of the extract was found to be above 2000 mg/kg. The leaves extract was administered orally to a group of five Wistar rats at a single dose of 2000 mg/kg of body weight and the control group was treated with 0.85% sodium chloride (NaCl). The results showed that there were signs of mild toxicity, there were no changes in body weight or mortality in the animals, and gross pathology studies showed no abnormalities in the brain, stomach, liver, lungs, and kidneys (Zavala et al., 2010).

10.2.3.2 ANTIFUNGAL PROPERTIES

The efficacy of the bioactive compounds of *F. cernua* leaves in the inhibition of the growth of phytopathogenic fungi, that attack fruit crops of commercial interest has been demonstrated (Table 10.1).

Bioactive compounds from *F. cernua* leaves extracts obtained with hexane, ethanol, and a mixture of methanol–chloroform have been shown to have greater inhibitory activity against *C. gloeosporioides* (Guerrero-Rodríguez et al., 2007).

Higher inhibition values against *F. oxysporum* have also been obtained using ethanol as a solvent. It is noteworthy that there are few studies to evaluate the antifungal power of bioactive compounds from *F. cernua* leaves against *B. cinerea* and *R. stolonifer* (Jasso-De Rodríguez et al., 2007).

However, it is worth noting the evaluation of tannins using lanolin and cocoa butter as solvents, which had approximately 70% inhibition against *R. solani* at a concentration of 500 mg/L of tannins (Castillo et al., 2010).

The resin of an aqueous extract of *F. cernua* leaves showed inhibition of mycelial growth at a minimum inhibitory concentration (MIC) of 1362 mg/L against phytopathogenic fungi of *F. oxysporum, B. cinerea, Penicillium* sp., *A. alternata, R. stolonifer, Mucor* sp., *S. sclerotorium,* and *C. gloeosporioides*, thus representing a viable alternative to counteract the use of chemical compounds in agriculture. The great antifungal effectiveness of bioactive compounds from *F. cernua* leaves is attributable to the phenolic compounds present as hydroxyl groups of hydrolyzable tannins equivalent to gallic acid, flavonoids, terpenes, among others (Álvarez-Pérez et al., 2020).

TABLE 10.1 Antifungal Activity of *F. cernua* Extracts Rich in Bioactive Compounds.

Extracts	Dose	Phytopathogenic fungi	Reference
Leaves ethanolic extract	10 µl/L–	*Alternia* sp., *Rhizoctonia solani*, and *Fusarium oxysporum*	Jasso-De Rodríguez et al., 2007
	100 µl/L		Guerrero-Rodríguez et al., 2007
	2000 mg/L	*Colletotrichum gloeosporioides* and *Penicillum digitatum*	(Jasso de Rodríguez et al., 2017)
	2163 mg/L	*Fusarium oxysporum and Rhizopus stolonifer*	
Leaves methanol:chloroform extract	4000 mg/L	*Colletotrichum gloeosporioides*	Guerrero-Rodríguez et al., 2007)
Leaves hexanic extract	4000 mg/L	*Colletotrichum gloeosporioides*	Guerrero-Rodríguez et al., 2007)
Leaf extract with lanolin as the solvent	2000 mg/L	*Rhizoctonia solani*	Castillo et al., 2010
Leaf extract with cocoa butter as the solvent	1000 mg/L	*Rhizoctonia solani*	Castillo et al., 2010
Powdered resin from leaves aqueous extract	1.519–3.310 mg/L	*Rhizopus stolonifer, Botritis cinerea, Fusarium oxysporum*, and *Colletotrichum gloeosporioides*	De León-Zapata et al., 2016
Branches ethanolic extract of	4240 mg/L	*Fusarium oxysporum and Rhizopus stolonifer*	Jasso de Rodríguez et al., 2017
Leaves and branches ethanol extract	1692 mg/L	*Rhizopus stolonifer*	Jasso de Rodríguez et al., 2017
Powdered resin from leaves aqueous extract	125–1000 mg/L	*Fusarium oxysporum, Botrytis cinerea, Penicillium* sp., *Alternaria alternata, Rhizopus stolonifer, Mucor* sp., *Sclerotinia sclerotorium, and Colletotrichum. gloeosporioides*	Álvarez-Pérez et al., 2020

In the postharvest issue, extracts with a significant amount of bioactive compounds of tarbush have been used to apply to fruits and vegetables to increase the shelf life and avoid microbial contamination during storage. For example, microemulsions with candelilla wax encapsulate a fermented extract of polyphenols from *F. cernua* leaves to increase the shelf life and improve the quality of *Golden delicious* apples (De León-Zapata et al., 2015).

A nanoemulsion has also been formulated based on candelilla wax with the addition of a crude extract of *F. cernua* leaves as an antioxidant and antimicrobial additive. The nanoemulsion was made by high shear hot agitation and was able to increase the nutritional quality of apples at the laboratory and industrial level (De León-Zapata et al., 2017; De León-Zapata et al., 2018).

In addition, the effects of nanolaminate coatings made from nanolayers of alginate and chitosan, incorporated with extracts of *F. cernua*, have been investigated for their application and evaluation in tomatoes. Where, the nanolaminate with tarbush extract managed to extend the shelf life of the tomato by reducing weight loss and microbial growth, as well as reducing gas exchange and ethylene production, and maintaining firmness and color, so it is an alternative to prolong the shelf life of the tomato (Salas-Méndez et al., 2019). As well as the formulation and optimization of an edible coating based on candelilla wax, whey protein, glycerol, and *F. cernua* leaf extract, which was applied to tomato, managed to reduce weight loss and firmness in the fruit. Sensory evaluation showed that the product obtained is acceptable to consumers. The edible coating added with *F. cernua* extract was the most effective pathogenic fungi growth inhibitor and the visual appearance at the end of storage confirmed the beneficial effect of the edible coating (Ruiz-Martínez et al., 2020).

10.2.3.3 INSECTICIDAL AND BACTERICIDAL PROPERTIES

The insecticidal properties of bioactive compounds from *F. cernua* leaves have been confirmed in a few reported works, among which the study of the insecticidal activity of benzofuran 7-methoxy-2-isopropenyl-5-acetyl-2,3-dihydro benzofuran-3-ol-cinnamate from *F. cernua* leaves, which demonstrated its activity as a juvenile hormone causing anatomical malformations, retention of juvenile characteristics, and sterility in treated insects from their second to fourth development stages (Chan et al., 1975). In addition, the termiticidal activity of bioactive compound fractions of *F. cernua* leaves with hexane, diethyl ether, and ethanol has also been reported (Tellez et al., 2001).

And finally, a study where the insecticidal activity of crude extracts of *F. cernua* leaves extracted with solvents of different polarity on three pest insects with agronomic impacts, such as *Sitophilus oryzae*, *Phthorimaea operculella,* and *Brevicoryne brassicae* was evaluated, as well as the repellent or attraction effect on *Sitophilus oryzae*. Obtained results showed that the hexane fraction from the extract had the highest insecticidal effect against *B. brassicae* with 100% mortality at a concentration of 10,000 μlL^{-1} at 24 h. In addition, the hexane fraction showed an insecticidal effect by inciting repellency to *S. oryzae* at 5 and 45 days. It is worth mentioning that the repellency effect caused by the hexane fraction may be due to the volatile substances borneol and camphor that are present (García et al., 2007).

Few works have demonstrated the bactericidal effect of the bioactive compounds of *F. cernua* leaves, among which is reported that at a MIC of 50 and 25 μgL^{-1}, from the hexane extract of the aerial parts of *F. cernua* leaves, inhibits the growth of *Mycobacterium tuberculosis,* strains H37Rv and CIBIN: UMF: 15: 99, (Molina-Salinas et al., 2006). The bactericidal effect of extracts of *F. cernua* leaves obtained by the Soxhlet method using water, ethanol, and alternative organic solvents (lanolin and cocoa butter) against pathogenic bacteria that attack humans, such as *Enterobacter aerogenes, Escherichia coli, Salmonella typhi,* and *Staphylococcus aureus* has also been demonstrated (Méndez et al., 2012).

10.3 FOUQUIERIA SPLENDENS

10.3.1 MORPHOLOGICAL DESCRIPTION

Fouquieria splendens is characterized by having a candelabra-like shape, with branches emerging from a very short stem (Bowers, 2006) covered in resin and thorns. The maximum height it reaches is 4.5 m and adult individuals can resist up to 200 years (Ebert and Eberth, 2006).

The bark is laminated and has a layer of chlorophyll; therefore, the light that reaches this layer first passed through the translucent crust and long-term exposure to desert sunlight allows the transformation of the plant's phytochemicals (Warnhoff and Halls, 1965), achieving a net CO_2 uptake of more than 13 $\mu mol\ m^{-2}s^{-1}$ (Nobel and Zutta, 2005).

Branches on a single plant typically vary in length from 1 to 3 m; the number of branches increases with age to a maximum of 75–100 (Bowers et al., 2004). Lateral branches are few because most shoots are used for leaf production on short shoots (Bowers, 2006).

The plants can produce between 6 and 8 leaf crops annually, depending on the amount of rain that falls; but the minimum amounts of rainfall for germination are 18 mm because they have a continuous network anastomosis of cells in the cortex for water storage (Bowers, 2006) and the bush germinates in March, July, August, and October (Bowers et al., 2004); Although leaves were produced multiple times per year in response to rainfall, the nodal extension of steams does not occur every year and was highly correlated with the number of stem rings (Nobel and Zutta, 2005).

Panicles arise at or near the tip, usually one to several per branch; the basic components of the inflorescence are 1, 2, or 3 shoots known as monads, dyads, and triads. These basic units are combined in different patterns or types; the most common include two triads and one monad on a 7-flowered crest and 3 triads, 2 dyads, and 3 monads on a 16-flowered crest (Bowers, 2006).

The primary pollinators are hummingbirds and carpenter bees. The annual average of reproductive outputs per plant is 2500 flowers and 400 fruits. Flower production increases linearly with plant size, that is, the number of branches longer than 1 m, but fruit production reaches an asymptote (Bowers, 2005).

Short branches of *F. splendens* spend most of their "food reserves" on growth, secondary thickening, and maturation of the branches and can divert substantial reserves to reproduction (Darrow, 1943).

Long branches have an average of three times more flowers than short branches regardless of plant size, while long and short branches also differ in the number of panicles per branch, being more complex on long branches (Bowers, 2006).

The rings of woody trunks provide the opportunity to know the age of the plant and its sporadic growth, so common in this type of plant (Killingbeck, 2017). And it is considered a shrub, with an average lifespan of 75 years (Bowers et al., 2004).

10.3.2 BIOACTIVE COMPOUNDS

In the leaves, the inner bark and the outer bark polyphenols have been found, such as leukocyanidin and ellagic acid; flavonols like kaempferol and quercetin; cinnamic acids, such as p-coumaric, caffeic, and low concentration ferulic acids; coumarin, and scopoletin (Bate-Smith, 1964), as well as iridoid glucosides (Jensen and Nielsen, 1982). Specifically in the leaf, among angiosperms, isoquercitrin, rutin, caffeic acid, and scopolin can also be found (Scogin, 1978).

During acid hydrolysis of the leaves and bark, it has been observed that the blue color of the aqueous extract is rapidly replaced by a dark precipitate, characteristic of asperuloside (Bate-Smith, 1964).

Ocotillol (Figure 10.1) is a triterpene that can be isolated from the plant *F. splendens* Engelm, it is characterized by the union of a five-membered epoxy ring to a C-20 (Zhang et al., 2020). In addition, it has been reported that a possibility of synthesis of this compound in the plant is by the derivation of the biogenetic equivalent in the 1,2 epoxidation in the $C_{24}=C_{25}$ bond with the subsequent oxidation that opens C-24 for the hydroxyl group of the C-20 (Warnhoff and Halls, 1965). Regarding the mineral content, high variability in resorption of N, Zn, and P has been reported between individuals, growth sites, and years and can be considered an attribute of drought-deciduous plants that inhabit the desert (Killingbeck, 2008).

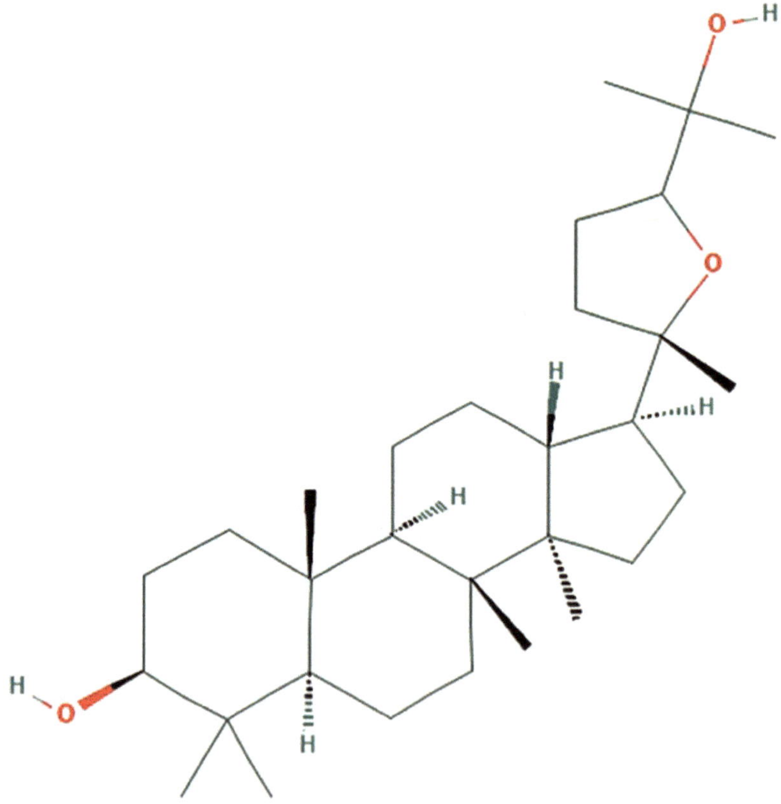

FIGURE 10.1 Chemical structure of ocotillol. Molecule isolated from the plant *Fouquieria splendens* with various biological activities.

10.3.3 BIOLOGICAL PROPERTIES

The structural modifications of the molecule were mainly focused on the positions of C-3 and C-12, resulting in compounds that have a varied pharmacological activity, including antitumor, anti-myocardial ischemia, and antibacterial (Zhang et al., 2020). It has been reported that methanolic extracts of leaves are effective as an antibiotic against *Staphylococcus aureus* at a concentration of 25 μg/mL and *Escherichia coli* at a concentration of 27.1 μg/mL (Sharma et al., 2017). As it has been shown that the plant is not toxic during the in vivo tests against the brine shrimp lethality test ((Sharma et al., 2017), it can be explored for other topical or medicinal applications due to the quantities of antioxidants it contains.

10.4 AGAVE LECHUGUILLA

10.4.1 MORPHOLOGICAL DESCRIPTION

Agave Lechuguilla (Figure 10.2) is a small, rosetophilous shrub with an average of 11–30 leaves per plant, it can measure between 30 and 40 cm wide and 20–70 cm tall. Its roots are fibrous, thin, and long, it is distributed at a depth between 8 and 12 cm of the soil. The leaves are formed from a considerably developed apical bud (called the bud) which is a cone formed by young leaves rolled around on a central axis. The newly unfolded young leaves are whitish-green or bluish-green, the mature ones are from 25 to 50 cm high and 2.5 to 6 cm wide, they have a linear–lanceolate shape, with a color that goes from light green to greenish-yellow, they are between diffuse and erect, sometimes curved and rarely falcate, concave above and convex below; sometimes they have a lighter green longitudinal stripe on the upper side, and narrow, dark-green stripes, 30–60 cm long on the underside. The leaf margin is entire, light brown to gray, easily separated on dry leaves, with 8–20 detached teeth (spines) per side; these teeth measure from 2 to 7 mm, brown or more frequently light gray, and separated from 1.5 to 4 cm. The apex is a strong, conical, subulate, greyish spine of 1.5–4 cm. The inflorescence receives the names of quiote, mequiote (from the Nahuatl metl, maguey, and quiyotl or quiotl, stem), Garrocha or bohordo; it is a spiky scape 2.5–3.6 or up to 6 m tall, with a glaucous surface and deltoid bracts, 10–20 cm long, with a spine at the apex (Reyes et al., 2000). The flowers have pedicels measuring from 20 to 50 mm and are grouped in pairs or triads. Fruit develops a brown or black oblong or pyriform dehiscent capsule, 18–25 mm long and 11–18 mm wide, short pedicelled or sessile, round and with a short break at the apex, glaucous; Seeds 4.5–6 mm long and 3.5–4.5 mm wide with a small thread or notch and few winged striae around the curved side (Reyes et al., 2000).

FIGURE 10.2 *Agave Lechuguilla* plant.

Source: The World Checklist of Vascular Plants (WCVP). https://www.gbif.org/es/species/2766505

10.4.2 BIOACTIVE COMPOUNDS

The chemical composition of *A. lechuguilla* corresponds to nonstructural components, such as sucrose, nitrate/nitrites, protein, ash, chlorophyll, and fatty compounds typical of the plant (Carmona et al., 2017). The relative humidity of the material, after the drying process, was 4.33%. The cellulose fraction was calculated based on glucose detected by HPL (0.55 gL^{-1}), hemicellulose was determined from the concentration of xylose (0.22 gL^{-1}), galactose (0.18 gL^{-1}), arabinose (0.068 gL^{-1}), and mannose (0.08 gL^{-1}). The cellulose, hemicellulose, and total lignin contents (dry basis) in the samples were 17.72%, 17.15%, and 7.32%, respectively (Table 10.2).

TABLE 10.2 Chemical Composition of *A. Lechuguilla.*

Chemical component	Composition (% dry basis)
Cellulose	17.72 ± 0.68
Hemicellulose	17.15 ± 0.91
Soluble lignin	3.64 ± 0.34
Insoluble lignin	3.68 ± 0.21
Ashes	12.45 ± 0.88
Extractives	45.34 ± 1.2

Source: Carmona et al., 2017.

In addition, the concentration of antioxidant compounds was evaluated, from which it was concluded that using different solvents, such as methanol and ethanol, considerable amounts of possible commercial interest were extracted.

In lechuguilla extracts, the presence of smilagenin (steroidal sapogenin) has also been reported in addition to 8 more saponins: yucagenin, gitogenin, hecogenin, tigogenin, diosgenin, gentrogenin, chlorogenin and ruizgenin, and even estrogen and progesterone. The content of saponins in the lechuguilla extract is approximately 0.1–0.3% on a dry basis (Hernández et al., 2005).

10.4.3 BIOLOGICAL PROPERTIES

In the literature, it has been reported that the lechuguilla extract is rich in saponins, which have surfactant properties that can be used as surfactants and/or natural detergents (Guclu-Ustundag and Mazza, 2007).

10.5 POTENTIAL APPLICATION IN ORGANIC AGRICULTURE

In general, the natural extracts and resins of *F. cernua* have great potential as biofungicides, bioinsecticides, and biobactericides for application in organic agriculture during harvest and postharvest. Extracts of *F. splendens* have great application potential as a biobactericide as well as protector and stabilizer of emulsions with bioactive compounds in organic agriculture due to their antibacterial and antioxidant properties.

Until now, different applications to agro-industrial residues have been investigated and in the specific case of *A. Lechuguilla,* it is possible to find that the guishe residue resulting from carving leaves to obtain the fiber, is rich in chemical compounds of industrial interest. An example of this may be that some reports show that Agave may contain saponins which can function as a potential herbicide due to allelopathic activity, preventing seed germination (Hussain et al., 2018).

Similarly, studies have also reported the biotechnological activities developed to obtain and purify antioxidant compounds from *A. lechuguilla* residues, specifically flavonoids. The processing of agricultural residues can be a great source of raw material for molecules of industrial interest (Morreeuw et al., 2021).

10.6 REGULATIONS AND RESTRICTIONS

10.6.1 INTERNATIONAL REGULATORY FRAMEWORK

Pesticides with the highest persistence in the environment are organochlorines. The production and use of this type of compound have decreased worldwide; however, regulations and restrictions have not been able to prevent many of these compounds from continuing to be used. The restrictions applied to these compounds are described through treaties, conventions, and agreements, such as the Montreal Protocol, the Stockholm Convention, and the Rotterdam Convention (Table 10.3).

TABLE 10.3 International Treaties and Conventions on the Use of Biopesticides.

Document	Description
Montreal Protocol	It is an international treaty designed to protect the ozone layer by reducing the production and consumption of numerous substances
Stockholm Convention	Is an International Treaty promoted in 2001 by the United Nations Environment Program (UNEP), which has the objective of protecting human health and the environment
Rotterdam Convention	Entered into force on February 24, 2004, and aims to guarantee the protection of the population and the environment of all countries from possible dangers related to the trade in pesticides and highly dangerous chemicals
Food and Agriculture Organization of the United Nations (FAO)	United Nations agency leading the international effort to end hunger and achieve food safety

Source: Programa de las Naciones Unidas para el Desarrollo, 2021; SEMARNAT, 2021; FAO, 2021.

TABLE 10.4 Mexican Laws and Regulations.

Federal laws and national regulations	Description
Federal Law of Ecological Balance and Environmental Protection (LGEEPA) (DOF, 2011)	It was published on January 28, 1988, and its last modification was issued on January 28, 2011. This law groups pesticides with fertilizers and toxic substances, granting the federation regulation on the use and management of these chemical substances (Art. 120). SEMARNAT grants authorizations for the manufacture, import, and for all kinds of activities related to pesticides (Art. 135). It establishes the authorization for the importation of products (Art.144). Finally, the LGEEPA establishes that the particular activities related to the management of pesticides are subject to the Official Mexican Norms (Art. 143)
General Law for Integrated Waste Management (LGPGIR)	It was published in 2003 and its last modification corresponds to June 19, 2007. Regulates the management of waste, defining it as products that are discarded. Article 31 establishes that the containers as pesticide residues must be subject to a management plan. And the Art. 67 indicates that the incineration of organochlorine pesticides is prohibited due to the generation of pollutants such as dioxins and furans (DOF, 2007)
Federal Plant Health Law (LFSV)	It was published in January 1994 and the last modification was published in July 2007. The main objectives are to promote plant health, as well as the application, verification, and certification of risk reduction systems for physical, chemical, and microbiological in vegetable production (Art. 1). Points out that it is the responsibility of the Federation, through SAGARPA, to regulate the field studies to carry out the establishment of the limits of pesticide residues (Art. 7 and 42). SAGARPA collaborates with the Social Development to monitor compliance with the regulations to pesticides (Art. 10). On the other hand, this Secretariat is obliged to carry out a National Program for the Monitoring of Pesticide Residues and to disseminate the information generated (Art. 42 bis) (DOF, 2007)
General Health Law (LGS)	Regulates the right to health protection that every person has. According to the LGS, it is the responsibility of the SSA to authorize new products that may be toxic, including pesticides, as well as coordinate the conditions of manufacturing, formulation, packaging, and labeling with other agencies (Art. 279 and 281). Likewise, due to its toxic characteristics, any type of pesticide requires a sanitary registration with the SSA (Art. 376). Another of its attributions that correspond to the SSA is the issuance of Official Norms regarding the application and protection against pesticides and the authorizations for their importation (Art. 298). In Article 17 of this law, all the attributions and obligations regarding the regulation and control of pesticides are indicated, assigning attributions to the Federal Commission for the Protection against sanitary risks (COFEPRIS)

TABLE 10.4 *(Continued)*

Federal laws and national regulations	Description
Regulations on Registration, Import and Export Authorizations and Export Certificates of Pesticides, Plant Nutrients, and Toxic or Dangerous Substances and Materials	It was issued on December 28, 2004, by the SSA, and Article 3 indicates the procedure for the application of the regulation itself is the responsibility of the following organizations COFEPRIS, SEMARNAT, and SAGARPA. This regulation also establishes the requirements to ensure that pesticides that reach the Mexican market are safe during handling. Among these requirements, a series of toxicological studies stands out, which include acute, sub-chronic, chronic, and specific effects toxicity (neurotoxicity, teratogenicity)

Source: DOF, 2021.

TABLE 10.5 Official Mexican Regulations.

	Official Mexican standards
Sampling	NOM-AA-105-1988. Pesticide determination of residues in soil – method of sampling
Ecological (SEMARNAT)	NOM-090-ECOL-1994, which establishes the requirements for the design and construction of agrochemical receptors
	NOM-052-ECOL-1993, which establishes the characteristics of hazardous wastes, their listing, and the limits that make a hazardous waste due to its toxicity to the environment
	NOM-052-SEMARNAT-2005 (DOF, 2005) catalogs pesticides in the "Part 6, pesticides and herbicides." Due to the structural complexity of these compounds and their intrinsic characteristics, some pesticides can be classified as dangerous substances
Sanitary (SSA)	NOM-044-SSA1-1993, which establishes the requirements to contain pesticides. Packing
	NOM-045-SSA1-1993, which establishes the labeling of pesticides. Products for agricultural, forestry, livestock, gardening, urban, and industrial use
	NOM-046-SSA1-1993, which establishes the labeling of pesticides. Products for domestic use
	Draft NOM-058-SSA1-1993, establishing the sanitary requirements for establishments that manufacture and formulate pesticides and fertilizers and that process dangerous substances
	Draft NOM-043-SSA1-1993, regarding the storage of pesticides
	NOM-232-SSA1-2009, which establishes the requirements, indications and characteristics that must be met by the packaging and labeling of pesticides, both technical and formulated and in their different presentations, in order to minimize risks to the health of workers occupationally exposed and the general population, during storage, transport, handling, and application
Zoosanitary (SAGARPA)	NOM-021-ZOO-1994, which establishes the analysis of residues of organochlorine pesticides and polychlorinated biphenyls in fat from cattle, horses, pigs, sheep, and birds by gas chromatography
	NOM-028-ZOO-1995 for organophosphate pesticide residues, in liver and muscle of cattle, horses, pigs, sheep, goats, cervids, and birds
	NOM-004-ZOO-1994, establishes the maximum limit of toxic residues in food of animal origin

TABLE 10.5 *(Continued)*

	Official Mexican standards
Phytosanitary (SAGARPA)	NOM-032-FITO-1995, which establishes the phytosanitary requirements and specifications for conducting biological effectiveness studies of agricultural pesticides and its technical opinion
	NOM-033-FITO-1995, which establishes the phytosanitary requirements and specifications for the notice of start of operation that must be met by individuals and companies interested in marketing agricultural pesticides
	NOM-034-FITO-1995, which establishes the phytosanitary requirements and specifications for the notice of start of operation that must be met by natural and legal persons interested in the manufacture, formulation by assembly plant, formulation and/or assembly plant and import of agricultural pesticides
	NOM-036-FITO-1995. It establishes the criteria to serve as diagnostic laboratories.
	NOM-050-FITO-1995, which establishes the phytosanitary requirements and specifications to carry out field tests to establish maximum residue limits for pesticides in agricultural products
	NOM-052-FITO-1995, establishes the criteria to carry out an adequate application of pesticides by aerial means
	NOM-051-FITO-1995, which establishes the phytosanitary requirements and specifications for the management of agricultural pesticides whose acquisition and application is subject to the written recommendation of a phytosanitary professional
	NOM-053-FITO-1995, which establishes the phytosanitary requirements and specifications to carry out the diffusion of the advertising of phytosanitary supplies
	NOM-057-FITO-1995, which establishes the phytosanitary requirements and specifications to issue the opinion on the analysis of pesticide residues
Industrial Hygiene and Safety (STPS)	NOM-003-STPS-1999, regulates hygiene and safety conditions to prevent risks to workers who carry out agricultural activities due to the handling and use of pesticides
	NOM-010-STPS-1993, regarding safety and hygiene conditions in workplaces where chemical substances capable of generating contamination in the work environment are produced, stored or handled

Source: DOF, 2021.

TABLE 10.6 Mexican Institutions and Associations are Implicated in the Use and Application of Biopesticides.

Government institutions and civil associations	Description
COFEPRIS (Government institution)	It is a decentralized administrative agency of the Ministry of Health, with technical, administrative, and operational autonomy, which oversees exercising the attributions in matters of health regulation, control, and promotion (DOF, 2004). It is responsible for authorizing the registration and issuance of certificates of free sale for the export of pesticides and plant nutrients, as well as granting import permits for pesticides, plant nutrients, and toxic or dangerous substances, after analysis, evaluation, and judgment of technical information, toxicological and safety. For evaluation purposes, it corresponds to classifying pesticides into five categories, chemical, biochemical, microbiological, botanical, and miscellaneous
CICOPLAFEST (Government institution)	It is a Commission that was created in 1987 and is empowered to regulate the exploration, elaboration, manufacture, formulation, mixing, conditioning, packaging, handling, transport, distribution, application, storage, commercialization, possession, use, and final disposal of pesticides, fertilizers, and toxic substances. In response to the need for a regulatory framework for pesticides, SSA, SAGARPA, SEMARNAT, and the Ministry of the Economy (SE) published in 2004 the "Catalog of Pesticides" (CICOPLAFEST, 2004)
INECC (Government institution)	It is a decentralized agency of SEMARNAT, whose functions include generating, integrating, and disseminating knowledge and information through applied scientific research and capacity building, to support the formulation of environmental policy and decision-making that promote sustainable development. As part of INECC is the National Center for Environmental Research and Training (CENICA), which for several years has carried out studies related to the evaluation of persistent toxic substances and has also focused efforts on the analysis of pesticides in various environmental matrices
CNRPyC (Government institution)	It was created in 1991 in Matamoros, Tamaulipas city, and is currently located in Tecamac, State of Mexico. Its function is to assist in monitoring to comply with the maximum limits of pesticide residues in agricultural products. This National Center is part of the General Directorate of Food, Aquaculture, and Fisheries Safety (DGIAAP) of SAGARPA
SENASICA (Government institution)	This organism implements an annual monitoring program for pesticide residues in agricultural fields. The presence of unauthorized compounds may reflect malpractice and misinformation of personnel who purchase and apply these products, and perhaps the use of illegally distributed substances

TABLE 10.6 *(Continued)*

Government institutions and civil associations	Description
PROCCYT (Civil association)	Provides technical advice, carries out technical training activities, collaborates in education, training, and prevention in health aspects. This institution orients its activities toward comprehensive pest management and sustainable agriculture, care for the environment, and the health of workers and the community

Source: COFEPRIS, 2021; CICOPLAFEST, 2021; INECC, 2021; CNRPyC, 2021; SENASICA, 2021; PROCCYT, 2021.

10.6.2 MEXICO REGULATORY FRAMEWORK OF BIOPESTICIDES

Several agencies are involved in regulations related to the production and use of pesticides in Mexico. Health aspects are regulated by the Department of Health and Assistance (SSA); the impact on the environment by the Department of Environmental and Natural Resources (SEMARNAT); the biological efficacy of products for agricultural use by the Department of Agriculture and Rural Development (SAGARPA); and the Ministry of Labor and Social Security (STPS) issues the criteria of industrial hygiene and safety. In Table 10.4, the laws and regulations issued by these agencies are discussed.

As mentioned above, the regulation of the management and application of pesticides involves various federal agencies. Table 10.5 presents a description of the different Mexican Official Standards according to the agency that issues them.

In Mexico, the institutions involved (Table 10.6) in the use and application of pesticides are the Federal Commission for the Protection against Sanitary Risks (COFEPRIS), the Inter-ministerial Commission for the Control of the Process and use of Pesticides, Fertilizers, and Toxic Substances (CICOPLAFEST), the National Institute of Ecology and Climate Change (INECC), the National Reference Center for Pesticides and Contaminants (CNRPyC), the National Service for Health, Safety and Agro-Food Quality (SENASICA), and the civil associations such as the Mexican Association for the Crop Protection, Science and Technology (PROCCYT).

10.7 CONCLUSIONS

The plants *Flourensia cernua*, *Fouquieria splendens,* and *Agave lechuguilla* grow extensively in the arid zones of northern Mexico, and are neither threatened in these ecosystems nor are they currently exploited. So the rational use of non-timber forest resources would adhere to the Forest Law. The controlled use of these plants would only imply harvesting or pruning the upper third of the foliage, which would serve to promote greater vigor and sprouting of the young buds that would later produce new branches, leaves, and more foliage, which could be used every 2 or 3 years. There is a clear need to develop new natural alternatives for their application in organic agriculture, as well as the necessary regulations. Several plants belonging to the arid zones with biocidal and antioxidant properties, such as *Flourensia cernua, Fouquieria splendens,* and *Agave lechuguilla* may represent good potential options as biofungicides, bioinsecticides, and biobactericides to deal with pests of significant economic importance in crops; in addition to having great potential as protectors and emulsion stabilizers, as they are a primary source of bioactive phytochemicals, they are also biodegradable and relatively nontoxic to humans, animals, and the surrounding environment.

KEYWORDS

- **biological properties**
- **natural extracts**
- **bioactive compounds**
- **arid plants**
- **sustainable agriculture**
- **pesticide regulation**

REFERENCES

Abd El-Ghani, M. M.; Huerta-Martínez, F. M.; Hongyan, L.; Qureshi, R. The Deserts of Mexico. In *Plant Responses to Hyperarid Desert Environments*; Springer: Cham, 2017. https://doi.org/10.1007/978-3-319-59135-3_9.

Álvarez, R. E.; Jimenez, G. O. J.; Posada, A. C. M.; Rojano, B. A.; Gil, G. G. H.; García, P. C. M.; Durango, R. D. L. Actividad antioxidante y contenido fenólico de los extractos provenientes de las bayas de dos especies del género Vismia (Guttiferae). *Vitae, Revista de la Facultad de Química Farmacéutica* **2008,** *15*, 165–172.

Álvarez-Pérez, O. B.; Ventura-Sobrevilla, J. M.; Ascacio-Valdés, J. A.; Rojas, R.; Verma, D. K.; Aguilar, C. N. Valorization of *Flourensia cernua* DC as Source of Antioxidants and Antifungal Bioactives. *Ind. Crops Prod.* **2020,** *152*, 112422. https://doi.org/10.1016/j.indcrop.2020.112422.

Bate-Smith, E. C. Chemistry and Taxonomy of *Fouquieria splendens* Engelm: A New Member of the Asperuloside Group. *Phytochemistry* **1964,** *3*, 623–625. https://doi.org/10.1016/S0031-9422(00)82938-6.

Belmares, R.; Garza, Y.; Rodríguez, R.; Contreras-Esquivel, J. C.; Aguilar, C. N. Composition and Fungal Degradation of Tannins Present in Semiarid Plants. *Elec. J. Environ. Agric. Food Chem.* **2009,** *8*, 312–318.

Bowers, J. E. Influence of Plant Size and Climatic Variability on the Floral Biology of *Fouquieria splendens* (ocotillo). *Madroño* **2005,** *52*, 158–165. https://doi.org/10.3120/0024-9637 (2005)52[158:IOPSAC]2.0.CO;2.

Bowers, J. E. Branch Length Mediates Flower Production and Inflorescence Architecture of *Fouquieria Splendens* (ocotillo). *Plant Ecol.* **2006,** *186*, 87–95. https://doi.org/10.1007/s11258-006-9114-7.

Bowers, J. E.; Turner, R. M.; Burgess, T. L. Temporal and Spatial Patterns in Emergence and Early Survival of Perennial Plants in the Sonoran Desert. *Plant Ecol.* **2004,** *172*, 107–119. https://doi.org/10.1023/B:VEGE.0000026026.34760.1b.

Carmona, E.; Morales, T.; Mussatto, S.; Castillo, D.; Ríos, L. Propiedades químicas, estructurales y funcionales de la lechuguilla (*Agave lechuguilla* Torr.). *Rev. Mex. Cienc. Forestales* **2017,** *8*, 100–122. http://dx.doi.org/10.29298/rmcf.v8i42.21.

Castillo, F.; Hernández, D.; Gallegos, G.; Méndez, M.; Rodríguez, R.; Reyes, A.; Aguilar, C. N. In Vitro Antifungal Activity of Plant Extracts Obtained With Alternative Organic Solvents Against *Rhizoctonia solani* Kühn. *Ind. Crops Prod.* **2010,** *32*, 324–328. https://doi.org/10.1016/j.indcrop.2010.05.013.

Castillo, Q. D.; Mares, A. O.; Villavicencio, G. E. E. Lechuguilla (Agave lechuguilla Torr.) planta suculenta de importancia económica y social de las zonas áridas y semiáridas de Mexico. *Boletín de la Sociedad Latinoamericana y del Caribe de Cactáceas y otras Suculentas* **2012,** *8*, 6–9.

CICOPLAFEST, Comisión Intersecretarial para el Control del Proceso y Uso de Plaguicidas y Sustancias Tóxicas, Gobierno De Mexico [Online] 2021. http://www.ordenjuridico.gob.mx/listDependencia.php?idEst=386&poder=ejecutivo&liberado=no#gsc.tab=0 (accessed June 8, 2021)

Chan, G. F. Q.; Towers, G. H. N.; Mitchell, J. C. Ultraviolet Mediated Antibiotic Activity of Thiophene Compounds of Tagetes. *Phytochemistry* **1975,** *14*, 2295–2296. https://doi.org/10.1016/S0031-9422(00)91121-X.

Chelinho, S.; Maleita, C. M. N.; Francisco, R.; Braga, M. E. M.; Da Cunha, M. J. M.; Abrantes, I.; De Sousa, H. C.; Morais, P. V.; Sousa, J. P. Toxicity of the Bionematicide 1,4-naphthoquinone on Non-Target Soil Organisms. *Chemosphere* **2017,** *181*, 579–588. https://doi.org/10.1016/j.chemosphere.2017.04.092.

Cheung, L. M.; Cheung, H. P. C. K.; Ooi, V. E. C. Antioxidant Activity and Total Phenolics of Edible Mushroom Extracts. *Food Chem.* **2003,** *81*, 249–255. https://doi.org/10.1016/S0308-8146(02)00419-3.

CNRPyC, Centro Nacional de Referencia de Plaguicidas y Contaminantes, Servicio Nacional de Sanidad, Inocuidad y Calidad Agroalimentaria [Online] 2021. https://www.gob.mx/senasica/acciones-y-programas/centro-nacional-de-referencia-de-plaguicidas-y-contaminantes-cnrpyc. (accessed June 08, 2021).

COFEPRIS, Comisión Federal Para la Protección contra Riesgos Sanitarios, Gobierno De Mexico y Secretaría de Salud [Online] 2021. https://www.gob.mx/cofepris (accessed June 08, 2021).

Darrow, R. A. Vegetative and Floral Growth of *Fouquieria splendens*. *Ecology* **1943,** *24*, 310–322. https://doi.org/10.2307/1930533.

De Ancos, B.; Gonzáles, E. M.; Cano, M. P. Ellagic Acid, Vitamin C, and Total Phenolic Contents and Radical Scavenging Capacity Affected by Freezing and Frozen Storage in Raspberry Fruit. *J. Agric. Food Chem.* **2000,** *48*, 4565–4570. https://doi.org/10.1021/jf0001684.

De León, M. A.; Sáenz, A.; Jasso-Cantu, D.; Rodríguez, R.; Pandey, A.; Aguilar, C. N. Fermented *Flourensia cernua* Extracts and Their In Vitro Assay Against *Penicillium expansum* and *Fusarium oxysporum*. *Food Technol. Biotechnol.* **2013,** *51*, 233–239.

De León-Zapata, M. A.; Pastrana-Castro, L.; Barbosa-Pereira, L.; Rua-Rodríguez, M. L.; Saucedo, S.; Ventura-Sobrevilla, J. M.; Salinas-Jasso, T. A.; Rodríguez-Herrera, R.; Aguilar,

C. N. Nanocoating With Extract of Tarbush to Retard Fuji Apples Senescence. *Postharvest Biol. Technol.* **2017,** *134*, 67–75. https://doi.org/10.1016/j.postharvbio.2017.08.010.
De León-Zapata, M. A.; Pastrana-Castro, L.; Rua-Rodríguez, M. L.; Alvarez-Pérez, O. B.; Rodríguez-Herrera, R.; Aguilar, C. N. Experimental Protocol for the Recovery and Evaluation of Bioactive Compounds of Tarbush Against Postharvest Fruit Fungi. *Food Chem.* **2016,** *198*, 62–67. https://doi.org/10.1016/j.foodchem.2015.11.034.
De León-Zapata, M. A.; Sáenz-Galindo, A.; Rojas-Molina, R.; Rodríguez-Herrera, R.; Jasso-Cantú, D.; Aguilar, C. N. Edible Candelilla Wax Coating With Fermented Extract of Tarbush Improves the Shelf Life and Quality of Apples. *Food Packag. Shelf Life* **2015,** *3*, 70–75. https://doi.org/10.1016/j.fpsl.2015.01.001.
De León-Zapata, M. A.; Ventura-Sobrevilla, J. M.; Salinas-Jasso, T. A.; Flores-Gallegos, A. C.; Rodríguez-Herrera, R.; Pastrana-Castro, L.; Rua-Rodríguez, M. L.; Aguilar, C. N. Changes of the Shelf Life of Candelilla Wax/Tarbush Bioactive Based-Nanocoated Apples at Industrial Level Conditions. *Sci. Hortic.* **2018,** *231*, 43–48. https://doi.org/10.1016/j.scienta.2017.12.005.
Ebert, T. A.; Eberth, T. A. Decomposition Rate of Ocotillo (*Fouquieria splendens*) Wood in the Desert of Southern California and its Use in Estimating Adult Survival by Life-Cycle Graph Analysis. *Plant Ecol.* **2006,** *186*, 177–187. https://doi.org/10.1007/s11258-006-9121-8.
Estell, R. E.; James, D. K.; Fredrickson, E. L.; Anderson, D. M. Within-Plant Distribution of Volatile Compounds on the Leaf Surface of *Flourensia cernua. Biochem. Syst. Ecol.* **2013,** *48*, 144–150. https://doi.org/10.1016/j.bse.2012.11.020.
FAO, Organización de las Naciones Unidas para la Alimentación y la Agricultura, ONU [Online] 2021. http://www.fao.org/about/es/.
DOF, Diario Oficial de la Federación. Secretaría de Gobernación [Online] 2021. https://www.dof.gob.mx/ (accessed June 08, 2021)
Henrickson, J. Anatomy of Periderm and Cortex of Fouquieriaceae. *A J. Syst. Evolut. Bot.* **1969,** *7*, 97–126.
García, M.; Gonzalez-Coloma, A.; Donadel, O. J.; Ardanaz, C. E.; Tonn, C. E.; Sosa, M. E. Insecticidal Effects of *Flourensia oolepis* Blake (Asteraceae) Essential Oil. *Biochem. Syst. Ecol.* **2007,** *35*, 181–187. https://doi.org/10.1016/j.bse.2006.10.009.
Guclu-Ustundag, O; Mazza, G. Saponins: Properties, Applications and Processing. *Crit. Rev. Food Sci. Nutr.* **2007,** *47*, 231–258. https://doi.org/10.1080/10408390600698197.
Guerrero-Rodríguez, E.; Solis-Gaona, S.; Hernández-Castillo, F. D.; Flores-Olivas, A.; Sándoval-López, V.; Jasso-Cantú, D. In Vitro Biological Activity of Extracts from *Flourensia cernua* D.C. in Post-Harvest Pathogens: *Alternaria Alternata* (Fr.:Fr.) Keissl., *Colletotrichum gloeosporioides* (Penz.) Penz. and Sacc. and *Penicillium digitatum* (Pers.:Fr.) Sacc. *Revista Mexicana de Fitopatologia* **2007,** *25*, 48–53.
Hernández, R.; Lugo, E. C.; Díaz, L.; Villanueva, S. Extraccion y cuantificación indirecta de las saponinas de *Agave lechuguilla* Torrey. *eGnosis* **2005,** *3*, 1–9.
Holst, C. M.; Oredsson, S. M. Comparison of Three Cytotoxicity Tests in the Evaluation of the Cytotoxicity of a Spermine Analogue on Human Breast Cancer Cell Lines. *Toxicol. In Vitro* **2005,** *19*, 379–387. https://doi.org/10.1016/j.tiv.2004.10.005.
Hussain, M.; Debnath, B.; Qasim, M.; Bamisile, B.; Islam, W.; Hameed, M. Role of Saponins in Plant Defense Against Specialist Herbivores. *Molecules* **2018,** *24*, 1–21. https://doi.org/10.3390/molecules24112067.
Hyder, P. W.; Fredrickson, E. L.; Estell, R. E.; Lucero, M. E.; Remmenga, M. D. Loss of Phenolic Compounds From Leaf Litter of Creosote Bush [*Larrea tridentata* (Sess. & Moc.

Ex DC.) Cov.] and Tarbush (*Flourensia cernua* DC.). *J. Arid Environ.* **2005,** *61*, 79–91. http://dx.doi.org/10.1016/j.jaridenv.2004.08.001.

INECC, Instituto Nacional De Ecología y Cambio Climático, Gobierno De Mexico [Online] 2021. https://www.gob.mx/inecc (accessed June 08, 2021).

Innes, R. J. Flourensia cernua. In Fire Effects Information System; U.S. Department of Agriculture, Forest Service, Rocky Mountain Research Station, Fire Sciences Laboratory (Producer); [Online] 2022. https://www.fs.fed.us/database/feis/plants/shrub/flocer/all.html (accessed Feb 08, 2022).

Jasso de Rodríguez, D.; De Angulo-Sánchez, J. L.; Hernández-Castillo, F. D. Chapter 14 An Overview of the Antimicrobial Properties of Mexican Medicinal Plants. *Adv. Phytomed.* **2006,** *3*, 325–377. https://doi.org/10.1016/S1572-557X(06)03014-5.

Jasso-De Rodríguez, D.; Hernández, C. D.; Angulo, S. J. L.; Rodríguez, G. R.; Villarreal, Q. J. A.; Lira, S. R. H. Antifungal Activity In Vitro of *F. cernua* Extracts on *Alternaria* sp., *Rhizoctonia solani*, and *Fusarium oxysporum*. *Ind. Crops Prod.* **2007,** *25*, 111–116. http://dx.doi.org/10.1016/j.indcrop.2006.08.007.

Jasso de Rodríguez, D.; Puente-Romero, G. N.; Díaz-Jiménez, L.; Rodríguez-García, R.; Ramírez-Rodríguez, H.; Villarreal-Quintanilla, J. A.; Flores-López, M. L.; Carrillo-Lomelí, D. A.; Genisheva, Z. A. In Vitro Gastrointestinal Digestion of Microencapsulated Extracts of *Flourensia cernua*, *F. microphylla*, and *F. Retinophylla*. *Ind. Crops Prod.* **2019,** *138*, 111444. https://doi.org/10.1016/J.INDCROP.2019.06.007.

Jasso-De Rodríguez, D.; Rodríguez-García, R.; Hernández-Castillo, F. D.; Aguilar-González, C.N.; Sáenz-Galindo, A.; Villarreal-Quintanilla, J. A.; Moreno-Zuccolotto, L. E. In Vitro Antifungal Activity of Extracts of Mexican Chihuahuan Desert Plants Against Postharvest Fruit Fungi. *Ind. Crops Prod.* **2011,** *34*, 960–966. https://doi.org/10.1016/j.indcrop.2011.03.001.

Jasso de Rodríguez, D.; Salas-Méndez, E.; De, J.; Rodríguez-García, R.; Hernández-Castillo, F. D.; Díaz-Jiménez, M. L. V.; Sáenz-Galindo, A.; González-Morales, S.; Flores-López, M. L.; Villarreal-Quintanilla, J. A.; Peña-Ramos, F. M.; Carrillo-Lomelí, D. A. Antifungal Activity In Vitro of Ethanol and Aqueous Extracts of Leaves and Branches of *Flourensia* spp. Against Postharvest Fungi. *Ind. Crops Prod.* **2017,** *107*, 499–508. https://doi.org/10.1016/j.indcrop.2017.04.054.

Jasso de Rodríguez, D.; Trejo-González, F. A.; Rodríguez-García, R.; Díaz-Jimenez, M. L. V.; Sáenz-Galindo, A.; Hernández-Castillo, F. D.; Villarreal-Quintanilla, J. A.; Peña-Ramos, F. M. Antifungal Activity In Vitro of *Rhus muelleri* against *Fusarium oxysporum* f. sp. lycopersici. *Ind. Crops Prod.* **2015,** *75*, 150–158. https://doi.org/10.1016/j.indcrop.2015.05.048.

Jasso-Rodríguez, D.; Hernández-Castillo, F. D.; Solís-Gaona, S.; Rodríguez-García. D.; Rodríguez-Jasso, R. M. *Flourensia cernua* DC: A Plant from Mexican Semiarid Regions With a Broad Spectrum of Action for Disease Control. In *Integrated Pest Management and Pest Control-Current and Future Tactics*, 2012; pp 639–650. https://doi.org/10.5772/31783.

Jensen, R. S.; Nielsen, B. J. Iridoid Glucosides in Fouquieriaceae. *Phytochemistry* **1982,** *21*, 1623–1629. https://doi.org/10.1016/S0031-9422(82)85028-0.

Killingbeck, K. T. Can Zinc Influence Nutrient Resorption? A Test With the Drought-Deciduous Desert Shrub *Fouquieria splendens* (ocotillo). *Plant Soil* **2008,** *304*, 145–155. http://dx.doi.org/10.1007/s11104-007-9530-z.

Killingbeck, K. Are Growth Rings Accurate Fingerprints of Plants Age in the Stem-Succulent, Drought-Deciduous Shrub Growing in the Chihuahuan Desert?. *J. Arid Environ.* **2017,** *114*, 220–223. https://doi.org/10.1016/j.jaridenv.2017.03.016.

Krochmal, A.; Paur, S.; Duisberg, P. Useful Native Plants in the American Southwestern Deserts. *Econ. Bot.* **1954,** *8*, 3–20. https://doi.org/10.1007/BF02898875.

Martínez-Flórez, S.; González-Gallegos, J.; Culebras, J. M.; Tuñón, M. J. Los Flavonoides: Propiedades y Acciones Antioxidantes. *Nutr. Hosp.* **2002,** *17*, 271–278.

Mata, R.; Bye, R.; Linares. E.; Macías, M.; Rivero-Cruz, I.; Pérez, O.; Timmermann. B. N. Phytotoxic Compounds from *Flourensia Cernua. Phytochemistry* **2003,** *64*, 285–291. https://doi.org/10.1016/S0031-9422(03)00217-6.

Meinhold, K.; Darr, D. The Processing of Non-Timber Forest Products Through Small and Medium Enterprises-A Review of Enabling and Constraining Factors. *Forests* **2019,** *10*, 1026. https://doi.org/10.3390/f10111026.

Méndez, M.; Rodríguez, R.; Ruiz, J.; Morales-Adame, D.; Hernández-Castillo, F. D.; Aguilar, C. N. Antibacterial Activity of Plant Extracts Obtained With Alternative Organic Solvents Against Food-Borne Pathogen Bacteria. *Ind. Crops Prod.* **2012,** *37*, 445–450. http://dx.doi.org/10.1016/j.indcrop.2011.07.017.

Molina-Salinas, G. M.; Ramos-Guerra, M. C.; Vargas-Villarreal, J.; Mata-Cárdenas, B. D.; Becerril-Montes, P.; Said-Fernández, S. Bactericidal Activity of Organic Extracts from *Flourensia cernua* DC Against Strains of *Mycobacterium Tuberculosis*. *Arch. Med. Res.* **2006,** *37*, 45–49. https://doi.org/10.1016/j.arcmed.2005.04.010.

Morreeuw, Z.; Castillo, D.; Ríos, L.; Martínez, R.; Estrada, N.; Melchor, E.; Iqbal, H.; Parra, R.; Reyes, A. High Throughput Profiling of Flavonoid Abundance in *Agave lechuguilla* Residue-Valorizing Under Explored Mexican Plant. *Plants* **2021,** *10*, 695. https://doi.org/10.3390/plants10040695.

Narcia, M.; Castillo, D.; Vazquez, J.; Berlanga, R. Nota de investigación turno técnico de la lechuguilla (Agave lechuguilla Torr.) en el noreste de Mexico. *Revista Mexicana de Ciencias Forestales* **2012,** *3*, 81–88. http://dx.doi.org/10.29298/rmcf.v3i9.531.

Neeraj, G. S.; Kumar, A.; Ram, S.; Kumar, V. Evaluation of Nematicidal Activity of Ethanolic Extracts of Medicinal Plants to *Meloidogyne Incognita* (Kofoid and White) Chitwood Under Lab Conditions. *Int. J. Pure Appl. Biosci.* **2017,** *1*, 827–831. http://dx.doi.org/10.18782/2320-7051.2525.

Nobel, P. S.; Zutta, B. R. Morphology, Ecophysiology, and Seedling Establishment for *Fouqueria splendens* in the Northwestern Sonoran Desert. *J. Arid Environ.* **2005,** *62*, 251–265. https://doi.org/10.1016/j.jaridenv.2004.11.002.

Pando, M.; Eufracio, O.; Jurado, E.; Estrada, E. Post Harvest Growth of Lechuguilla (*Agave lecheguilla*, Torr.) in North-Eastern Mexico. *Econ. Bot.* **2004,** *58*, 78–82. https://doi.org/10.1663/0013-0001(2004)058[0078:PGOLAL]2.0.CO;2.

Peinado, M.; Macías, M. A.; Ocaña-Peinado, F. M.; Aguirre, J. L.; Delgadillo, J. Bioclimates and Vegetation Along the Pacific Basin of Northwestern Mexico. *Plant Ecol.* **2011,** *212*, 263–281. https://doi.org/10.1007/s11258-010-9820-z.

Pérez, R. M.; Vargas, R.; Martinez, F. J.; Garcia, E. V.; Hernández, B. Actividad antioxidante de los alcaloides de Bocconia arborea. Estudio sobre seis métodos de análisis. *Ars Pharm.* **2003,** *44*, 5–21.

PROCCYT, Protección de Cultivos, Ciencia y Tecnología. Asociación Líder Representante de la Industria de la Ciencia de la Protección de los Cultivos en Mexico, accessed 08/06/2021, http://www.proccyt.org.mx.

Programa de las Naciones Unidas para el Desarrollo, Protocolo de Montreal, ONU [Online] 2021. https://www.undp.org/content/undp/es/home/sustainable-development/environment-and-natural-capital/montreal-protocol.html#:~:text=El%20Protocolo%20de%20Montreal%20es,capa%20de%20ozono%20(SAO) (accessed June 18, 2021).

Reyes, J.; Aguirre, J.; Peña, C. Biología y aprovechamiento de Agave lechuguilla Torrey. *Boletín de la Sociedad Botánica de Mexico* **2000,** *067*, 75–88. https://doi.org/10.17129/botsci.1626.

Ruiz-Martínez, J.; Aguirre-Joya, J. A.; Rojas, R.; Vicente, A.; Aguilar-González, M. A.; Rodríguez-Herrera, R.; Alvarez-Perez, O. B.; Torres-León, C.; Aguilar, C. N. Candelilla Wax Edible Coating With *Flourensia cernua* Bioactives to Prolong the Quality of Tomato Fruits. *Foods* **2020,** *9*, 1303. https://doi.org/10.3390/foods9091303.

Salas-Méndez, E. D. J.; Vicente, A.; Pinheiro, A. C.; Ballesteros, L. F.; Silva, P.; Rodríguez-García, R.; Hernández-Castillo, F. D.; Díaz-Jiménez, M. D. L. V.; Flores-López, M. L.; Villarreal-Quintanilla, J. A.; Peña-Ramos, F. M.; Carrillo-Lomelí, D. A.; De Rodríguez, D. J. Application of Edible Nanolaminate Coatings With Antimicrobial Extract of Flourensia Cernua to Extend the Shelf-Life of Tomato (*Solanum lycopersicum* L.) Fruit. *Postharvest Biol. Technol.* **2019,** *150*, 19–27. https://doi.org/10.1016/j.postharvbio.2018.12.008.

Scogin, R. Leaf Phenolics of the Fouquieriaceae. *Biochem. Syst. Ecol.* **1978,** *6*, 297–298. https://doi.org/10.1016/0305-1978(78)90049-2.

SEMARNAT, Convenio de Estocolmo, Secretaría de Medio Ambiente y Recursos Naturales [Online] 2021. https://www.gob.mx/semarnat/acciones-y-programas/convenio-de-estocolmo#:~:text=El%20Convenio%20de%20Estocolmo%20tiene,nuevos%20COP%20a%20trav%C3%A9s%20del (accessed June 08, 2021)

SENASICA, Servicio Nacional De Sanidad, Inocuidad y Calidad Agroalimentaria, Gobierno De Mexico [Online] 2021. https://www.gob.mx/senasica (accessed June 08, 2021).

Sharma, A.; Flores-Vallejo, R. C.; Cardoso-Taketa, A.; Villarreal, M. L. Antibacterial Activities of Medicinal Plants Used in Mexican Traditional Medicine. *J. Ethnopharmacol.* **2017,** *208*, 264–329. https://doi.org/10.1016/j.jep.2016.04.045.

Tellez, M.; Estelle, R.; Frederickson, E.; Powell, J.; Wedge, D.; Schrader, K.; Kobaisy, M. Extracts of *Flourensia cernua* (L.), Volatile Constituents and Antifungal, Antialgal, Antitermite Bioactivities. *J. Chem. Ecol.* **2001,** *27*, 2263–2273. https://doi.org/10.1023/A: 1012283005014.

Towers, G. H. N.; Macrae, W. D.; Irwin, D. A. J.; Bisalputra, T. Membrane Lesions in Human Erytrocytes Induced the Naturally Occurring Compounds Alfa-terthienyl and Phenylheptatriyne. *Photobiochem. Photobiophys.* **1980,** *1*, 309–318.

Valdez-Villavicencio, J. H.; Peralta-García, A.; González-Gutiérrez, N. S.; Hernández-Morlán, X. I.; Hinojosa-Huerta, O. Diversity of the Herpetofauna in Restored and Disturbed Sites in the Colorado River Delta, Baja California, Mexico. *Rev. Mex. Biodivers.* **2021,** *92*, e923763. https://doi.org/10.22201/ib.20078706e.2021.92.3763.

Warnhoff, E. W.; Halls, C. M. M. Desert Plant Constituents. II. Ocotillol: An Intermediate in the Oxidation of Hydroxyl Isoöctenyl Side Chains. *Can. J. Chem.* **1965,** *43*, 3311–3321. https://doi.org/10.1139/v65-461.

Weng-Kong, K.; Mat-Junit, S.; Aminudin, N.; Ismail, A.; Abdul-Aziz, A. Antioxidant Activities and Polyphenolics from the Shoots of *Barringtonia racemosa* (L.) Spreng in a Polar to Apolar Medium System. *Food Chem.* **2012,** *134*, 324–332. https://doi.org/10.1016/j.foodchem.2012.02.150.

Zavala, D.; Carrillo, M. L.; Alvarado, B.; Sánchez, A. O. Evaluación de la toxicidad aguda de un extracto alcohólico de hojas de hojasén (*Flourensia cernua*). *Revista Mexicana de Ciencias Farmacéuticas* **2010,** *41*, 50–54.

Zhang, Z.; Chen, Z.; Zhang, S.; Shao, X.; Zhou, Z. Antibacterial Activity of the Structurally Novel Ocotillol-Type Lactone and its Analogues. *Fitoterapia* **2020,** *144*, 104597. https://doi.org/10.1016/j.fitote.2020.104597.

CHAPTER 11

Microalgae-Based Feed: A Sustainable Bioproduct for the Livestock Sector in Arid Zones

DAVID FRANCISCO LAFUENTE-RINCON,
JAVIER ULISES HERNÁNDEZ-BELTRÁN,
FERNANDO HERNÁNDEZ-TERÁN, and NAGAMANI BALAGURUSAMY

Laboratory of Bioremediation, Faculty of Biological Sciences, Autonomous University of Coahuila, Torreón, Mexico

ABSTRACT

Drylands are ecosystems commonly called deserts where water scarcity is a particular feature. Moreover, the production and supply of food are one of the main challenges for the economic development, due to soil deterioration. The use of natural resources and the integration of a cultivation and livestock production system suitable for arid areas is one of the current efforts to guarantee the food and nutritional security of arid regions and make them more productive, sustainable, and resilient to climate change. Goat, sheep, and other small ruminant breeding are an important economic activity for drylands in reason are naturally adapted to heat stress common among founding in livestock production in this region.

The main nutritional factor facing livestock production in arid areas is to supply the high dietary demand for protein. Specifically, microalgae (green algae and cyanobacteria) are a rich source of proteins, that in addition to complementing animal feed, provide an important amount of carbohydrates, lipids, and vitamins essential for the growth of livestock.

The availability of sunlight and soil in arid regions has enormous advantages for the implementation of bioprocesses in the production of microalgal

Exploration and Valorization of Natural Resources from Arid Zones. Cristobal Noé Aguilar, Norma Paola Meléndez-Renteria, & Karen Nathiely Ramirez-Guzman (Eds.)

biomass. Photoautotrophy is the most common method to produce high-value products from microalgal biomass, but we can find also reports using photoheterotrophic and mixotrophic production.

The reuse of industrial and urban water makes it possible to increase water resources in regions with water scarcity and uses that for the indoor and outdoor microalgal production. The application of microalgal biomass for the manufacture of livestock feed involves procedures and techniques for its recovery from biological reactors and its subsequent as feed; the microalgae-based feedstock is subjected to several toxicological studies before its release as a safe product for dietary and therapeutic uses in livestock production. Finally with the increase of the global population, the improvement of bioprocesses to produce microalgae as sources of proteins and nutrients to sustain the livestock sector are in the eye of biotechnologists.

11.1 INTRODUCTION

The arid and semi-arid lands dominate almost half the land use of earth's surface, where the role of soil and water conservation along the productive process of food represents a key constraint to the economic development in this region (Golla, 2021). The economic development of arid regions is vulnerable to climate change, as rapid population growth and anthropogenic water demand increase the current water resource scarcity (Lian et al., 2021). The water scarcity impact in urban zones is projected to increase from 933 million in 2016–2.373 billion people in 2050 (He et al., 2021). According to the FAO (2021), the world's total meat production is approximately around 337.2 million tons among all the poultry meat, pig meat, bovine meat, and ovine meat are the major types. The breeding of small ruminants in arid and semi-arid regions is an important economic activity that includes goats and sheep naturally adapted to arid conditions worldwide (Leite et al., 2021). The microalgae are a polyphyletic group of eukaryotes microorganisms, and some descriptions include the prokaryotes group of cyanobacteria. The large-scale cultivation of microalgae is a promising alternative feedstock for livestock and poultry production (Saadaoui et al., 2021). The production of microalgal biomass in photobioreactors has become a viable option for feedstock in animals mainly considering the estimations of 22.66 g/m^2 of biomass production (Valdovinos-García et al., 2021). In this sense, microalgae have important contributions to boosting the economic growth (Fernandez et al., 2020). Also, microalgal biomass in the animal feed had well-descripted health benefits of cattle, swine, poultry, lamb, and goat (Nagarajan et al., 2021). The bioeconomy is a strategic system to reduce the world dependence on natural resources such as water and provide sustainable development of microalgal production is by using wastewater from urban and food industry effluents that contain a considerable resource of nutrients and growth factors, such as nitrogen, phosphorus, potassium, amino acids, proteins, ammonium, and fatty acids (Ma et al., 2018; Ali et al., 2021). Microalgae utilization in the food industry including the integration to the treatment of high volumes of produced wastewater to the food industry as effective and sustainable methods to treat these effluents, and the increased livestock productivities as protein alternative source to avoid the competition between crops for agriculture and feed production (Esteves et al., 2021). Microalgae are versatile and adaptable to grow in photobioreactors (PBRs) in controlled or not controlled conditions, indoor, and outdoor conditions using urban wastewater, for example, *Chlorella vulgaris* (Sarker and Salam, 2019). Their use on large scale for commercial purposes remains to use only *Arthrospira platensis* (Spirulina) and *Chlorella* spp. as nutritional supplements related to the opportunities use microalgae as feed for livestock production.

11.2 WATER SCARCITY AND DESERTIFICATION IN DRYLANDS

Drylands are ecosystems commonly called deserts where water scarcity is a particular feature (Safriel, 2009). Water scarcity continues to intensify in urban and agricultural regions in arid zones in direct relation to anthropogenic activities and climate change (Qadir and Sato, 2016). This phenomenon is generally associated with desertification, as a human process of reducing soil ecosystem services in non-desert areas (Safriel, 2009; Adeel et al., 2005), this concept does not exclude the exploitation of water resources in hyper-arid lands (Martínez-Valderrama et al., 2020). Until 2005, deserts reached about 45% of the earth's surface including dry sub-humid, semi-arid, arid, and hyper-arid biomes (Adeel et al., 2005). The aridity index (AI) at these sites can range from 0.2 to 0.5 mm mm^{-1} dry sub-humid 0.5–0.65 mm mm^{-1} to less than 0.03 mm mm^{-1} in hyper-arid lands (Plaza et al., 2018). The deserts are inhabited by more than 2.8 billion people which represents about 40% of the world's population (Plaza et al., 2018), of which, close to 101 million people live in hyper-arid regions where the productivity of ecosystems and the availability of water resources are severe (Martínez-Valderrama et al., 2020). Moreover, the production and supply of food are one of the main challenges for the economic development, due to soil deterioration (Golla, 2021). More than 74% of the world's rangelands and half of the world' croplands are in semi-arid areas, paradoxically, only 11% and 30% of land use in drylands is used for cropland and grazing, respectively (Plaza et al., 2018). The reason looks obvious, in the sense the profitability and sustainability of dryland livelihoods are directly linked to increased aridity, limiting farming to semi-arid areas, and leaving grazing, aquaculture, solar energy, and tourism as the only economic development alternatives for arid and hyper-arid regions (Safriel, 2009; Adeel et al, 2005). Recent predictions show a progressive increase in the conversion of forests, pastures, and cropland into drylands throughout the 21st century, with adverse consequences for the lifestyles of more than 500 million people living mainly in South America, Africa, and crops in the northern hemisphere. Here, the worst-case scenario shows an increase of global temperature of up to 4°C with hotspots in the border with the United States and Mexico (Sonoran and Chihuahuan desert), North-east Brazil, South of Argentina, the Mediterranean countries, western and southern countries on Africa, Central Asia, China, and Australia (Spinoni et al., 2021).

11.3 FOOD SECURITY AND LIVESTOCK PRODUCTION IN ARID REGIONS

The food security of a country or region represents its ability to produce or import enough quality food to meet the nutritional and well-being needs of its inhabitants (FAO, 2006). One of the factors with greater impact in this area is the growing demand for food of animal origin around the world (Robinson et al., 2014). However, its production faces impediments, such as water scarcity and overexploitation of water sources derived from this activity (Richter et al., 2020). Specifically, in arid areas, the high vulnerability of soils to environmental pressure, erosion, and desertification is added to this challenge as the main limiting factor (FAO, 2006). Dairy production systems in Saudi Arabia are an example of a production not suitable for arid sites, which is sustained by importing inputs such as alfalfa meal and pellets to feed more than 368,000 head of cattle with a water demand of 74–110 L per head of cow a day. In addition, the water is used as a cooling system for the herd and cattle pens (Martínez-Valderrama et al., 2020). Consequently, the use of natural resources and the integration of a cultivation and livestock production system suitable for arid areas is one of the current efforts to guarantee the food and nutritional security of arid regions and make them more productive, sustainable, and resilient to climate change

(Sekaran et al., 2021). The climate is also a factor against it, only between 1981 and 2010, there have been dangerous events for livestock production in countries such as Kenya, where heat stress affects thousands of heads of dairy cattle, cattle, sheep, goats, pigs, and poultry (Rahimi et al., 2021). Goat, sheep, and other small ruminant breeding are an important economic activity for drylands in reason are naturally adapted to heat stress common among founding in livestock production in this region (Leite et al., 2021). Heat stress can lead to important metabolic disorders, oxidative stress, and immunosuppression, and contribute to the animal's death. Some strategies for mitigation of heat stress effect on animal livestock include dietary supplementation with vitamin C and E, selenium, zinc, and electrolytes (Leite et al., 2021). Currently, the livestock sector includes the production of cattle, poultry, pigs, sheep, goats, and fishing (Sekaran et al., 2021; Robinson et al., 2014). The total world meat production is estimated at 337.2 million tons in 2020 (FAO, 2021). In 2020, the world's leading meat exporters were the European Union, the United States of America, Brazil, Canada, Mexico, and the Russian Federation (FAO, 2021), while Brazil and China are the largest consumers of animal products, such as meat, milk, and eggs (Robinson et al., 2014). The world poultry meat production was about 133.3 million tons in 2020. In the same year, the world bovine meat output was estimated to be 71.4 million tons, while pig meat output was estimated to be 109.2 million tons; on the other hand, ovine meat output was estimated at 16 million tons (FAO, 2021). The global poultry meat production reached 133.3 million tons in 2020. The global meat production from countries of arid regions is given in Table 11.1.

11.4 MICROALGAE BIOMASS USE IN ANIMAL FEED

The main nutritional factor facing livestock production in arid areas is to supply the high dietary demand for protein, which is traditionally obtained by the production or import of protein-rich biomass, such as soybeans, corn, and wheat (Golla, 2021). An alternative that proposes the reduction of the environmental impact of the use of arable land is the production of microalgal biomass as an excellent prospect to meet this demand (Vigani et al., 2015). Specifically, microalgae (green algae and cyanobacteria) are a rich source of proteins (Borowitzka, 2013; Barkia et al., 2019; Smetana et al., 2017) with the content of microalgae varies between 18 and 46% of dry weight (Tibbetts et al., 2015), that in addition to complementing animal feed, provide an important amount of carbohydrates, lipids, and vitamins essential for the growth of livestock (Valdovinos-García et al., 2021; Benemann, 2013). The overview of the main species of microalgal biomass content is given in Table 11.2.

TABLE 11.1 Global Meat Production From Countries of Arid Regions in 2020.

Country	Meat production (Thousand tons/carcass weight equivalent) in 2020			
	Poultry	**Bovine**	**Pig**	**Ovine**
Algeria	nd	137	nd	352
Argentina	2261	3243	665	nd
Australia	1313	2123	436	656
Bangladesh	nd	nd	nd	234
Brazil	14,363	9894	4418	136
Chile	782	223	574	Nd
China	22,321	6739	42,044	4926
Colombia	nd	736	443	Nd
Egypt	nd	735	nd	nd
India	3821	2156	355	826
Iran	2336	350	nd	250
Mexico	3617	2079	1649	106
Nigeria	nd	nd	269	417
Pakistan	nd	2225	nd	748
Saudi Arabia	930	nd	nd	124
South Africa	1816	963	287	174
United States of America	23,131	12,353	12,837	74
World	133,266	71,408	109,200	16,276

Source: Adapted from FAO (2021).

11.4.1 HIGH VALUE AND BIOACTIVE COMPOUNDS FOR ANIMAL FEED

Other biomolecules obtained from microalgae with application in animal feed include pigments, such as astaxanthin, lutein beta-carotene, chlorophyll, phycobiliprotein (Borowitzka, 2013), polyunsaturated fatty acids (PUFAs), and polysaccharides, such as beta-1,3 glucan (Borowitzka, 2013; Yaakob et al., 2014; Kusmayadi et al., 2021) found in microalgae but have supported the growth and health in animals which feed them including

TABLE 11.2 Nutritional Content of the Biomass of the Main Microalgae and Cyanobacteria Species Cultivated for Animal Feed.

Group	Species	Total biomass content (g/100g)			Reference
		Protein	Lipids	Carbohydrates	
Microalgae	*Chlamydomonas reinhardtii*	48	17	21	Becker, 2007
	Chlorella vulgaris	42.8	8.7	Nd	Alfaia et al., 2021
	Chlorella vulgaris	51–58	nd	14–22	Becker, 2007
	Dunaliella salina	23.4	43.3	6.9	Colusse et al., 2020
	Nannochloropsis spp.	28.8	35.9	18.4	Rebolloso-Fuentes et al., 2021
	Nannochloropsis gaditana	47	16	21.7	Molino et al., 2019
	Scenedesmus obliquus	50–56	nd	Nd	Becker, 2007
	Scenedesmus obliquus	40.4	5.6	28	Silva et al., 2020
	Tetraselmis sp. CTP4	31.2	7	18.6	Pereira et al., 2019
Cyanobacteria	*Arthrospira platensis*	61	14	6	Habib et al., 2008
	Arthrospira platensis	63.9	10.7	12.8	Niccolai et al., 2019
	Spirulina platensis	61.3	15.1–15.8	7.1–8.0	Tokuşoglu and Uunal, 2019
	Spirulina platensis	76.6	2.5	6.5	Seghiri et al., 2019

feed for livestock, poultry, and aquaculture (Yaakob et al., 2014). The main benefits of animal feed with the microalgae-based product include increased growth and improved health, which are mainly reflected in increased weight gain and growth rate, greater digestibility, and an increase in the survival index. In addition to the antilipidemic, antiglycemic, cardioprotective, antioxidant, and hypertensive activity, they can increase the quality of the animals where they are applied (Nagarajan et al., 2021). Just as proteins are essential nutrients for muscle growth and gain in livestock, fatty acids, and lipids are an important source of energy and a primary source of fat that promotes weight gain in livestock (Nagarajan et al., 2021). Particularly, the use of microalgae supplementation enhancement of PUFA's in oil can improve the lipid metabolisms in farm animals (Nagarajan et al., 2021). Studies in the broiler chicken diet show the benefit of supplementation with *Spirulina plantensis* to increase the fatty acid profile in meat, especially in the PUFA eicosapentaenoic (EPA) and docosahexaenoic (DHA) (Bonos et al., 2016). The use of algae as a food supplement shows a marked increase in DHA in beef loin with 10–20 mg DHA/100 g meat, and in ham with 25–56 mg DHA/100 g meat compared with soybean oil (2.1–4.5 mg DHA/100 g meat) and flaxseed oil (4.3–9.1 mg DHA/100 g meat), (Bonos et al., 2016). According to Xu et al. (2021), the use of *Schizochytrium* sp. seaweed dietary supplement on bull improves the antioxidant status, fatty acid profile, and volatile compounds in beef meat. Except in the case of cyanobacteria, microalgae contain cellulosic cell walls that must be evaluated for use in non-ruminant livestock as they represent a digestibility problem (Becker, 2013). The carbohydrates and polysaccharides represent the major constituent of net energy and metabolizable energy on feed formulations. Although most livestock represent ruminants such as cattle and sheep where the digestibility of cellulosic material containing the cell walls of microalgae (except the case of cyanobacteria) is not a problem, treatment of microalgae biomass for use in non-ruminant or monogastric livestock should be considered (Nagarajan et al., 2021). Pigments such as chlorophylls, carotenoids, and phycobilins can be easily identified in photosynthetic organisms such as microalgae. Astaxanthin, beta-carotene chlorophylls, lutein, and phycobilins are microalgal pigments that play a remarkable role in nutraceuticals and functional foods (Borowitzka, 2013). Primarily, carotenoids are important nutraceuticals in animal feed, principally for their potent antioxidant activities, and have been widely used as high-end health supplements (Borowitzka, 2013; Hu et al., 2018). In addition, carotenoids are the most extracted microalgae biomolecules in terms of production volume (Hu et al., 2018). Currently, the most commonly produced carotenoid is astaxanthin, which is extracted mainly from the microalgae *Haematococcus pluvialis* (Borowitzka, 2013; Hu et al., 2018).

11.4.2 MICROALGAL GLOBAL MARKET AND COMMERCIALIZATION

Microalgae-based food/feed represents a market opportunity but still are significantly smaller compared with cereals and other commodity crops, but there has been impressive growth (Vigani et al., 2015). Also, biomass supplementation improves the nutritional and sensorial quality of meat (Nagarajan et al., 2021). Globally, the private companies producing commercial food and feed products derived from microalgae and in meat production in arid regions, are distributed in the USA (Food, dried algae, Astaxanthin, EPA, DHA), Mexico (Feed, dried algae), India (Food, dried algae, astaxanthin, beta-carotene), and China (Food, dried algae, EPA/DHA, beta-carotene). The main microalgae-based products in the market are whole-dried microalgae of *Spirulina* and *Chlorella* and high-value molecules, such as astaxanthin, phycobiliprotein, omega-3 fatty acids, and beta-carotene (Park et al., 2021).

11.5 MICROALGAE TROPHIC REQUIREMENTS

The availability of sunlight and soil in arid regions has enormous advantages for the implementation of bioprocesses in the production of microalgal biomass. The common system to produce microalgae is the photoautotrophic way that transforms carbon dioxide into organic molecules useful as an energy source for animal feed. Significantly, the cultivation of microalgae is simple and does not require harmful chemicals or a disproportionate amount of water and fertile land in comparison with other terrestrial crops (Veerabadhran et al., 2021). Microalgae are grown in culture media for autotrophic growth, except for the addition of a source of organic carbon in the media (Hillebrand and Sommer, 2000). Commonly, a culture medium for algae contains salts ubiquitous in the habitat of origin of the algae, certain ionic components, such as K^+, Mg^{2+}, Na^+, Ca^{2+}, SO_4^-, and Cl^-, the sources of nitrogen, especially nitrate, ammonium, and urea, the source of carbon, CO_2 or HCO_3^-, pH, trace elements and a chelating agent such as ethylenediaminetetraacetic acid (EDTA), and vitamins. In addition, organic carbon sources, such as glucose, fructose, and acetic acid are used in heterotrophic cultures (Grobbelaar, 2013). Nutrients are an important expense for large-scale cultivation, and their source and composition should be considered once you want to expand the cultivation system (Torzillo and Chini Zittelli, 2015). All microalgae are photoautotrophic, with autotrophic cultivation as the main method for biomass production. However, they have a versatile and flexible metabolism in terms of obtaining energy (Morales-Sánchez et al., 2016), with the ability to intercalate autotrophy and/or heterotrophy as the best option for its development (Grobbelaar, 2013). An example of the adaptability of microalgae is presented in oleaginous strains belonging to the genus *Microchloropsis*; these organisms adapt to different trophic conditions, thriving under strict photoautotrophic conditions, as well as in the concomitant presence of light and mixotrophy as they reduced external carbon as energy sources (Bo et al., 2021).

11.5.1 PHOTOAUTOTROPHIC PRODUCTION OF MICROALGAE BIOMASS

Photoautotrophy is the most common method to produce high-value products from microalgal biomass (Benemann et al., 2018). In the case of PUFA production in autotrophic microalgae cultures, no carbon source is required to grow as it is necessary to obtain a high-value compound in yeasts or bacteria; however, factors, such as radiation, temperature, state of nutritional deprivation, pH, salinity, stirring, and mixing factors must be controlled to maintain their optimal production (Saxena et al., 2020). Generally, open pounds are the most optimal systems to maintain autotrophic productivity because they allow greater diffusion of CO_2 from the atmosphere to the crop; however, atmospheric CO_2 cannot satisfy the carbon requirements of high-yield autotrophic algae production systems (Grobbelaar, 2013).

11.5.2 PHOTOHETEROTROPHIC PRODUCTION OF MICROALGAE BIOMASS

The main feature of a heterotrophic microalgae culture is the use of organic carbon sources to improve microalgae growth through aerobic respiratory metabolism using oxygen as the final electron acceptor. As a general rule, microalgae growing in the dark should have a mass ratio (C:N:P) of 46.1 parts carbon to 7.7 parts nitrogen and 1 part of phosphorus (Hillebrand and Sommer, 2000). In addition, this system can use solar energy for the metabolic processing of organic carbon sources (without CO_2), becoming a process photoheterotrophic (Yang et al., 2000). Instead, heterotrophic systems have the highest theoretical biomass yield in ATP compared with autotrophic cultivation, as microalgae expend a large amount of ATP for the cell maintenance process, and this system offers an organotrophic alternative for energy obtaining (Yang et al., 2000). In addition, the supplementation of microalgal cultures with carbon sources increases the content of lipids and carbohydrates in the cells (Perez-Garcia et al., 2011) and increases the biomass concentration up to 25 times compared with phototrophic conditions (Morales-Sánchez et al., 2016). Under well-controlled conditions in biorecators, heterotrophic cultivation of microalgae for pigment production has been the attention center for commercial applications due to its overcoming the difficulties associated with CO_2 and light supply as well as avoiding contamination issues and soil requirements in open autotrophic cultivation systems (Hu et al., 2018). According to Smetana et al. (2017), incorporating the energy sources in the production of heterotrophic microalgae has the potential to reduce the environmental impact by 40%, while changing the carbon source could reduce the impact by 25–40%. Park et al. (2021) analyzed a photoheterotrophic culture of *Chlorella sorokiniana* and revealed that the heterotrophic metabolism becomes dominant when the microorganism begins to consume organic carbon (24–72 h). However, as the organic carbon source was reduced, autotrophy became completely dominant. Regarding food development, Smetana et al. (2017) describe that protein biomass obtained from heterotrophic *Chlorella pyrenoidosa* crops have greater nutritional benefits than pork and beef. In this case, it is proposed that a further optimization of production with *Chlorella pyrenoidose* with hydrolyzed food residues could provide a more beneficial protein concentrate with 2.25 kg CO_2 eq, the energy demand of 28.4 Mj, and land occupation of 0.31 m^2 per kg (Smetana et al., 2017). Currently, these types of cultures are economically viable only for high-value products, including PUFAs, pigments, antioxidants, polysaccharides, food, and feed for aquaculture from carbon sources, such as glucose, acetate, or glycerol (Morales-Sánchez et al., 2016).

11.5.3 MIXOTROPHIC PRODUCTION OF MICROALGAE BIOMASS

The mixotrophic culture uses inorganic and organic carbon sources in the presence of light, called the simultaneous autotrophy and heterotrophy system. In this process, CO_2 is fixed by photosynthesis, while organic compounds are oxidized by aerobic respiration (Bo et al., 2021). Zhang et al. (2011) reported that the addition of low concentrations of glucose can shorten the lag phase of the culture mainly due to its strong cellular assimilation in microalgae during their growth. Yang et al. (2000) agrees that the simultaneous supply of light, CO_2, and a source of organic carbon in the medium can obtain a simultaneous absorption of photosynthetic carbon and organic carbon because of higher yields in biomass production. Mixotrophic cultivation is a good option to produce microalgae at a low cost because culture media that use agro-industrial by-products as supplements have higher biomass yields than heterotrophic and autotrophic media. Therefore, some complex substrates such as the use of beet molasses for the culture of *Tetradesmus obliquus* cells work better in mixotrophic culture conditions than that in photoheterotrophic in terms of higher biomass yield and specific growth rate, as well as shorter biomass doubling time (Piasecka et al., 2020). Table 11.3 presents an example of the differences between biochemical composition and biomass yield in microalgae cultures in photoautotrophic, photoheterotrophic, and mixotrophic systems using the cultivation regime of *Tetradesmus obliquus* to produce essential fatty acid as a reference.

TABLE 11.3 Comparative Analysis of Microalgal Trophic Regimen (Piasecka et al., (2020).

	Tetradesmus obliquus			
	Biomass yield (gL^{-1})	**Protein content (% w/w)**	**Lipid content (% w/w)**	**Carbohydrates content (% w/w)**
Photoautotrophy	2.12 ± 0.15	24.13 ± 5.02	14.13 ± 5.02	61.82 ± 0.98
Photoheterotrophy	2.86 ± 0.56	35.36 ± 7.05	13.74 ± 2.09	32.62 ± 3.54
Mixotrophy	4.21 ± 0.74	45.15 ± 7.68	13.48 ± 2.95	24.97 ± 3.2

Tetradesmus obliquus as culture Piasecka et al., (2020).

11.6 URBAN AND FOOD INDUSTRY EFFLUENTS USE FOR MICROALGAE PRODUCTION

Water scarcity is a major challenge for industrial and urban development. Since these developments are based on a safe water supply, strategies must be put in place to ensure the necessary quantities of water. Concerning the above, the reuse of industrial and urban water makes it possible to increase water resources in regions with water scarcity (Bauer et al., 2020). Many industrial centers around the world are in some regions classified as arid (sub-humid to hyper-arid), where large quantities of urban effluents and effluents from the food industry are produced. Some municipal wastewaters also classified as raw wastewater, are nutrient-rich effluent with secondary effluents, and centrates are obtained, which are by-products of sludge. Centrate is a nutrient-rich effluent with an approximate content of 124.2, 208.3, 2320 mg/L nitrogen, phosphorus, and chemical oxygen demand (COD), respectively (Li et al., 2019). In untreated municipal wastewater, the concentrations of 59.9, 7.3, and 460.6 mg/L for total nitrogen, total phosphorus, and COD respectively are evident (Kong et al., 2021), while in agricultural wastewater containing some dairy manure effluents, the concentration of total nitrogen, total phosphorus, and DOC reaches to 305, 266, and 38,230 mg/L, respectively (Dai et al., 2015). Several studies demonstrate an important application of microalgae, attributed to the viability as an advanced wastewater treatment due to their high efficiency in the extraction of contaminants from different sources such as domestic, industrial, and agricultural wastewater (Acién-Fernández et al., 2018). The recovery of resources from industrial effluents through the cultivation of microalgae in addition to the here discussed utility to produce supplements for animal feed allows the emission reduction of gaseous pollutants due to its ability to remove volatile organic compounds in conjunction with bacterial cultures (Ali et al., 2021). According to Yu et al. (2019), the use of wastewater as a rich nutrient medium can improve the lipid yield when used during the starvation of nitrogen sources during *Chlorella* culture. Despite its high nutrient content, the use of municipal wastewater is not an option for feed production. The use of animal waste to produce animal feed has appeal but is challenged by issues of scale, waste composition, and the use of specific algae species to allow food approval (Benemann et al., 2018). Instead, food waste as a main source of carbon is considered the most suitable option for the heterotrophic production of microalgae (Smetana et al., 2017). However, to ensure the viability of production on an industrial scale, the most economical substrates, such as glucose, glycerol, and acetates are chosen (Barclay and Apt, 2013).

11.7 INDOOR AND OUTDOOR PRODUCTION OF MICROALGAE

Due to the differences in the geographical distribution of food and feed obtained from microalgae are conditioned to the different climatic conditions of the sites where they are produced, so their production is carried out in open and closed systems (Enzing et al., 2014). Open or semi-closed systems are more common in places with high temperatures and sunlight intensity (Terry and Raymond, 1985). In contrast, the use of closed and indoor systems such as PBRs is more common in places with cold or extreme temperatures. More than 98% of commercial microalgae biomass production uses open ponds, including raceway-type designs mixed with paddle wheels (Benemann et al., 2018). The most common open systems are outdoor ponds, highly dependent on climatic and environmental conditions, but with considerably lower operating costs than the closed systems when used for large biomass production (Vigani et al., 2015). Although open ponds have a more favorable energy balance than closed systems, they have a considerable water ecological footprint when used on a large scale, and it is difficult to maintain a viable crop for long periods because of frequent microbial contamination (Zittelli et al., 2013). The indoor production of microalgae biomass is performed mainly on PBRs. Specifically, the industrial exploitation of microalgae to produce feed and biofuels is mainly limited to the high costs related to large-scale cultivation. PBRs, including closed systems for algae cultivation, suffer high energy expenses to operate the mixing, cooling systems, and the energy incorporated into the process (Zittelli et al., 2013). The main operational variables for the cultivation of microalgae in PBRs are irradiation intensity, temperature, optical depth, turbulence, acclimatization to the light of microalgae, nutrient composition, and complementation of the degassing of carbon dioxide and oxygen (Masojídek et al., 2015). Classic PBR designs include flat and tubular panels and some hybrid systems that combine open ponds and PBRs (Zittelli et al., 2013). Among the semi-closed cultivation systems is the vertical PBR used to produce algal biofertilizer for rice production (Jochum et al., 2018). PBR has a system of precisely controlled growing conditions, reduced ecological water footprint due to the prevention of water losses caused by evaporation, and area-dependent productivity (Troschl et al., 2017). PBR takes up less space compared with open tank systems; most PBR is made in indoor systems use artificial lighting or natural lighting in greenhouses. During microalgae cultivation in vertical PBRs, light-emitting diode (LED) is used as lighting (Gouveia et al., 2016). These systems achieve higher biomass yields than open systems, but their direct and indirect costs to build and maintain them is high, hence, they are considered specifically to produce high-value metabolites that cushion the cost of sterile and controlled conditions sustained (Masojídek et al., 2015). The culture of microalgae in tubular PBRs presents a spatial configuration that favors the surface available for lighting. In these systems, a greater possible relationship between area and volume exposed to light irradiation is achieved, and this is favorable to the photosynthetic activity and mixing patterns (Troschl et al., 2017). Among the main disadvantages of the use of tubular bioreactors are the photosynthetic limitations that occur in outdoor microalgal cultivation (Veerabadhran et al., 2021). Currently, global climate challenges make it difficult for photobioprocesses to scale from the laboratory level to free-field production, likewise, the choice of a cultivation system depends on the optimal working regime for the growth of microalgal strain and its potential application (Masojídek et al., 2015). In industrial-scale microalgae production systems, light conversion efficiency ranges from 1 to 2% on a solar basis, well below the theoretical maximum of about 10% due to energy loss at all stages of the processes. In this sense, the availability of light is the most important variable for choosing the working regime (Torzillo and Chini Zittelli, 2015). The cultivation of cyanobacteria is an example of large-scale microalgal biomass production, which occupies large extensions of land for the installation of open raceway ponds or sophisticated closed production systems, such as PBR (Troschl et al., 2017). Smetana et al. (2017) show us a scenario where the annual dry matter production of *A. platensis* and *C. vulgaris* at a pilot industrial scale of 1.4–1.8 tons using autotrophic (tubular PBR, open channel pond) and heterotrophic (fermenters) systems had a better environmental impact than the production of most traditional protein sources for food and feed. This is related to the use of heat and energy (autotrophic culture) and glucose for feeding microalgae (heterotrophic). Figure 11.1 shows the overall view systems settling for microalgal biomass production in arid and semi-arid regions.

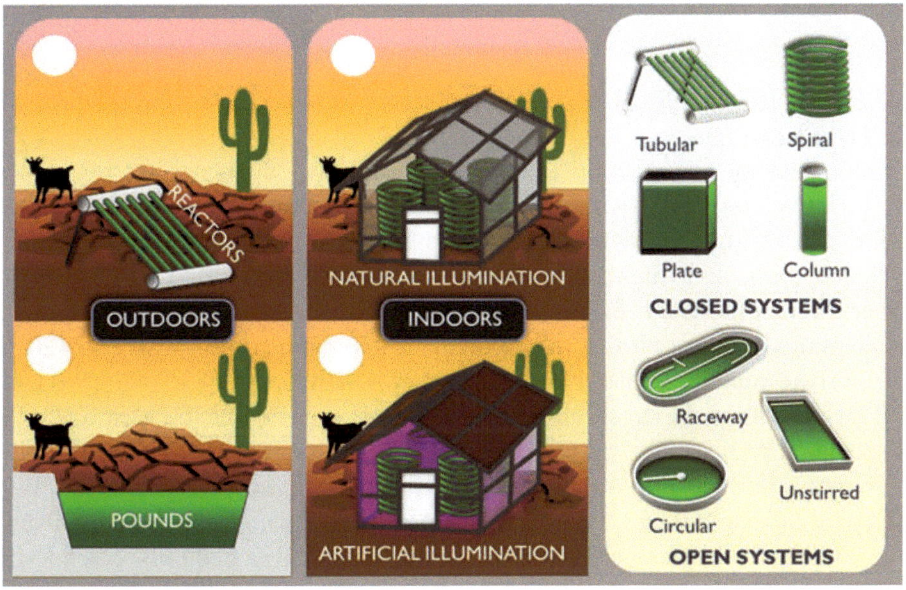

FIGURE 11.1 Culture systems of microalgae in arid zones.

11.8 DOWNSTREAM PROCESS FOR MICROALGAL FEED PRODUCTION

The application of microalgal biomass for the manufacture of livestock feed involves procedures and techniques for its recovery from biological reactors and its subsequent treatment before release as feed. The downstream processes in a bioprocess of algae biomass as feedstock consists of three stages: biomass harvesting (including dewatering), biomass concentration (dewatering and drying), and biomass processing for high value and bioactive compounds recovery (cell disruption, metabolites extraction, and purification technologies). For reasons of feasibility, the operations and processes for the treatment of microalgal biomass must be scalable with low energy costs and with great usability (Perez-Garcia et al., 2011). Biomass harvesting alone costs about 30% of the total costs of microalgae biomass production and biomass sedimentation in large-scale crops (Darpito et al., 2015; Deprá et al., 2018). Only, in vertical PBR cultivation, air compressor systems are used for both harvesting and drying of microalgal biomass (Gouveia et al., 2016). To produce 1 kg of biomass of *Scenedesmus obliquus* in vertical photoreactors, the energy consumption of about 117.4 kWh is used for cultivation, 25 kWh for harvesting by centrifuge, and 0.2 kWh for dehydration in the oven (Gouveia et al., 2016).

11.8.1 BIOMASS HARVESTING BY CENTRIFUGATION AND FLOCCULATION

The harvesting and dewatering systems are considered single-unit operations for processing of microalgal biomass mainly performed by combining flocculation, centrifugation, or filtration in a sequence to improve the biomass harvesting efficiency (Fasaei et al., 2018). The self-settling properties in some microalgal cultures allow a reduction in the harvesting process. Yu et al. (2019) report the improvement of biomass harvesting efficiency of 83.5–93.1% during 24 h cultivation due to automatic sedimentation properties found in *Chlorella* species (FAO, 2006). The collection of microalgae by centrifugation is more common in obtaining high value-added products because it requires a high capital investment and a high energy cost. Therefore, its use in the industrial production of products of lower commercial value requires considering the steps of preconcentration of biomass such as flocculation for buffering the product's final cost (Kratzer and Murkovic, 2021). The flocculants are chemicals added to the microalgal culture to accelerate the sedimentation rate. Zhu and Hiltunen (2018) reports that adding aluminum sulfate (2.5 g/L) or chitosan (2.25 g/L) as flocculants achieves more than 90% of biomass recovery of *Chlorella vulgaris* culture, although the use of chitosan had no effects on the final lipid yield; the use of aluminum sulfate caused a reduction of 4.4%.

11.8.2 BIOMASS CONCENTRATION BY FLOCCULATION AND FILTRATION

Biomass concentration consists of dewatering and drying operations. Specifically, the dewatering process of microalgal biomass accounts for 20–30% of the product cost that consists of the flocculation and filtration process (Musa et al., 2020). The processing of dewatering is performed by flocculation and or filtration (Fasaei et al., 2018). One process for harvesting microalgal biomass is filtration, which has a lower energy cost compared with centrifugation. Filtration allows the separation of flocculated cells using micromembranes with pore sizes greater than 10 μm (Monte et al., 2018). One of the limitations of filtration is the frequent fouling and clogging of the membrane. According to Monte et al. (2018), the usefulness of the crossflow ultrafiltration and microfiltration procedure was proven to alleviate membrane fouling, intrapore fouling, and port blockages during the process of concentration of *Dunaliella salina* cells with a lighter permeate with ultrafiltration. For the extraction of oils and pigments from *Dunaliella salina*, membrane filtration is required; subsequently, the cellular filtration is dehydrated and dried from the biomass, cell breakdown, centrifugation of cell lysate, and extraction of bioactive substances with solvents (Monte et al., 2018).

11.8.3 BIOMASS DRYING

After harvesting the biomass, the next step in the downstream process is drying microalgal biomass. Drying consists of dehydrating biomass performed on a drum dryer and spray dryer (Fasaei et al., 2018). The downstream process for the recovery of microalgal biomass must be carried out under conditions that allow the conservation of the bioactive and nutritional properties of compounds, such as carbohydrates, pigments, omega-3 fatty acids, vitamins, amino acids, fatty acids, and proteins (Perez-Garcia et al., 2011). In the case of open ponds, the collection of biomass is carried out with the help of a paddle, and for the drying of the biomass, solar dryers are used (Gouveia et al., 2016).

11.8.4 HIGH VALUE AND BIOACTIVE COMPOUNDS RECOVERY

For animal feed purposes, dry biomass must undergo mechanic, enzymatic or chemical pretreatments to improve its digestibility because only ruminants can assimilate nutrients trapped between recalcitrant cell wall cellulose in microalgae (Becker, 2013). Instead of recovering high-value and bioactive compounds, the microalgal biomass is subjected to chemical or enzymatic treatments to disrupt the cell wall and cellular membrane for releasing the target molecules. For protein and carbohydrates, cell disruption, and extraction are employed. These methods include high-pressure homogenization, osmotic shock, sonication, autoclave, microwave, and bead-beating. For lipid extraction, the most common method is based on solvent extraction (using water or ethanol), because of its low cost and relatively high efficiency. Furthermore, these methods usually have a heating step that can cause thermodegradation of the compounds of interest (Coudert et al., 2020).

11.8.5 TOXICOLOGICAL ASSESSMENT

For use in animal feed, the microalgae-based feedstock is subjected to several toxicological studies before its release as a safe product for dietary and therapeutic uses in livestock production (Becker, Microalgae for Human and Animal Production). The first stage of evaluation consists of the determination of the proximal chemical composition, which includes the identification of biogenic toxic substances such as phycotoxins or xenobiotics, such as heavy metals, harvest, and processing residues (Becker, 2013). The EU Food safety regulation, as with other some food and feed sources, requires the assessment of harmful compounds, such as toxins, allergens, or other potentially dangerous compounds produced by the algae before their commercial release (EC, 2002). Protein quality evaluation includes sanitary analyses of microbiological spoilage, toxicological, and safety evaluation for feeding trials with an experimental animal. To use further, the harvested and concentrated algal biomass, a water content of less than 10% is required in the final product due to the moisture-promoting spoilage of the dried algal product by the bacteria, mold, or fungi growth (Becker, 2013). The product safety for their launching on market requirements delays their commercialization (Vigani et al., 2015). Currently, only *Chlorella vulgaris*, *Chlamydomonas reinhardtii*, *Arthrospira platensis*, *Auzenchlorella protothecoides*. *Euglena gracilis*, and *Dunaliella bardawil* are given GRAS status (Generally Recognized as Safe) by the United States Food and Drug Administration (Food and Drug Administration, 2022; Torres-Tiji et al., 2020). A schematic view of general bioprocess steps for microalgal biomass production for animal feed in drylands is shown in Figure 11.2.

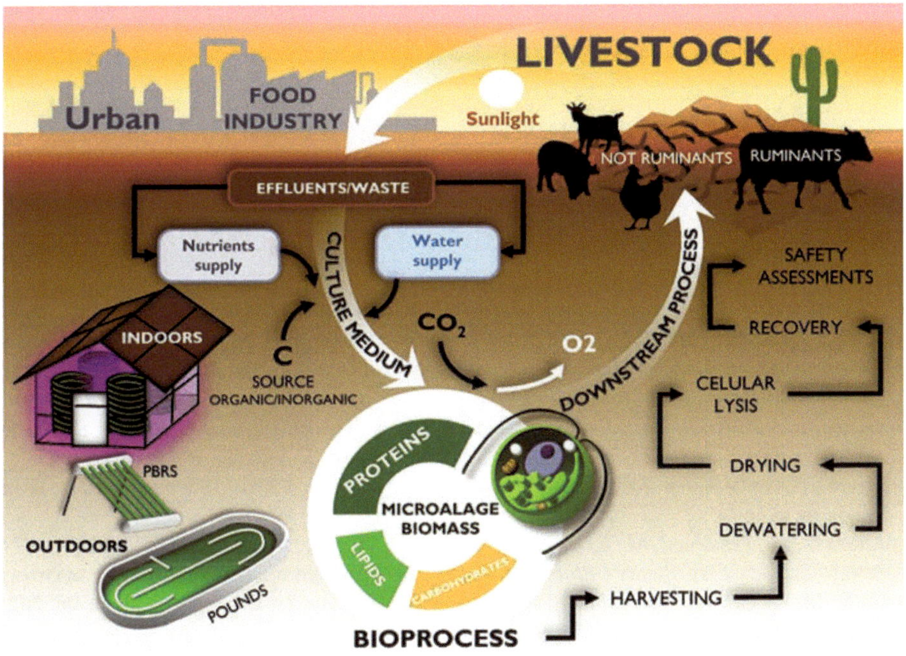

FIGURE 11.2 A schematic view of general bioprocess steps for microalgal biomass production for animal feed in drylands.

11.9 FUTURE TRENDS

Future scenarios for global food security and the global population will reach 9 billion by 2050. These effects will be more severe for the inhabitants of the arid regions of the world; therefore, the improvement of bioprocesses to produce microalgae as sources of proteins and nutrients to sustain the livestock sector are in the eye of biotechnologists. Challenges include harnessing the local natural resources, water scarcity, and reducing production costs to implement suitable and sustainable production of microalgae-based feed.

KEYWORDS

- **microalgae-based feed**
- **wastewater treatment**
- **food source**
- **microalgae culture systems in drylands**
- **novel nutrients sources**
- **climate change challenges**

REFERENCES

Acién-Fernández, F. G.; Gómez-Serrano, C.; Fernández-Sevilla, J. M. Recovery of Nutrients From Wastewaters Using Microalgae. *Front. Sust. Food Sys.* **2018,** *2*. DOI: 10.3389/fsufs.2018.00059

Adeel, Z.; Safriel, U.; Niemeijer, D.; White, R.; de Kalbermatten, G.; Glantz, M.; Salem, B.; Scholes, B.; Niamir-Fuller, M.; Ehui, S.; Yapi-Gnaore, V. Ecosystems and Human Well-Being: Desertification Synthesis; A Report of the Millennium Ecosystem Assessment, World Resources Institute, Washington, 2005.

Alfaia, C. M.; Pestana, J. M.; Rodrigues, M.; Coelho, D.; Aires, M. J.; Ribeiro, D. M.; Prates, J. A. M. Influence of Dietary *Chlorella vulgaris* and Carbohydrate-Active Enzymes on Growth Performance, Meat Quality, and Lipid Composition of Broiler Chickens. P*oultry Sci.* **2021,** *100* (2), 926–937. DOI: 10.1016/j.psj.2020.11.034

Ali, S.; Peter, A. P.; Chew, K. W.; Munawaroh, H. S. H.; Pau Loke Show, P. Resource Recovery from Industrial Effluents Through the Cultivation of Microalgae: A Review. *Bioresour. Technol.* **2021,** *337*, 125461. https://doi.org/10.1016/j.biortech.2021.125461

Barclay, W.; Apt, K. Strategies for Bioprospecting Microalgae for Potential Commercial Applications. In *Handbook of Microalgal Culture*; Richmond, A., Hu, Q., Eds.; 2013. https://doi.org/10.1002/9781118567166.ch4

Barkia, I.; Saari, N.; Manning, S. R. Microalgae for High-Value Products Towards Human Health and Nutrition. *Mar. Drugs* **2019,** *17* (5), 304. DOI: 10.3390/md17050304.

Bauer, S.; Linke, H. J.; Wagner, M. Combining Industrial and Urban Water-Reuse Concepts for Increasing the Water Resources in Water-Scarce Regions. *Water Environ. Res.* **2020,** *92* (7), 1027–1041. DOI: 10.1002/wer.1298

Becker, E. W. Micro-algae as a Source of Protein. *Biotechnol. Adv.* **2007,** *25* (2), 207–210. DOI: 10.1016/j.biotechadv.2006.11.

Becker, E. W. Microalgae for Human and Animal Nutrition. In *Handbook of Microalgal Culture: Applied Phycology and Biotechnology,* 2013; pp 461–503. DOI: 10.1002/9781118567166.ch25

Benemann, J. Microalgae for Biofuels and Animal Feeds. *Energies* **2013,** *6* (11), 5869–5886. DOI: 10.3390/en6115869

Benemann, J. R.; Woertz, I.; Lundquist, T. Autotrophic Microalgae Biomass Production: From Niche Markets to Commodities. *Ind. Biotechnol.* **2018,** *14* (1), 3–10.

Bo, D. D.; Magneschi, L.; Bedhomme, M.; Billey, E.; Deragon, E.; Storti, M.; Menneteau, M.; Richard, C.; Rak, C.; Lapeyre, M.; Lembrouk, M.; Conte, M.; Gros, V.; Tourcier, G.; Giustini, C.; Falconet, D.; Curien, G.; Allorent, G.; Petroutsos, D.; Laeuffer, F.; Fourage, L.; Jouhet, J.; Marechal, E.; Finazzi, G.; Collin, S. Consequences of Mixotrophy on Cell Energetic Metabolism in Microchloropsis gaditana Revealed by Genetic Engineering and Metabolic Approaches. *Front Plant Sci.* **2021,** *12.* https://doi.org/10.3389/fpls.2021.628684

Bonos, E.; Kasapidou, E.; Kargopoulos, A.; Karampampas, A.; Christaki, E.; Florou-Paneri, P.; Nikolakakis, I. Spirulina as a Functional Ingredient in Broiler Chicken Diets. *S. Afr. J. Animal Sci.* **2016,** *46* (1), 94. DOI: 10.4314/sajas.v46i1.12

Borowitzka, M. A. High-Value Products from Microalgae—Their Development and Commercialization. *J. Appl. Phycol.* **2013,** *25*, 743–756. https://doi.org/10.1007/s10811-013-9983-9

Colusse, G. A.; Mendes, C. R. B.; Duarte, M. E. R.; Carvalho, J. C.; Noseda, M. D. Effects of Different Culture Media on Physiological Features and Laboratory-Scale Production Cost of Dunaliella salina. *Biotechnol. Rep.* **2020,** *27*, e00508. DOI: 10.1016/j.btre.2020.e00508

Coudert, E.; Baéza, E; Berri, C. Use of Algae in Poultry Production: A Review, *World's Poult. Sci. J.* **2020,** *76* (4), 767–786. DOI: 10.1080/00439339.2020.1830012

Dai, X. R.; Saha, C. K.; Ni, J. Q.; Heber, A. J.; Blanes-Vidal, V.; Dunn, J. L. Characteristics of Pollutant Gas Releases from Swine, Dairy, Beef, and Layer Manure, and Municipal Wastewater. *Water Res.* **2015,** *76*, 110–119. DOI: 10.1016/j.watres.2015.02.050

Darpito, C.; Shin, W. S.; Jeon, S.; Lee, H.; Nam, K.; Kwon, J. H.; Yang, J. W. Cultivation of Chlorella Protothecoides in Anaerobically Treated Brewery Wastewater for Cost-Effective Biodiesel Production. *Bioprocess Biosyst. Eng.* **2015,** *38* (3), 523–530. DOI: 10.1007/s00449-014-1292-4.

Deprá, M. C.; dos Santos, A. M.; Severo, I. A.; Santos, A. B.; Zepka, L. Q.; Jacob-Lopes, E. Microalgal Biorefineries for Bioenergy Production: Can We Move from Concept to Industrial Reality?. *Bioenerg. Res.* **2018,** *11*, 727–747. https://doi.org/10.1007/s12155-018-9934-z

EC. Regulation (EC) No 178/2002 of the European Parliament and of the Council of January 2002 Laying Down the General Principles and Requirements of Food Law, Establishing the European Food Safety Authority and Laying Down Procedures in Matters of Food Safety. Official Journal of the European Communities, 2002; vol L 31, pp 1–24.

Enzing, C. M.; Ploeg, M.; Barbosa, M. J., Sijtsma, L. Microalgae-Based Products for the Food and Feed Sector: An Outlook for Europe. EUR 26255, Publications Office of the European Union; JRC85709, 2014. DOI: 10.2791/3339

Esteves, A. F.; Pires, J. C. M.; Gonçalves, A. L.; Current Utilization of Microalgae in the Food Industry Beyond Direct Human Consumption. *Cultured Microalgae Food Ind.* **2021,** 199–248. DOI: 10.1016/b978-0-12-821080-2.00005-8

FAO. Policy Brief: Food Security - Issue 2, June 2006.

FAO. Meat Market Review: Overview of Global Meat Market Developments in 2020, Rome, **2021.**

Fasaei, F.; Bitter, J. H.; Slegers, P. M.; van Boxtel, A. J. B. Techno-Economic Evaluation of Microalgae Harvesting and Dewatering Systems. *Algal Res.* **2018,** *31*, 347–362. DOI: 10.1016/j.algal.2017.11.038

Fernandez, F. G. A.; Reis, A.; Wijffels, R. H.; Barbosa, M.; Verdelho, V.; Llamas, B. The Role of Microalgae in the Bio-Economy. *New Biotechnol.* **2020,** *61*, 99–107. DOI: https://doi.org/10.1016/j.nbt.2020.11.011

Food and Drug Administration, USA. Substances Generally Recognized as Safe [Online] 2022. https://www.federalregister.gov/d/2016-19164/p-1 (accessed Feb 22, 2022)

Golla, B. Agricultural Production System in Arid and Semi-Arid Regions. *J. Agric. Sci. Food Technol.* **2021,** *7* (2), 234–244. DOI: https://dx.doi.org/10.17352/2455-815X.000113

Gouveia, L.; Graça, S.; Sousa, C.; Ambrosano, L.; Ribeiro, B.; Botrel, E. P.; Silva, C. M. Microalgae Biomass Production Using Wastewater: Treatment and Costs. *Algal Res.* **2016,** *16*, 167–176. DOI: 10.1016/j.algal.2016.03.010

Grobbelaar, J. U. Inorganic Algal Nutrition. In *Handbook of Microalgal Culture;* 2013; pp. 123–133. DOI: 10.1002/9781118567166.ch8

Habib, M. A. B.; Parvin, M.; Huntington, T. C.; Hasan, M. R. A Review on Culture, Production and Use of Spirulina as Food for Humans and Feeds for Domestic Animals FAO Fisheries and Aquaculture Circular, No. 1034; FAO: Rome, 2008; p 33.

He, C.; Liu, Z.; Wu, J; Pan, X.; Fang, Z., Li, J.; Bryan, B. A.; Future Global Urban Water Scarcity and Potential Solutions. *Nat. Commun.* **2021,** *12*, 4667. https://doi.org/10.1038/s41467-021-25026-3

Hillebrand, H.; Sommer, U. Effect of Continuous Nutrient Enrichment on Microalgae Colonizing Hard Substrates. *Hydrobiologia* **2000,** *426*, 185–192. https://doi.org/10.1023/A:1003943201178

Hu, J.; Nagarajan, D.; Zhang, Q.; Chang, J.S.; Lee, D. J. Heterotrophic Cultivation of Microalgae for Pigment Production: A Review. *Biotechnol. Adv.* **2018,** *36* (1), 54–67. DOI: 10.1016/j.biotechadv.2017.09.

Jochum, M.; Moncayo, L. P.; Jo, Y. K. Microalgal Cultivation for Biofertilization in Rice Plants Using a Vertical Semi-Closed Airlift Photobioreactor. *PLOS ONE* **2018,** *13* (9), e0203456. DOI: 10.1371/journal.pone.0203456

Kong, Z.; Wu, J.; Rong, C.; Wang, T.; Li, L.; Luo, Z.; Li, Y. Y. Large Pilot-Scale Submerged Anaerobic Membrane Bioreactor for the Treatment of Municipal Wastewater and Biogas Production at 25°C. *Bioresour. Technol.* **2021,** *319*, 124123. DOI: 10.1016/j.biortech.2020.124123

Kratzer, R.; Murkovic, M. Food Ingredients and Nutraceuticals from Microalgae: Main Product Classes and Biotechnological Production. *Foods* **2021,** *10* (7), 1626. DOI: 10.3390/foods10071626.

Kusmayadi, A.; Leong, Y. K.; Yen, H. W.; Huang, C. Y.; Chang, J. Microalgae as Sustainable Food and Feed Sources for Animals and Humans – Biotechnological and Environmental Aspects. *Chemosphere* **2021,** *271*, 129800. DOI: 10.1016/j.chemosphere.2021.12

Leite, J. H. G. M.; Façanha, D. A. E.; Bermejo, J. V. D.; Guilhermino, M. M.; Bermejo, L. A. Adaptive Assessment of Small Ruminants in Arid and Semi-Arid Regions. *Small Rumin. Res.* **2021,** *203*, 106497. DOI: 10.1016/j.smallrumres.2021.10

Li, K.; Liu, Q.; Fang, F.; Luo, R.; Lu, Q.; Zhou, W.; Ruan, R. Microalgae-Based Wastewater Treatment for Nutrients Recovery: A Review. *Bioresour. Technol.* **2019,** 121934. DOI: 10.1016/j.biortech.2019.121934

Lian, X.; Pia, S.; Chen, A.; Huntingford, C.; Fu, B.; Li, L. Z. X.; Huang, J.; Sheffield, J.; Berg, A. M.; Keenan, T. F.; McVicar T.R.; Wada, Y.; Wang, X., Wang, T.; Yang, Y.; Roderick, M. L. Multifaceted Characteristics of Dryland Aridity Changes in a Warming World. *Nat. Rev. Earth Environ.* **2021,** *2*, 232–250. https://doi.org/10.1038/s43017-021-00144-0

Ma, H.; Guo, Y.; Qin, Y.; Li, Y. Y. Nutrient Recovery Technologies Integrated with Energy Recovery by Waste Biomass Anaerobic Digestion. *Bioresour. Technol.* **2018,** *269*, 520–531. https://doi.org/10.1016/j.biortech.2018.08.114.

Martínez-Valderrama, J.; Guirado, E.; Maestre, F. T. Desertifying Deserts. *Nat. Sustain.* **2020,** *3*, 572–575. https://doi.org/10.1038/s41893-020-0561-2

Masojídek, J.; Sergejevová, M.; Malapascua, J. R.; Kopecký, J. Thin-layer Systems for Mass Cultivation of Microalgae: Flat Panels and Sloping Cascades. In *Algal Biorefineries*; Springer: EUA, 2015; pp 237–261.

Molino, A.; Martino, M.; Larocca, V.; Di Sanzo, G.; Spagnoletta, A.; Marino, T.; Musmarra, D. Eicosapentaenoic Acid Extraction from *Nannochloropsis gaditana* using Carbon Dioxide at Supercritical Conditions. *Marine Drugs* **2019,** *17* (2), 132. DOI: 10.3390/md17020132

Monte, J.; Sá, M.; Galinha, C. F.; Costa, L.; Hoekstra, H.; Brazinha, C.; Crespo, J. G. Harvesting of Dunaliella Salina by Membrane Filtration at Pilot Scale. *Sep. Purif. Technol.* **2018,** *190*, 252–260. DOI: 10.1016/j.seppur.2017.08.019

Morales-Sánchez, D.; Martinez-Rodriguez, O. A.; Martinez, A. Heterotrophic Cultivation of Microalgae: Production of Metabolites of Commercial Interest. *J. Chem. Technol. Biotechnol.* **2016,** *92* (5), 925–936. DOI: 10.1002/jctb.5115

Musa, M.; Ward, A.; Ayoko, G. A.; Rösch, C.; Brown, R.; Rainey, T. J. Single-Step Dynamic Dewatering of Microalgae from Dilute Suspensions Using Flocculant Assisted Filtration. *Microb. Cell Fact.* **2020,** *19*, 222. https://doi.org/10.1186/s12934-020-01472-4

Nagarajan, D.; Varjani, S.; Lee, D.; Chang, J. Sustainable Aquaculture and Animal Feed from Microalgae – Nutritive Value and Techno-Functional Components. *Renew. Sustain. Energy Rev.* **2021,** *150*, 111549. https://doi.org/10.1016/j.rser.2021.111549

Niccolai, A.; Chini Zittelli, G.; Rodolfi, L.; Biondi, N.; Tredici, M. R. Microalgae of Interest as Food Source: Biochemical Composition and Digestibility. *Algal Res.* **2019,** *42*, 101617. DOI: 10.1016/j.algal.2019.101617

Park, J. E.; Zhang, S.; Han, T. H.; Hwang, S. J. The Contribution Ratio of Autotrophic and Heterotrophic Metabolism during a Mixotrophic Culture of Chlorella sorokiniana. *Int. J. Environ. Res. Public Health* **2021,** *18* (3), 1353. DOI: 10.3390/ijerph18031353

Pereira, H.; Silva, J.; Santos, T.; Gangadhar, K. N.; Raposo, A.; Nunes, C.; Varela, J. Nutritional Potential and Toxicological Evaluation of Tetraselmis sp. CTP4 Microalgal Biomass Produced in Industrial Photobioreactors. *Molecules* **2019,** *24* (17), 3192. DOI: 10.3390/molecules24173192

Perez-Garcia, O.; Escalante, F. M.; de-Bashan, L. E.; Bashan, Y. Heterotrophic Cultures of Microalgae: Metabolism and Potential Products. *Water Res.* **2011,** *45* (1), 11–36. DOI: 10.1016/j.watres.2010.08.037.

Piasecka, A.; Nawrocka, A.; Wiącek, D.; Krzemińska, I. Agro-industrial by-Product in Photoheterotrophic and Mixotrophic Culture of Tetradesmus obliquus: Production of ω3 and ω6 Essential Fatty Acids with Biotechnological Importance. *Sci. Rep.* **2020,** *10*, 6411. DOI: 10.1038/s41598-020-63184-4

Plaza, C.; Zaccone, C.; Sawicka, K.; Mendez, A. M.; Tarquis, A., Gascó, G.; Heuvelink, G. B. M.; Schuur, E. A. G.; Maestre, F. T. Soil Resources and Element Stocks in Drylands to Face Global Issues. *Sci. Rep.* **2018,** *8*, 13788. https://doi.org/10.1038/s41598-018-32229-0

Qadir, M.; Sato, T. Water Reuse in Arid Zones. In *Urban Water Reuse Handbook*; CRC Press: USA, 2016.

Rahimi, J.; Mutua, J. Y.; Notenbaert, A. M. O.; Marshall, K.; Butterbach-Bahl, K. Heat Stress Will Detrimentally Impact Future Livestock Production in East Africa. *Nat. Food* **2021,** *2*, 88–96. https://doi.org/10.1038/s43016-021-00226-8

Rebolloso-Fuentes, M. M.; Navarro-Pérez, A.; García-Camacho, F.; Ramos-Miras, J. J.; Guil-Guerrero, J. L. Biomass Nutrient Profiles of the Microalga Nannochloropsis. *J. Agricul. Food Chem.* **2001,** *49* (6), 2966–2972. DOI: 10.1021/jf0010376

Richter, B. D.; Bartak, D.; Caldwell, P.; Davis, K. F.; Debaere, P.; Hoekstra, A. Y., Li, T.; Marston, L.; McManamay, R.; Mekonnen, M. M.; Rudell, B. L.; Rushforth, R. R.; Troy, T. J. Water Scarcity and Fish Imperilment Driven by Beef Production. *Nat Sustain.* **2020,** *3,* 319–328. https://doi.org/10.1038/s41893-020-0483-z

Robinson, T. P.; Wint, G. R. W.; Conchedda, G.; Van Boeckel, T. P.; Ercoli, V.; Palamara, E.; Cinardi, G.; D′Aietti, L.; Hay, S. I.; Gilbert, M. Mapping the Global Distribution of Livestock. *PLoS ONE* **2014,** *9* (5), e96084. https://doi.org/10.1371/journal.pone.0096084

Saadaoui, I.; Rasheed, R.; Cherif, M.; Al Jabri, H.; Sayadi, S.; Manning, S. R.; Microalgal-Based Feed: Promising Alternative Feedstocks for Livestock and Poultry Production. *J. Animal Sci. Biotech.* **2021,** *12,* 76. https://doi.org/10.1186/s40104-021-00593-z

Safriel, U. Deserts and Desertification: Challenges but also Opportunities. *Land Degrad. Dev.* **2009,** *20* (4), 353–366. DOI: 10.1002/ldr.935

Sarker, N. K.; Salam, P. A. Indoor and Outdoor Cultivation of *Chlorella vulgaris* and its Application in Wastewater Treatment in a Tropical City—Bangkok, Thailand, SN. *Appl. Sci.* **2019,** *1,* 1645. https://doi.org/10.1007/s42452-019-1704-9

Saxena, P.; Kumar, M.; Harish. Autotrophic Cultivation of Microalgae for the Production of Polyunsaturated Fatty Acid. In *Nutraceutical Fatty Acids,* 2020; pp 165–185. DOI: 10.1002/9781119631729.ch6

Seghiri, R.; Kharbach, M.; Essamri, A. Functional Composition, Nutritional Properties, and Biological Activities of Moroccan Spirulina Microalga. *J. Food Qual.* **2019,** 1–11. DOI: 10.1155/2019/3707219

Sekaran, U.; Lai, L.; Ussiri, D. A. N.; Kumar, S.; Clay, S. Role of Integrated Crop-Livestock Systems in Improving Agriculture Production and Addressing Food Security – A Review. *J. Agric. Food Res.* **2021,** *5,* 100190. DOI: 10.1016/j.jafr.2021.100190

Silva, M. E. T. da; Correa, K. de P.; Martins, M. A.; da Matta, S. L. P.; Martino, H. S. D.; Coimbra, J. S. dos R. Food Safety, Hypolipidemic and Hypoglycemic Activities, and In Vivo Protein Quality of Microalga Scenedesmus Obliquus in Wistar Rats. *J. Funct. Foods* **2020,** *65,* 103711. https://doi.org/10.1016/j.jff.201

Smetana, S.; Sandmann, M.; Rohn, S.; Pleissner, D.; Heinz, V. Autotrophic and Heterotrophic Microalgae and Cyanobacteria Cultivation for Food and Feed: Life Cycle Assessment. *Bioresour. Technol.* **2017,** *245,* 162–170. DOI: 10.1016/j.biortech.2017.08.11

Spinoni, J.; Barbosa, P.; Cherlet, M.; Forzieri, G.; McCormick, N.; Naumann, G.; Vogt, J. V.; Dosio, A. How Will the Progressive Global Increase of Arid Areas Affect Population and Land use in the 21st Century? *Global Planet Change* **2021,** *205,* 103597. https://doi.org/10.1016/j.gloplac

Terry, K. L.; Raymond, L. P. System Design for the Autotrophic Production of Microalgae. *Enzyme Microb. Technol.* **1985,** *7,* 474–487. http://dx.doi.org/10.1016/0141-0229(85)90148-6

Theegala, C. S. Algal Cell Disruption and Lipid Extraction: A Review on Current Technologies and Limitations. In *Algal Biorefineries*; Prokop, A., Bajpai, R., Zappi, M., Eds; Springer: EUA, 2015. https://doi.org/10.1007/978-3-319-20200-6_1

Tibbetts, S. M.; Milley, J. E.; Lall, S. P. Chemical Composition and Nutritional Properties of Freshwater and Marine Microalgal Biomass Cultured in Photobioreactors. *J. Appl. Phycol.* **2015,** *27,* 1109–1119. https://doi.org/10.1007/s10811-014-0428-x

Troschl, C.; Meixner, K.; Drosg, B. Cyanobacterial PHA Production-Review of Recent Advances and a Summary of Three Years' Working Experience Running a Pilot Plant. *Bioengineering* **2017,** *4* (2), 26. DOI: 10.3390/bioengineering4020026

Tokuşoglu, O.; Uunal, M. K. Biomass Nutrient Profiles of Three Microalgae: *Spirulina platensis, Chlorella vulgaris*, and *Isochrisis galbana. J. Food Sci.* **2019,** *68* (4), 1144–1148. DOI: 10.1111/j.1365-2621.2003.tb09615.x

Torres-Tiji, Y.; Fields, F. J.; Mayfield, S. P. Microalgae as a Future Food Source. *Biotechnol. Adv.* **2020,** 107536. DOI: 10.1016/j.biotechadv.2020.107

Torzillo G.; Chini Zittelli, G. Tubular Photobioreactors. In *Algal Biorefineries*; Prokop, A., Bajpai, R., Zappi, M., Eds.; Springer: EUA, 2015. https://doi.org/10.1007/978-3-319-20200-6_5

Valdovinos-García, E. M.; Petriz-Prieto, M. A.; Olán-Acosta, M. D. L. Á.; Barajas-Fernández, J.; Guzmán-López, A.; Bravo-Sánchez, M. G. Production of Microalgal Biomass in Photobioreactors as Feedstock for Bioenergy and Other Uses: A Techno-Economic Study of Harvesting Stage. *Appl. Sci.* **2021,** 114386. https://doi.org/10.3390/ app11104386

Veerabadhran, M.; Gnanasekaran, D.; Wei, J.; Yang, F. Anaerobic Digestion of Microalgal Biomass for Bioenergy Production, Removal of Nutrients and Microcystin: Current Status. *J. Appl. Microbiol.* **2021,** *131* (4), 1639–1651. DOI: 10.1111/jam.15000

Veerabadhran, M.; Natesan, S.; Mubarakali, D.; Xu, S.; Yang, F. Using Different Cultivation Strategies and Methods for the Production of Microalgal Biomass as a Raw Material for the Generation of Bioproducts. *Chemosphere* **2021,** *285*, 131436. DOI: 10.1016/j. chemosphere.2021.13

Vigani, M.; Parisi, C.; Rodríguez-Cerezo, E.; Barbosa, M. J.; Sijtsma, L.; Ploeg, M.; Enzing, C. Food and Feed Products from Micro-Algae: Market Opportunities and Challenges for the EU. *Trends Food Sci. Technol.* **2015,** *42* (1), 81–92. DOI: 10.1016/j.tifs.2014.12.004

Xu, C.; Zhang, S.; Sun, B.; Xie, P.; Liu, X.; Chang, L.; Lu, F.; Zhang, S. Dietary Supplementation with Microalgae (Schizochytrium sp.) Improves the Antioxidant Status, Fatty Acids Profiles and Volatile Compounds of Beef. *Animals* **2021,** *11*, 3517. https://doi.org/10.3390/ ani11123517

Yaakob, Z.; Ali, E.; Zainal, A.; Mohamad, M.; Takriff, M. S. An Overview: Biomolecules from Microalgae for Animal Feed and Aquaculture. *J. Biol. Res. Thessaloniki* **2014,** *21*, 6.

Yang, C.; Hua, Q.; Shimizu, K.; Energetics and Carbon Metabolism During Growth of Microalgal Cells Under Photoautotrophic, Mixotrophic and Cyclic Light-Autotrophic/ Dark-Heterotrophic Conditions. *Biochem. Eng. J.* **2000,** *6* (2), 87–102. DOI: 10.1016/ s1369-703x(00)00080-2

Yu, H.; Kim, J.; Lee, C. Nutrient Removal and Microalgal Biomass Production from Different Anaerobic Digestion Effluents with Chlorella Species. *Sci. Rep.* **2019,** *9* (1). DOI: 10.1038/ s41598-019-42521-2

Zittelli, G. C.; Biondi, N.; Rodolfi, L.; Tredici, M. R. Photobioreactors for Mass Production of Microalgae. In *Handbook of Microalgal Culture*; Richmond, A., Hu, Q., Eds.; 2013. https:// doi.org/10.1002/9781118567166.ch13

Zhang, H.; Wang, W.; Li, Y.; Yang, W.; Shen, G. Mixotrophic Cultivation of Botryococcus braunii. *Biomass Bioenergy* **2011,** *35* (5), 1710–1715. DOI: 10.1016/j.biombioe.2011.01.002

Zhu, L.; Li, Z.; Hiltunen, E. Microalgae Chlorella Vulgaris Biomass Harvesting by Natural Flocculant: Effects on Biomass Sedimentation, Spent Medium Recycling and Lipid Extraction. *Biotechnol. Biofuels* **2018,** *11*, 183. https://doi.org/10.1186/s13068-018-1183-z

Index

U

V